Birkhäuser

Science Networks. Historical Studies
Founded by Erwin Hiebert and Hans Wußing
Volume 50

More information about this series at http://www.springer.com/series/4883

Robert E. Bradley • Salvatore J. Petrilli
C. Edward Sandifer

L'Hôpital's Analyse des infiniments petits

An Annotated Translation with Source Material by Johann Bernoulli

Birkhäuser

Robert E. Bradley
Mathematics and Computer Science
Adelphi University
Garden City, NY, USA

Salvatore J. Petrilli
Mathematics and Computer Science
Adelphi University
Garden City, NY, USA

C. Edward Sandifer
Newtown, CT, USA

ISSN 1421-6329 ISSN 2296-6080 (electronic)
Science Networks. Historical Studies
ISBN 978-3-319-17114-2 ISBN 978-3-319-17115-9 (eBook)
DOI 10.1007/978-3-319-17115-9

Library of Congress Control Number: 2015935196

Mathematics Subject Classification (2010): 01A45, 01A75, 26-03

Springer Cham Heidelberg New York Dordrecht London

Cover illustration: From Waller Ms de-00215, August Beer: Über die Correction des Cosinusgesetzes bei der Anwendung des Nicol'schen Prismas in der Photometrie, after 1850. With friendly permission by The Waller Manuscript Collection (part of the Uppsala University Library Collections)

Printed on acid-free paper

Springer International Publishing AG Switzerland is part of Springer Science+Business Media (www.springer.com)

Translators' Preface

Analyse des infiniment petits, pour l'intelligence des lignes courbes was the first textbook of the differential calculus. The title translates as *Analysis of the infinitely small, for the understanding of curved lines*. It was published anonymously in Paris in 1696, although members of the French mathematical community were well aware that the author was Guillaume François Antoine de l'Hôpital,[1] the Marquis of Saint-Mesme (1661–1704). The textbook was successful, as evidenced by the appearance of a posthumous second edition (L'Hôpital 1716), which identified the author.[2] Pierre Varignon (1646–1722), who was professor of mathematics at Collège des Quatre-Nations in Paris and a friend of l'Hôpital, created a collection of clarifications and additions to the *Analyse*. These were published posthumously (Varignon 1725), a few years after the 1716 edition of the *Analyse*. Later editions of the *Analyse* included similar commentary and continued to appear throughout the 18th century (L'Hôpital 1768, 1781).

Differential and integral calculus are generally considered to have their origins in the works of Sir Isaac Newton (1642–1727) and Wilhelm Gottfried von Leibniz (1646–1716)[3] in the late 17th century, although the roots of the subject reach far back into that century and, arguably, even into antiquity. Leibniz first described his new calculus in a cryptic article more than a decade before the publication of the *Analyse* (Leibniz 1684). For all practical purposes, Leibniz' early papers were not

[1]L'Hôpital spelled his name "Hospital" and this was the spelling used in the posthumous second edition of the text in 1715/1716. Fontenelle used the spelling "Hôpital," which is the standard modern spelling of the name, in his Eulogy (p. 295). There is no consensus among English-language authors of the early 21st century as to which spelling ought to be used.

[2]Technically, this was the third edition, despite the words "Seconde Edition" on the title page. There was a "Seconde Edition" one year earlier (L'Hôpital 1715), with many typographical errors, of which the 1716 edition is a corrected version (Bernoulli 1955, pp. 499–500).

[3]For more on the priority dispute over the discovery of the calculus, which is not a matter of interest for this volume, see Hall (1980).

understood, until Jakob Bernoulli (1654–1705) and his younger brother Johann[4] (1667–1748) began studying them in about 1687 and making discoveries of their own using his techniques.

Bernard de Fontenelle (1657–1757) became the secretary of the *Académie des Sciences* in Paris in 1697 and wrote the eulogy of l'Hôpital for the academy's journal. He said that in 1696,

> ...the Geometry of the Infinitely small was still nothing but a kind of Mystery, and, so to speak, a Cabalistic Science shared among five or six people. They often gave their Solutions in the Journals without revealing the Method that produced them, and even when one could discover it, it was only a few feeble rays of this Science that had escaped, and the clouds immediately closed again.[5] (Fontenelle 1708, pp. 133–134)

Later on, Jean Etienne Montucla (1725–1799) went one step further and listed the only people that he believed understood Leibniz' calculus before 1696: Leibniz himself, Jakob and Johann Bernoulli, Pierre Varignon and l'Hôpital (Montucla 1799, p. 397). L'Hôpital's *Analyse* changed all of this and for much of the 18th century, his book served aspiring French mathematicians as their first introduction to the new calculus.

For all that the *Analyse* was a popular and successful introduction to the differential calculus, it's remarkable that there is no account of the integral calculus in the book. In his Preface, l'Hôpital explained why

> In all of this there is only the first part of Mr. Leibniz' calculus, ... The other part, which we call *integral Calculus*, consists in going back from these infinitely small quantities to the magnitudes or the wholes of which they are the differences, that is to say in finding their sums. I had also intended to present this. However, Mr. *Leibniz*, having written me that he is working on a Treatise titled *De Scientiâ infiniti*, I took care not to deprive the public of such a beautiful Work ... [p. liii].

Unfortunately, Leibniz never completed this book *On the Science of the Infinite*.

The *Analyse* consists of ten chapters, which L'Hôpital called "sections." We consider it to have three parts. The first part, an introduction to the differential calculus, consists of the first four chapters:

1. In which we give the Rules of this calculus.
2. Use of the differential calculus for finding the Tangents of all kinds of curved lines.
3. Use of the differential calculus for finding the greatest and the least ordinates, to which are reduced questions *De maximis & minimis*.
4. Use of the differential calculus for finding inflection points and cusps.

Taken together, these chapters provide a thorough introduction to the differential calculus in about 70 pages. The next five chapters are devoted to what can only be described as an advanced text on differential geometry, motivated in part by what were then cutting-edge research problems in optics and other fields. The final

[4] Often referred to as Johann (I) Bernoulli to distinguish him from his son Johann and grandson Johann, who were also successful mathematicians.

[5] This quotation from p. 299 of this text.

chapter is mildly polemical, demonstrating the superiority of Leibniz' new calculus, when compared to the methods of René Descartes (1596–1660) and Johann van Waveren Hudde (1628–1704).

The Role of Johann Bernoulli

Most biographical information known about the Marquis de l'Hôpital comes from Fontenelle's eulogy (Fontenelle 1708, pp. 116–146), a translation of which is included in Chapter 13 of this volume. However, Fontenelle knew little or nothing at the time of l'Hôpital's death about the role of Johann Bernoulli in the composition of the *Analyse*. L'Hôpital himself acknowledged a debt to Johann Bernoulli in his preface:

> I acknowledge having received much from the illuminations of Messrs. Bernoulli, particularly those of the younger, presently Professor at Groningen. I have made plain use of their discoveries and those of Mr. Leibniz. This is why I grant that they may claim as much of this as they may wish, being content with that which they are willing to leave for me [p. liv].

In light of this, it seems somewhat strange that Montucla would write "We may only find fault in that Mr. de l'Hôpital did not make well enough known the debt he owed to Mr. Bernoulli" (Montucla 1799, p. 397), but the record shows that Johann Bernoulli's influence on the structure and content of the *Analyse* was much more significant than these words of recognition would suggest.

Among the few details known about l'Hôpital's early life, Fontenelle recounted that he solved one of Pascal's problems involving the cycloid at the age of 15. The Marquis became a cavalry officer, but had only attained the rank of captain when he resigned his commission due to poor eyesight. He devoted all of his energy to mathematics from that point onward. Some time around 1690, he joined Nicolas Malebranche's (1638–1715) circle, which was engaged, among other things, in the study of higher mathematics. It was there in November 1691 that he met the 24-year-old Johann Bernoulli, who was visiting Paris and had been invited by Malebranche to present his construction of the catenary at the salon. Although Fontenelle made no mention of this meeting, it is documented by Spiess in his introduction to the Bernoulli-l'Hôpital correspondence, which contains what may be considered a definitive biography of the Marquis de l'Hôpital (Bernoulli 1955, pp. 123–130).

There is no contemporary account of this meeting. Bernoulli wrote of the encounter in his autobiography, which he composed in 1741, but Spiess considers an earlier account that he gave in a letter to Pierre Rémond de Montmort (1678–1719) to be more reliable; see Bernoulli (1955, p. 135–137). In May 21, 1718, Bernoulli told Montmort that upon meeting the Marquis, he soon found him to be a good enough mathematician with regard to ordinary mathematics, but that he knew nothing of the differential calculus, other than its name, and had not even heard of the integral calculus. L'Hôpital had apparently mastered Fermat's method of finding maxima and minima and told Bernoulli that he had used it to invent a rule for determining the radius of curvature for arbitrary curves. The method was unwieldy and actually could only be used at local extrema of algebraic curves. Bernoulli showed him the formula for the radius of curvature that he had developed with his brother Jakob, which employed second-order differentials. Apparently, this so impressed the Marquis that he visited Bernoulli the very next day and engaged him as his tutor in the differential and integral calculus.

The Marquis de L'Hôpital (1661–1704) – portrait in walnut by Susan Petry, 2014, 25 × 19 × 3 cm.; photograph by the artist

Bernoulli tutored the Marquis in his Paris apartment four times a week from late 1691 through the end of July 1692. We are fortunate that l'Hôpital insisted that Bernoulli commit his lessons to paper. Bernoulli typically composed each lesson, which he wrote in Latin, the night before he gave them to the Marquis. Fortunately, his friend and later colleague at the University of Basel, Johann Heinrich von

Johann Bernoulli (1667–1748) – portrait in cherry by Susan Petry, 2014, 26 × 19 × 4 cm.; photograph by the artist

Stähelin (1668–1721), was rooming with him in Paris. Stähelin made copies of the lessons before Bernoulli handed them over to the Marquis.

In the summer of 1692, Bernoulli accompanied the Marquis to his estate in Oucques, near the French city of Blois, where he continued giving him tutorials until some time in October. Bernoulli's lessons from this period have not survived, although he told Montmort that the Marquis' valet made copies of some of them, one of which appears to have survived (Bernoulli 1955, p. 137). It is also possible that some of the Bernoulli's lessons on the integral calculus, which he later reported to have been given in Paris, were actually given at Oucques.

In any case, Bernoulli kept copies of his lessons to the Marquis throughout his long and productive career. The first part, on the differential calculus, was incorporated by l'Hôpital into the first four chapters of the *Analyse*. Bernoulli himself published the much larger second part, concerning the integral calculus, in his collected works (Bernoulli 1742, pp. 385–558). Titled *Lectiones mathematicae de methodo integralium*, (Mathematical Lectures Concerning the Method of Integration), this treatise bears the subtitle "written for the use of the Illustrious Marquis de l'Hôpital while the author spent time in Paris in the years 1691 & 1692." The first sentence of this work makes reference to what was seen "in the preceding." A footnote explained that Bernoulli meant the lectures in differential calculus, which had preceded this but which he had omitted, because all of it had appeared in the *Analyse*, "which is in everyone's hands." (Bernoulli 1742, p. 387) That is, he left out the portion of his Paris lessons that l'Hôpital had incorporated into his introductory chapters. Because Bernoulli chose not to publish this part, it was impossible in the 18th century to say how closely l'Hôpital's textbook coincided with Bernoulli's lessons.

A comparison finally became possible when Paul Schafheitlin discovered a manuscript copy of the full set of lessons, on both the differential and integral calculus, in the library of the University of Basel in 1921. Schafheitlin published the first portion as *Lectiones de calculo differentialum* (Schafheitlin 1922) and argued in his introduction that the manuscript was a copy made in 1705 by Bernoulli's nephew Nikolaus (1687–1759), who had been living with him in Groningen. Because the latter part was a near-perfect match to what Bernoulli had published in 1741, he could be quite certain that the first part was essentially the same set of lessons l'Hôpital had used when composing the *Analyse*.

In this volume, we have brought together both l'Hôpital's *Analyse* and Bernoulli's *Lectiones* for the first time, in English translation. We have cross-referenced the texts in order to facilitate a comparison of Bernoulli's original contributions with l'Hôpital's final version. Since the appearance of the *Lectiones*, various authors have characterized the *Analyse* as having essentially been written by Bernoulli. Indeed, Bernoulli himself, in an angry letter to Varignon of February 26, 1707, said that "to speak frankly, Mr. de l'Hôpital had no other part in the production of this book than to have translated into French the material that I gave him, for the most part, in Latin ..." (Bernoulli 1992, p. 215). The truth is much more nuanced. The superstructure of l'Hôpital's first four chapters is certainly due to Bernoulli and many of the details are essentially the same in both texts. However, l'Hôpital added much, in both quantity and quality. For one thing, Bernoulli's *Lectiones* occupied 37 manuscript pages, compared to 70 typeset pages for the first four chapters of the *Analyse*, but the Marquis added much more than mere verbiage to Bernoulli's lesson. He was a very talented pedagogue. He organized his material very well, extracting general propositions where Bernoulli gave examples, and explained matters clearly and in some detail. Furthermore, he frequently included many illustrative examples, gradually increasing in difficulty, generally providing an appropriate level of detail, but always leaving some things for readers to work out for themselves.

To the best of our knowledge, our English translation of the *Lectiones* is the first to be published. Strictly speaking, our translation of the *Analyse* is the second to appear. L'Hôpital published only one other book, a posthumous textbook on conic sections (L'Hôpital 1707), which also went through multiple editions. Edmund Stone (ca. 1700–1768) published an English translation of this more elementary text (Stone 1723). The son of a gardener of the Duke of Argyll and a self-taught mathematician, Stone subsequently published an English translation of L'Hôpital (1696). However, it was a re-writing of l'Hôpital's book in the sense that Stone translated every statement about differentials into the language of Newton's fluxions. This translation made up the first part of his calculus textbook (Stone 1730), which concluded with an original text on the integral calculus.

The L'Hôpital-Bernoulli Correspondence

Bernoulli departed Oucques and returned to Basel in the fall of 1692. He began a correspondence with l'Hôpital in November 1692, which continued until 1702. He completed a doctorate in medicine, for which the thesis was really a work of applied mathematics, in 1694 and ascended to the Chair of Mathematics at the University of Groningen in Holland late in 1695. He returned to Basel in 1705 and took over the Chair of Mathematics at the University of Basel, which had been occupied by his brother Jakob until his death the same year. As father of Daniel Bernoulli (1700–1782) and mentor to Leonhard Euler (1707–1783), his influence on Continental mathematics was even more significant than his impressive list of publications would suggest.

Bernoulli's estate contained 60 letters from the Marquis, two from his wife, the Marquise de l'Hôpital, and copies Bernoulli had made of 25 of his letters to the Marquis. Spiess infers that Bernoulli wrote at least 28 other letters to the Marquis. None of Bernoulli's original letters are known to have survived. In fact, all of l'Hôpital's grandchildren died childless and none of the Marquis' mathematical papers seem to have been preserved. Bernoulli's papers, including the correspondence with the Marquis, went to his son and subsequently to his grandson Johann (III) Bernoulli (1744–1807), who found them a home in the archives of the Royal Swedish Academy of Sciences. The letters remained essentially unknown until they were rediscovered by Gustav Eneström (1852–1923) in 1879. Although

Eneström and others published bits and pieces in the late 19th and early 20th centuries, e.g. Eneström (1894), it was not until much later that a critical edition of the entire correspondence was eventually published (Bernoulli 1955) as part of the *Bernoulli Edition*.

We have included significant excerpts from this correspondence in Chapter 12 of this volume. The editors of the *Bernoulli Edition* numbered the letters in Johann Bernoulli's correspondence and we have used their numbering system to refer to the letters here. Numbers were only assigned to letters that are actually in the archives. For example, l'Hôpital's letters of December 8, 1692 and January 2, 1693 were numbered 6 and 7, respectively. In letter 7, l'Hôpital makes mention of Bernoulli's letter of December 18, and the symbol (6, 1) was assigned to this lost letter. In this volume, when we skip from letter 7 to letter 11, for example, you will know that we have skipped over three extant letters, which can be found in Bernoulli (1955), and that there are possibly some missing letters in between as well. The letters themselves were written in French, with occasional Latin phrases, and Spiess' accompanying modern editorial information in Bernoulli (1955) was written German.

We have included translations of some of the letters in their entirety, for some others we have only included certain portions, and many others have been omitted. The entire correspondence is quite large (about 225 pages in Bernoulli 1955), so we have needed to be selective. Some of the discussion in the correspondence has little or no relevance to the *Analyse* and its composition. The portions that we have chosen to include are generally either relevant to the contents of the book, or they shed light on some aspect of the personal or professional relationship between l'Hôpital and Bernoulli. We have also chosen to include letter 5, from l'Hôpital to Malebranche, which shows that the Marquis was writing about the "Arithmetic of the Infinities" before he met Bernoulli.

Between Bernoulli's first letter (6, 1) of November 1692 and his letter (15, 1) of late September 1693, letters between the two men were written in almost perfect alternation. Then a number of l'Hôpital's letters went unanswered. Bernoulli was apparently unhappy that l'Hôpital had published one of the results that he had given him in his lessons. L'Hôpital's paper,[6] which gave the solution to a problem that had originally been proposed by Florimund de Beaune, was published under the pseudonym "Mr. G***," but Bernoulli would undoubtedly have recognized his own work. Bernoulli resumed writing on January 26, 1694. That letter is lost, but Spiess infers that Bernoulli had requested that the Marquis provide him an honorarium for the services he was rendering him through their correspondence (Bernoulli 1955, p. 201). We do not know how much l'Hôpital paid Bernoulli for his lessons in 1691–92, but it appears to have been enough to support him during his year or so in Paris and Oucques and perhaps for some time afterwards in Basel. In January 1694, Bernoulli was newly engaged to be married and not gainfully employed while he

[6] *Journal des Sçavans*, **34**, p. 401–403.

was completing his doctorate of medicine. In his response, on March 17, 1694, the Marquis wrote

> I will happily to give you a *pension* of three hundred pounds, which will begin the first of January of this present year, ... I promise you to increase this stipend shortly, which I well understand to be very modest, and it will be as soon as my affairs are somewhat straightened out ... I am not so unreasonable as to demand all of your time for this, but I will ask you at intervals to give me a few hours of your time, to work on what I will ask you and also to communicate your discoveries to me, while asking you at the same time not to share any of them with others. I even ask you not to send here to Mr. Varignon, nor to others, any copies of the writings you have left with me; if they should become public I would not be at all pleased [p. 246].

Bernoulli's letter (20, 1) of late March is lost to posterity, but he clearly accepted the terms of what we will refer to here as "The Contract." A few words are in order. First of all, the French word *pension* here does not have the same sense as the modern English "pension," although the terms are related. L'Hôpital is offering an annual honorarium or retainer, but not necessarily in perpetuity. In fact, the duration of the Contract was a little over two years. L'Hôpital sent Bernoulli his last installment of 300 ℔[7] in June 1696, shortly after the *Analyse* appeared and some eight months after Bernoulli had taken up his position at Groningen. Bernoullii received a total of 800 ℔, owing to a slightly larger payment of 200 ℔ in July of 1695. For context, 300 ℔ has been described as half of the annual salary of a professor, but it was in fact only 21% of the salary that Bernoulli would earn at Groningen. Not a great fortune, but undoubtedly welcome to Bernoulli in his circumstances, with no other income and a new family to support (his first child was born in February 1695).

We note that it was only after agreeing to the terms of The Contract that Bernoulli began making copies of his letters to l'Hôpital. The first of these was letter 22, of April 22, 1694. From this point onwards, Bernoulli made copies of his communications to the Marquis, except in a few cases where there was no mathematical content in his letters.

We must also note that l'Hôpital was paying for more than just instruction and advice, he was essentially acquiring the rights to publish Bernoulli's discoveries. This became a bone of contention in early 1695. L'Hôpital sent Bernoulli his solution to a challenge problem that had appeared in *Acta Eruditorum*, asking him to translate it into Latin and send it to the journal. Bernoulli could not resist adding some remarks of his own, that greatly simplified the Marquis' solution. Not only did this have the effect of making the Marquis look bad in print, but it was also a violation of The Contract, to have sent his discovery concerning a matter of their discussion to a third party. In letter 41, Bernoulli reassured the Marquis that he would adhere strictly to The Contract in the future, offering as evidence that he had declined Leibniz' request that he submit material for his prospective book *de Scientiâ infiniti* (p. 272).

---

[7]This is the symbol used for French *livres* or pounds, as transcribed in Bernoulli (1955).

The best known of the results that l'Hôpital "purchased" from Bernoulli by means of The Contract is the rule that became known as L'Hôpital's Rule, after its appearance in the *Analyse*. Bernoulli had evidently worked out the simplest case of L'Hôpital's Rule, the case of the indeterminate form $\frac{0}{0}$ for a finite value of x, in June 1693 or perhaps earlier. In letter 11, l'Hôpital mentioned that he had learned that Varignon had been challenged by Bernoulli to evaluate

$$\frac{\sqrt{2a^3x - x^4} - a\sqrt[3]{a^2x}}{a - \sqrt[4]{ax^3}} = y$$

when $x = a$ (p. 239). We note that they did not speak of limits, but rather they considered y to actually take on the value the we would call the limit as x approaches a. Modern examples involving L'Hôpital's Rule frequently involve transcendental functions, but Leibniz' calculus at this time was only a calculus of algebraic expressions. Bernoulli's $\frac{0}{0}$ Challenge involved an algebraic function, but the problem could not be easily solved by factoring (as with $\frac{x^2-a^2}{x-a}$, for example), making it an excellent vehicle for demonstrating the power of L'Hôpital's Rule. In letter 11, l'Hôpital incorrectly gave the answer as $x = 2a$. Bernoulli apparently told him that he was wrong in (11, 1), but did not reveal his method. Between that point and The Contract, l'Hôpital asked Bernoulli for the solution on three occasions, but to no avail. Once Bernoulli had entered into The Contract, however, he had no choice but to reveal his Rule, which he did in letter 28 (p. 267). L'Hôpital presented the rule in §163 of the *Analyse* (p. 151), followed immediately by its application to Bernoulli's $\frac{0}{0}$ Challenge.

The correspondence remained very active for the next year or so. Bernoulli clarified many issues for l'Hôpital, concerning topics that appeared in the *Analyse* and many others as well. There was also much non-mathematical discussion, including births and deaths and other family matters, and the payment of Bernoulli's stipend. In letter 47 of March 12, 1695, l'Hôpital announced that 'I hope to be able to procure you a chair of mathematics in Holland" (p. 273). He was referring to the position at the University of Groningen that Bernoulli would take up later in the same year. In this letter and subsequently, l'Hôpital gives the impression that he is primarily responsible for securing the position for Bernoulli. It was Christiaan Huygens (1629–1695), in fact, who recommended Bernoulli for the position, but apparently only after asking l'Hôpital for a letter of recommendation. For the next few months, the position at Groningen and the rumored death of Huygens[8] dominated the non-mathematical portion of the correspondence.

In letter 56, of August 22, 1695, l'Hôpital first mentioned the *Analyse* (p.285). What he actually said was that he planned to print his treatise on conic sections very soon, a book on which he had apparently been working for some time, but as it was still unfinished upon his death in 1704, only appeared posthumously (L'Hôpital

[8]Huygens died on July 8, but on June 10, l'Hôpital heard a rumor that he had died. There was some confusion over this in the letters that followed.

1707). "I will add to it," he continued, "a small treatise on the differential calculus, in which I will give you all credit that you deserve." Bernoulli was by this time preoccupied with his move to Groningen and did not respond until after he was settled, on November 8. He made no copy of that letter, but presumably gave his blessing for the proposed treatise. However, some time around the end of the year, l'Hôpital became very ill and after a long silence, his wife wrote to Bernoulli on February 1, 1696, to explain why the Marquis had not written for so long. L'Hôpital returned to Paris on or about the beginning of March and completed the publication process of the *Analyse*. In letter 63 of June 15, 1696, he wrote to Bernoulli that his book would "appear any day now" (p. 288), but that his illness had kept him from including his work on the conic sections, as he had originally planned. As for the *Analyse* itself, he assured Bernoulli that "you will see that I give you the credit you deserve."

Bernoulli's response is included in this volume for at least three reasons. Bernoulli congratulated the Marquis on the publication of the *Analyse*, which he had not yet received himself, but about which he had heard from other correspondents. He also acknowledged the receipt of 300 ₶, which was to be his last payment under the terms of The Contract. Also of interest is that Bernoulli described the Brachistochrone Problem to the Marquis in some detail (p. 289, 296). This challenge problem occupies a place of great significance in the history of mathematics and especially in the development of the Calculus of Variations; see Bernoulli (1988, pp. 329–334) for more on l'Hôpital's solution of the problem and Katz (2009, pp. 586–588) for a general introduction to the Brachistochrone Problem.

It was many months before Bernoulli finally received the *Analyse*. In February 1697, he wrote

> I have finally received a copy of your book and I thank you most humbly for it. You have done me too great an honor in speaking so highly of me in the preface. When I compose something in my turn I will not fail to give you the same in return. You explain things most intelligibly; I also find a beautiful order there and the propositions well organized; everything is admirably well done, and a thousand times better than I could have done. Finally, I desire nothing more than that you had put your name on the cover of the book, which would have given a much greater glory, and lent more Authority to our new method. [p. 220]

L'Hôpital and Bernoulli continued in frequent correspondence until the end of 1697. They wrote less frequently over the next six years and the final letter of their correspondence was written by l'Hôpital on September 15, 1702, in response to a lost letter from Bernoulli four months earlier. L'Hôpital died on February 2, 1704.

In the years immediately following the publication of the *Analyse*, Bernoulli seems to have been content with his largely unacknowledged role in its composition. However, shortly after the Marquis' death, he became unsatisfied with the general mention he received in the preface and began making priority claims on the *Analyse*, especially with regard to L'Hôpital's Rule. Eneström, in analyzing Bernoulli's correspondence (Eneström 1894), found that his dissatisfaction had its first major expression in a letter he wrote to Varignon on July 18, 1705 (Bernoulli 1992, pp. 167–174). By quoting from the Marquis' many letters involving his $\frac{0}{0}$ Challenge, he tried to convince Varignon that the discovery of L'Hôpital's Rule was far beyond the Marquis' ability in the mid-1690s. Bernoulli became increasingly quarrelsome during his long career and his priority claims, made both publicly and in private correspondence, eventually became fodder for historians of mathematics. Montucla, for example, accepted Bernoulli's word and stated plainly that the Marquis did not give him due credit (Montucla 1799, p. 397), whereas Bossut argued vociferously for rejecting Bernoulli's claims of priority (Bossut 1810, pp. 49–52). With the fullness of time, both Bernoulli's *Lectiones* and his correspondence have come to light and modern scholars are now in a position to make their own assessment of the matter. We believe that readers will find that Bernoulli was indeed shortchanged in credit for the *Analyse*, but at the same time, we see the book as a very successful collaboration between a brilliant researcher and a talented expositor.

The Contents of the *Analyse*

The Preface

The *Analyse* opens with a Preface that traces the history of the "Analysis of the Infinitely Small" back to Archimedes and through the 17th century. It has been suggested that Fontenelle actually wrote the Preface, or at least this historical survey, which makes up the largest part of it. Costabel argues convincingly against this thesis (Bernoulli 1992, pp. 13–14). The suggestion that Fontenelle was the author appears to have originated with Fontenelle himself in the 1730s and then became widely known after his death through his eulogy. It seems that Fontenelle did indeed assist l'Hôpital in the process of getting his book published following his illness in early 1696, but Costabel makes a strong case against the possibility that he wrote the Preface.

Chapter 1: Notation and the Rules of Calculus

In Chapter 1, l'Hôpital gives the rules for differential calculus. Modern readers should not expect anything that looks like a calculus book of our time. Leibniz' calculus concerns equations and differentials, not functions and derivatives. Even the graphs and terminology will take a little getting used to. There is no x-y coordinate system. Following Descartes' analytic geometry, there is one axis, which we will refer to as the x-axis, even if the independent variable is not x. The axis is usually horizontal, but sometimes it is drawn vertically. The x-coordinates are usually referred to as *abscissas*, literally meaning that they are "cut off" on the axis at some distance x from the origin. In the place of a y-coordinate, there is an *ordinate* that is "applied" to the axis at the point corresponding to the abscissa x. This is a line segment that is usually perpendicular to the axis, although *oblique* ordinates were occasionally arranged at a different angle to the axis. Even the familiar parabola looks unfamiliar in the late 17th century. It is written as $ax = yy$.

As a typographical convention, squares were usually written by repeating the letter representing the variable, but higher powers were written with superscripts. Thus, the *second cubical parabola* was usually written $axx = y^3$.

Leibniz' calculus works by translating geometric problems into the language of algebraic expressions, performing operations on those expressions (taking differentials and integrals) and extracting solutions from the results. This is still largely true of calculus as it is practiced now, if one understands by algebraic expressions those functions that are composed of algebraic expressions involving both variables and a number of transcendental functions. Geometry has a very long history, having already reached a high degree of sophistication by the time of Euclid, Archimedes and Apollonius, whereas analytic geometry was less than half a century old when Leibniz discovered his calculus. Not surprisingly, then, the *Analyse* describes a calculus that has a much more geometrical flavor than the calculus of later centuries. For example, the parabola could be considered as a relationship between two-dimensional figures: given an abscissa x and a *parameter* a (this name for the constant goes back to Apollonius), one seeks a square with the same area as the rectangle with sides of length a and x. The equation $ax = yy$ expresses a relation between "planar numbers" representing two-dimensional areas, so it would not have been written as $x = cyy$, which would seem to violate the rules of geometry by comparing a one-dimensional length to a three-dimensional volume. Descartes had taught that it was not necessary to write equations in such a *homogeneous* form, involving only terms of the same dimension, but even in the 1690s l'Hôpital, and especially Bernoulli, usually did write them in this manner.

Leibniz' differential calculus tells us how to find the relations among infinitely small increments dx, dy, etc., among the variables x, y, etc., in an equation. L'Hôpital gives the definition on p. 2: "The infinitely small portion by which a variable quantity continually increases or decreases is called the *Differential*." This cryptic definition might be made somewhat clearer by considering his Figure 1. AMB is a curve, presumably a parabola, with axis AC and origin A. AD is not the y-axis, but the parameter for the parabola. The abscissa is AP, denoted x and the

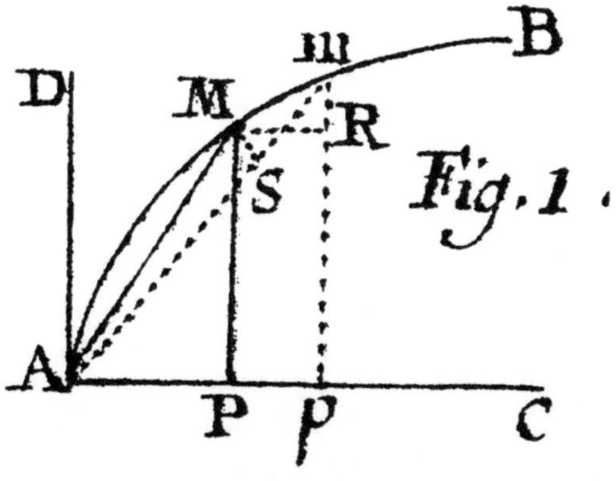

Fig. 1 Definition of the Differential

ordinate is PM, denoted y. We imagine that p is infinitely close to P and pm is the ordinate that is infinitely close to PM. Then Pp is the differential dx and Rm, where $MR \perp pm$, is the differential dy.

At this stage, we still don't know what "infinitely small" means, but perhaps Postulate I can clarify this. Bernoulli's first postulate is "Quantities that decrease or increase by an infinitely small quantity neither decrease nor increase." L'Hôpital is a little more comprehensible: "We suppose that two quantities that differ by an infinitely small quantity may be used interchangeably, or (what amounts to the same thing) that a quantity which is increased or decreased by another quantity that is infinitely smaller than it is, may be considered as remaining the same." Although this still does not give us much to go on, things become clearer when they begin to perform calculations. What one does is to replace the variables x, y, etc., in an equation with $x + dx$, $y + dy$, etc., and then cancel the original equation, much as finite differences are calculated. Then, because of Postulate I, terms involving two or more differentials multiplied together may be omitted, giving a relation among the variables and their differentials, such as $a\,dx = 2y\,dy$ in the case of the parabola.

L'Hôpital then gives the rules for the differential calculus:

- Constant Rule: The differential of a constant is 0.
- Sum and Difference Rule: the differential of $x \pm y$ is $dx \pm dy$.
- Product Rule: The differential of xy is $y\,dx + x\,dy$.
- Quotient Rule: The differential of $\frac{x}{y}$ is $\frac{y\,dx - x\,dy}{yy}$
- Power Rule: The differential of x^n is nx^{n-1} for any rational n.

We note that neither l'Hôpital nor Bernoulli used these names for their rules; we use them here because modern readers are familiar with these names and this will help them to see the correspondence between modern calculus and Leibniz' calculus. We also note that this is the ordering of the rules given by l'Hôpital. We find his exposition of the rules of calculus to be more elegant and satisfying that Bernoulli's and this is an example of a place where l'Hôpital has added value to Bernoulli's *Lectiones*. Finally, we note that there is no need for the Chain Rule in this calculus, because it is not restricted to functions. Variables may be freely introduced and when they are, the above rules are all that is needed. For example, if $y = \sqrt{ax - x^2}$, then we may let $u = ax - x^2$. Then the above rules give $dy = \frac{1}{2}u^{-\frac{1}{2}}\,du$ and $du = a\,dx - 2x\,dx$, so

$$dy = \frac{a\,dx - 2x\,dx}{2\sqrt{ax - x^2}}.$$

With a little thought, modern readers can see that whenever $y = f(x)$ for some algebraic function f, these rules will give $dy = f'(x)\,dy$, which explains the origin of our notation $\frac{dy}{dx}$ for the derivative $f'(x)$. We stress that functions and derivatives are nowhere to be found in the *Analyse* or the *Lectiones*. However, the calculations on equations that take place in their pages are reminiscent of calculations in modern Related Rates and Implicit Differentiation problems.

L'Hôpital and Bernoulli both made extensive use of proportions, another legacy of classical Greek mathematics. The proportional relation $a : b :: x : y$ means "as a is to b, so x is to y." To modern readers, this is just the relation

$$\frac{a}{b} = \frac{x}{y},$$

which may be solved to give $y = \frac{b}{a}x$, but to the ancient Greeks, there was a distinction made between the number $\frac{a}{b}$ and the proportion $a : b$. There was a large set of rules for manipulating proportions, which l'Hôpital's readers would have learned from Book V of Euclid's *Elements*. We note that Bernoulli wrote proportional relations in the form $a \cdot b :: x \cdot y$ and l'Hôpital wrote them in the form $a \,.\, b :: x \,.\, y$; we will write them in the form $a : b :: x : y$ in both treatises.

Chapter 2: Finding Tangents

Bernoulli's Postulate II is "Any Curved line consists of infinitely many straight lines, each of which is infinitely small." L'Hôpital's is much the same, explained a little more fully and with illustrations. This postulate is used for the first time to find tangents. This was not a question of calculating a slope, but rather the geometric question of how to *draw* the tangent at M. This reduces to the question of finding the point T where the tangent line intersects the axis and joining it to M. The line segment TP is called the *subtangent* so the problem is reduced to finding $t = TP$. By Postulate II, when p is infinitely close to P, then Pp is a straight line and by similar triangles, we have $mR : RM :: MP : PT$ (see Fig. 3). Bernoulli expresses this in the form $dy : dx :: y : t$, a relation that he repeats many times in the *Lectiones*. In his Proposition I on p. 11, L'Hôpital gives it in the form

$$mR(dy) : RM(dx) :: MP(y) : PT = \frac{y\,dx}{dy}.$$

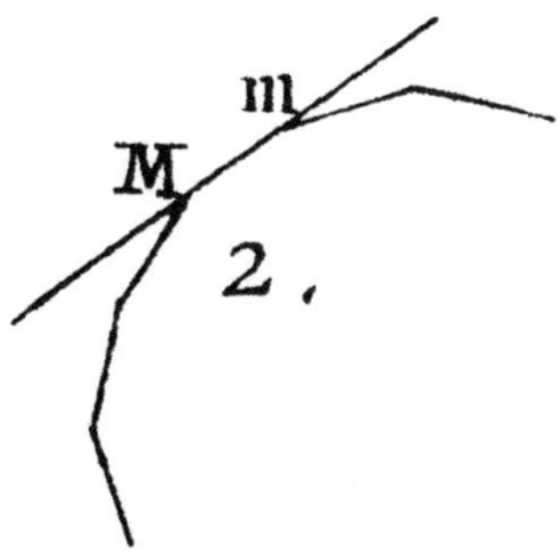

Fig. 2 Postulate II

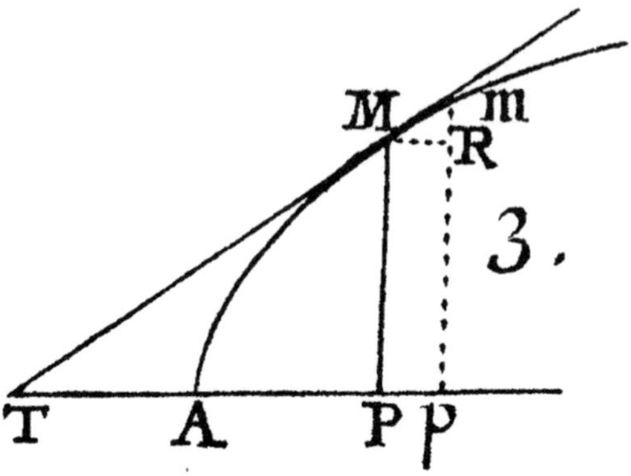

Fig. 3 The Differential Triangle

This is typical of the way that l'Hôpital uses proportional relations: The relation $a : b :: c : d$ can always be used to solve for d when a, b, and c are known, so l'Hôpital puts the corresponding values of the first three terms in parentheses and solves for the fourth term at the end of the expression. In this case, he has found a general formula for the length of the subtangent that can be used freely throughout the remainder of the *Analyse*. Modern readers will note that in the case where $y = f(x)$, this is equivalent to saying the length of the subtangent at (x_0, y_0) is $\frac{y_0}{f'(x_0)}$.

In both the *Analyse* and the *Lectiones*, what follows is the calculation of tangents for a large collection of curves that were known to mathematicians in the 1690s. The collection is especially extensive in the *Analyse* and is sort of a catalog of the curves that were known to mathematicians of that time. Some of these curves were already known to the ancient Greeks. L'Hôpital and Bernoulli presented almost all of them geometrically, rather than algebraically. Admittedly, in many cases, algebraic expressions are not possible without the use of transcendental functions.

Conic Sections

The conic sections – Ellipse, Hyperbola, and Parabola – were known to the ancient Greeks and were still the object of intense study in the 17th century. In fact, l'Hôpital originally conceived of the *Analyse* as a chapter to be included as a part of a larger treatise on conic sections; see letter 56 on p. 285. L'Hôpital's treatise on conic sections did not appear until after his death (L'Hôpital 1707).

We assume that the reader is familiar with the conic sections as they are now treated in analytic geometry. Of course, they were defined by the ancient Greeks as sections of a double-napped cone, but this treatment is not given in the *Analyse*. Instead, l'Hôpital gives them through equations in x and y that may be unfamiliar to modern readers. The first mention of the ellipse is in §12 (see p. 13), where it is given by the equation

$$\frac{ayy}{b} = ax - xx. \tag{1}$$

Rearranging terms and completing the square, this may be written as

$$\frac{\left(x - \frac{a}{2}\right)^2}{\left(\frac{a}{2}\right)^2} + \frac{y^2}{\left(\frac{\sqrt{ab}}{2}\right)^2} = 1,$$

which modern readers will recognize as an ellipse with center $\left(\frac{a}{2}, 0\right)$, horizontal axis a and vertical axis $\sqrt{ab}$. Thus, the origin is at the left end of the horizontal axis, rather than the center of the ellipse. When $b = a$, this is the equation of a circle of diameter a, with the origin on the circumference of the circle. Equation (1) reduces in this case to $yy = ax - xx$, i.e. $y^2 = x(a - x)$, an equation that is used freely throughout the *Analyse*.

Specifying an ellipse with a given center requires two parameters: in the modern treatment, those parameters are the horizontal and vertical semi-axes, whereas in l'Hôpital's treatment, the horizontal axis a is clearly a natural choice for one parameter. The other parameter, b, is also a natural choice for someone steeped in the classical theory of proportions. An ellipse is a curve in which the ratio of the rectangle on x and $a - x$ to the square on y is always the same, or $x(a - x) : y^2 :: a : b$, for some magnitude b. Because the magnitude a is the length of a line segment, the classical theory of proportions demands that b be a magnitude of the same kind. Thus in Figure 1.4 on page 14, a and b are the lengths of the segments *AB* and *AD*. As in Figure 1 on page xvii for the case of the parabola, *AD* is not the y-axis, but simply a representation of the second parameter of the ellipse.

Equation (1) generalizes naturally to higher order: in the last paragraph of §12, l'Hôpital gives

$$\frac{ay^{m+n}}{b} = x^m(a - x)^n \tag{2}$$

as "the general equation of all ellipses up to infinity." L'Hôpital writes the term $(a-x)^n$ as $\overline{a - x}^n$, because the overline is his way of grouping terms. In analogy to the classical case, the quantities $x^m(a - x)^n$ and y^{m+n} here are always in the same proportion as a is to b. In §13, l'Hôpital gives

$$\frac{ay^{m+n}}{b} = x^m(a + x)^n \tag{3}$$

as the generalized equation of hyperbolas. Similarly to the ellipse, the quantities $x^m(a + x)^n$ and y^{m+n} here are always in the same proportion as a is to b. In the classical case, where $m = n = 1$, we may rearrange terms and complete the square to obtain:

$$\frac{\left(x + \frac{a}{2}\right)^2}{\left(\frac{a}{2}\right)^2} - \frac{y^2}{\left(\frac{\sqrt{ab}}{2}\right)^2} = 1,$$

which is a hyperbola with center at $\left(-\frac{a}{2}, 0\right)$, so that the origin is once again a point on the curve, in this case the vertex of the right-hand branch. In analogy to the ellipse, a represents the major or transverse axis of the hyperbola, while the conjugate axis (called the conjugate diameter in the last paragraph of §13) is the quantity $\sqrt{ab}$. The major axis of a hyperbola is the distance between its vertices. If one erects a perpendicular to the major axis at a vertex, then the conjugate axis is the segment of that line contained between the points where it intersects the asymptotes of the hyperbola. In Figure 6 on page 15, only the upper half of the right-hand branch of the hyperbola is given, but nevertheless the point B is the vertex of the left-hand branch and a is the length of the line segment AB. The second parameter, b, is the length of the line segment AD, not to be confused with AE the semi-conjugate axis.

L'Hôpital treats parabolas in §11 on page 13. The classical case is $ax = yy$, which may strike a modern reader as strange, since it does not express y as a function of x, but is actually quite natural in that the axis of symmetry of the parabola is the same as the x-axis. This parabola has only one parameter, a line segment of length a so that the square on y is always equal to the rectangle on a and x.

In the third part of §11, l'Hôpital gives the equation of the generalized parabola as $y^m = x$ when m denotes a positive rational number. We note that the parameter is suppressed in this expression, but in the particular example of $m = \frac{3}{2}$, l'Hôpital writes the equation as $y^3 = axx$. Bernoulli writes this *semi-cubical* or second cubic parabola the same way in his Problem I on page 193, and gives the first cubical parabola as $aax = y^3$. Similarly, Bernoulli give the *biquadratic* parabolas as $a^3x = y^4$, $aaxx = y^4$ and $ax^3 = y^4$. In all of these equations, the parameter has the appropriate order so that the equations are all *homogenous*; that is, they involve terms all of the same total degree.

In §11 l'Hôpital also considers the case where $y^m = x$ and m denotes a negative rational number. These are the hyperbolas "between the asymptotes"; i.e. hyperbolas where the asymptotes are the axis and its perpendicular. The cases $aa = xy$ and $a^3 = xyy$ are both mentioned in this article.

Construction of Conic Sections

In Proposition IV on page 19, l'Hôpital gives a general construction of a new curve from two given curves that intersect. This is a generalization of a method for constructing the conic sections as the geometric means of the ordinates of two straight lines. In Figure 8, the given curves are AQC and BCN, which intersect at the point C. The ordinates PM of the new curve MC, denoted by y, are defined through

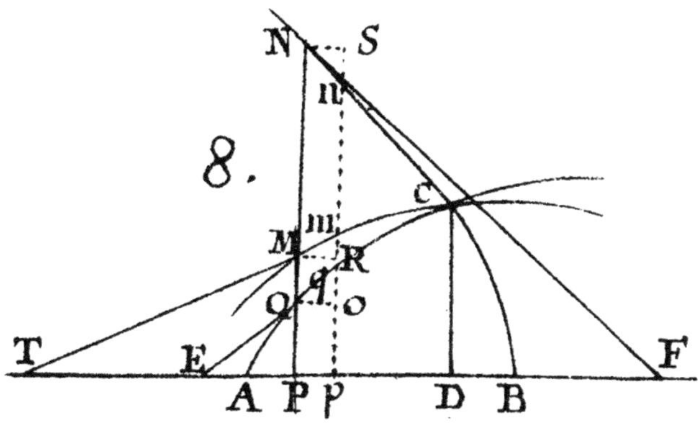

Fig. 8 Conic Sections and Related Examples

an equation relating them to the ordinates PQ of AQC and PN of BCN, denoted, respectively, by x and z. In the particular case where AQC and BCN are straight lines and the equation is $yy = xz$, then the new curve MC is a conic section, depending on the relationship between the lines and their point of intersection. L'Hôpital presents a generalization of this construction in §21. We will use modern ideas and notation to explain the construction.

We take the point A as the origin of the abscissas, which we denote by u, and determine x and z as linear functions of u. The first case, which is more or less illustrated in Figure 8, is the case where the ordinate DC of the point C meets the axis between A and B. Let the coordinates of C be denoted (d, c) and the distance AB be a, then we have

$$x = \frac{c}{d}u \quad \text{and} \quad z = \frac{c}{a-d}(a-u).$$

If $y^{m+n} = x^m z^n$, we then have

$$\frac{d^m(a-d)^n}{c^{m+n}}y^{m+n} = u^m(a-u)^n,$$

which has the form of equation (2) with horizontal axis a and parameter

$$b = \frac{ac^{m+n}}{d^m(a-d)^n}.$$

Thus, the curve MC is a generalized ellipse and in the classical case of $m = n = 1$, the ordinate PM can be constructed with a compass and straightedge, because $y = \sqrt{xz}$ is the geometric mean of the ordinates PQ and PN.

In the second case considered by l'Hôpital, the ordinate DC meets the axis outside of the line segment AB. Without loss of generality, A may still be considered as the origin and the point B is to its left. In this case we have the generalized

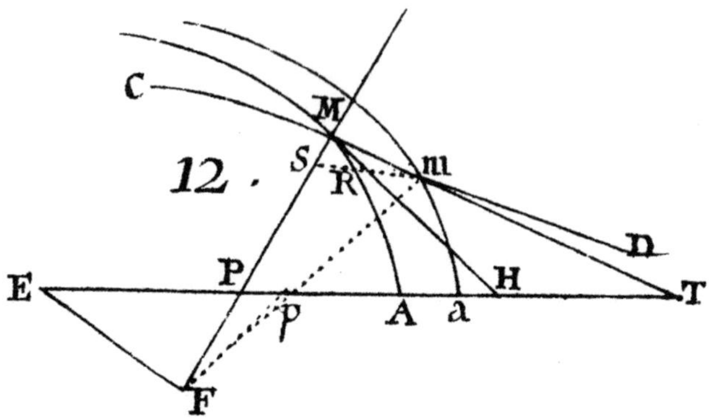

Fig. 12 Hyperbola and Related Examples

hyperbola, because

$$x = \frac{c}{d}u \quad \text{and} \quad z = \frac{c}{a+d}(a+u),$$

so that $y^{m+n} = x^m z^n$ yields

$$\frac{d^m(a+d)^n}{c^{m+n}} y^{m+n} = u^m(a+u)^n,$$

which has the form of equation (3) with major axis a and parameter

$$b = \frac{ac^{m+n}}{d^m(a+d)^n}.$$

Finally, we consider the case where one of the straight lines is parallel to the axis. Without loss of generality, the line BC is replaced by the horizontal line passing through the point C. Thus $x = \frac{c}{d}u$ and $z = c$, so that $y^{m+n} = a^n x^m$, which is a generalized parabola with parameter

$$a = \frac{c^{1+\frac{m}{n}}}{d^{\frac{m}{n}}}.$$

Alternate Construction of the Hyperbola

In Proposition VII on page 23, l'Hôpital gives a general method for constructing a new curve by sliding a given curve *MRA* along the axis. In figure 12, the line *FP* rotates about a fixed point *F* below the axis. As the line *FP* rotates, the curve *MRA* moves rigidly in such a way as to keep the length of the segment *PA* fixed. The new curve *CMD* is then the locus of all points *M* where the line *FP* meets the curve

MRA. In the particular case where the curve *MRA* is actually the straight line *MH*, then *CMD* is a hyperbola. L'Hôpital mentions that his hyperbola has the axis *ET* as one of its asymptotes, but gives no further details.

We give here a derivation of the equation of this hyperbola using modern methods and notation. We suppose *ET* to be the x-axis and the perpendicular from the fixed point F to the axis to be the y-axis. Let b denote the distance from F to *ET*, so that F has coordinates $(0, -b)$. If we let $(z, 0)$ be the coordinates of the point P, then H has coordinates $(0, z + a)$ for a fixed a. We suppose that the line *MH* has slope v, where v may be positive, negative, or infinite (i.e., the line *MH* may be vertical). There is no need to consider the case $v = 0$, i.e. the case where *MH* is horizontal, because in this case the curve *CMD* coincides with the line *MH*.

Using this notation, the lines *MH* and *FP* have equations

$$y = vx - v(z + a) \quad \text{and} \quad y = \frac{b}{z}x - b,$$

respectively. The point M is found by solving these equations simultaneously, giving

$$x = \frac{v(z + a) - b}{vz - b}z \quad \text{and} \quad y = \frac{abv}{vz - b}.$$

We eliminate the parameter z by solving the second equation to get

$$z = \frac{ab}{y} + \frac{b}{v}, \quad \text{so that } x = (y + b)\left(\frac{a}{y} + \frac{1}{v}\right).$$

If the line *MH* is vertical, so that $\frac{1}{v} = 0$, we have $y = \frac{ab}{x - a}$, which gives a hyperbola with asymptotes $y = 0$ and $x = a$. Otherwise, we have

$$x = \frac{y^2 + (av + b)y + abv}{vy},$$

which is a second degree equation in x and y, whose general form is

$$-vxy + y^2 + (av + b)y + abv = 0.$$

The discriminant of this equation is $v^2 > 0$, so the curve is a hyperbola. It is a rotated, skewed hyperbola whose equation can be written as

$$-\frac{v^2}{4}x^2 + \left(-\frac{v}{2}x + \left(y + \frac{av + b}{2}\right)\right)^2 = -\frac{v(av + b)}{2}x + \frac{(av - b)^2}{4}.$$

From this form, we can determine that the asymptotes are $y = 0$ and $y = vx$.

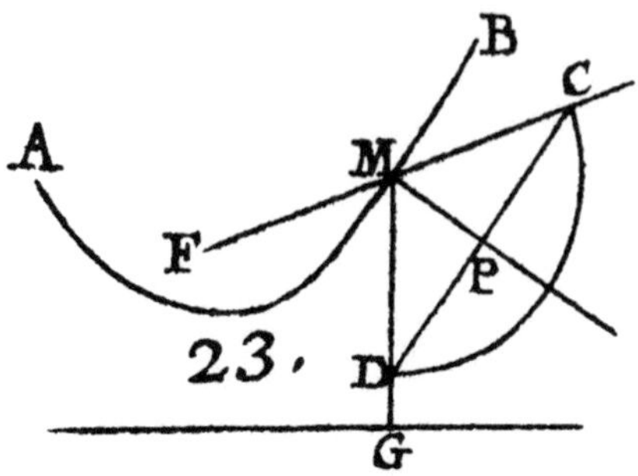

Fig. 23 Conic Sections Defined by Focus and Directrix

Focus-Directrix Construction of the Conic Sections

In Proposition X on page 29, l'Hôpital considers curves that are defined by means of two or more foci. This is a generalization of the conic section, because ellipses and hyperbolas may be defined by means of two foci: the ellipse is the locus of all points such that the sum of their distances to two given foci is constant, whereas for the hyperbola, it is the difference of the distances that is constant. Furthermore, in this portion of the text that is more than six pages long, l'Hôpital extends to the notion of focus to include both the case of a straight line and that of a curved line. When the focus is a straight line, then we understand the distance between a point and the focus to mean the perpendicular distance from the point to the line. Such a focus is usually called a *directrix*. When the focus is a curved line, then we understand the distance to mean the length of the portion of the tangent from the focal curve to the point on the curve being constructed; see Figure 2.20 on page 33.

In §34 on page 35, l'Hôpital considers curves defined by means of two foci, one being a point and one being a directrix. Modern readers are probably familiar with the focus-directrix definition of the parabola, but l'Hôpital uses the focus and directrix to construct all of the conic sections. In Figure 23, F is the focus, and the line marked simply as G is the directrix. For a point M on the curve being defined, we require that $MF : MG :: a : b$ for two positive constants a and b. L'Hôpital asserts that the curve is a parabola if $a = b$, a hyperbola if $a > b$, and an ellipse if $a < b$. To verify this, we set up a coordinate system as follows. We drop a perpendicular from F to the directrix G and make that line the y-axis. The vertex V of the curve is the point where the y-axis intersects the curve. We take V to be the origin and draw the x-axis parallel to G. Because V is one of the points M on the curve, we have $VF : VG :: a : b$, so without loss of generality, we may take the coordinates of F and G to be $(0, a)$ and $(0, -b)$, respectively. The proportional relation $MF : MG :: a : b$ then becomes

$$\frac{\sqrt{x^2 + (y - a)^2}}{y + b} = \frac{a}{b},$$

so that

$$b^2\left(x^2+(y-a)^2\right)=a^2(y+b)^2.$$

Simplifying, we have

$$b^2x^2+(b^2-a^2)y^2=2ab(a+b)y.$$

If $a = b$, this reduces to $x^2 = 4ay$, the parabola with focal distance a. Otherwise, we have

$$\frac{x^2}{kc^2}+\frac{(y-c)^2}{c^2}=1,$$

where

$$k=1-\frac{a^2}{b^2}\quad\text{and}\quad c=\frac{ab}{b-a}.$$

This is indeed an ellipse if $a < b$, so that $k > 0$, whereas if $a > b$, the coefficient of x^2 is negative and the curve is hyperbola.

Cycloid

The cycloid is given parametrically in modern textbooks by

$$\begin{aligned} x &= b\theta - a\sin\theta \\ y &= b - a\cos\theta, \end{aligned} \tag{4}$$

where a and b are positive constants. When $a = b$, the cycloid is called *simple*. The curve can be generated by rolling a circle of radius b along the x-axis and tracing out the path of the point on its circumference that is at the origin when $\theta = 0$. If $b > a$, the cycloid is called *curtate* and the curve is traced out by an interior point of the circle that is in the interval $(0, b)$ on the y-axis when $\theta = 0$. The cycloid is *prolate* if $b < a$, in which case the tracing point is outside the circle, on the negative y-axis, when $\theta = 0$.

The cycloid is an example of a *roulette*, a very general construction of a curve that is produced by rolling one curve, called the *mobile* curve, along a different, fixed curve. In the case of the cycloid, the mobile curve is a circle and the fixed curve is a line. If one rolls a circle around inside another circle, a point on the circumference of the mobile circle traces out a *hypocycloid*. If the mobile circle rolls around the outside of the fixed circle, an *epicycloid* is traced out.

In later chapters of the *Analyse*, l'Hôpital considers cycloids, hypocycloids, and epicycloids as roulettes, which he usually calls "half-roulettes" because he is

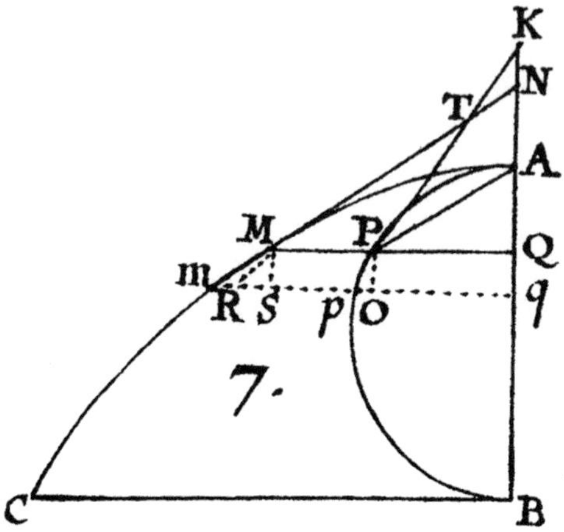

Fig. 7 The Cycloid

frequently only interested in a half-turn of the mobile circle, which in the case of the parametric definition of the cycloid means $0 \leq \theta \leq \pi$.

However, in Chapter 2, l'Hôpital does not define the cycloid as a roulette. Rather, in Proposition II on p. 17 he takes the axis to be a curved line *APB* (see fig. 7) and the abscissas s to be given by the arc length on this curve, originating at A. The ordinates t, which are perpendicular to the line segment *AB*, are applied to the curve *APB*. When *APB* is a semi-circle and the ordinates are determined by the equation $s = \frac{at}{b}$ (l'Hôpital uses x and y, respectively, where we use s and t), then the curve *AMC* a cycloid, simple when $b = a$, curtate when $b > a$ and otherwise prolate. To see that these two definitions are equivalent, first consider the simple cycloid, the case $a = b$. Let *CB* be the x-axis, using ordinary Cartesian coordinates, and let the origin be at the point C. The diameter *AB* has length b and is on the vertical line $x = \pi b$. The parametric equations (4) can then be derived by letting θ parameterize the angle, in radians, along the semi-circle *APB*, with $\theta = 0$ corresponding to the point B and $\theta = \pi$ to the point A. For the other cases, the semicircle has radius a, but the center of the diameter *AB* is still at the point $(\pi b, b)$. In this case, the segment *CB* is parallel to the x-axis, lying above the in the curtate case, or below in the prolate case.

Bernoulli also treated the cycloid (p. 198). However, he only considered the case of the simple cycloid and did not use curvilinear coordinates as l'Hôpital did. Given l'Hôpital's reported interest in the cycloid at the age of 15, it is not surprising that he devoted relatively more of his work to this curve than Bernoulli did.

The cycloid has many fascinating properties, including the fact that it gives the solution to the Brachistochrone Problem. It was also studied by Huygens in his work on the pendulum clock (Huygens 1673). For more on the cycloid, see Lockwood (1971, pp. 80–89). For roulettes, see Lockwood (1971, pp. 138–151).

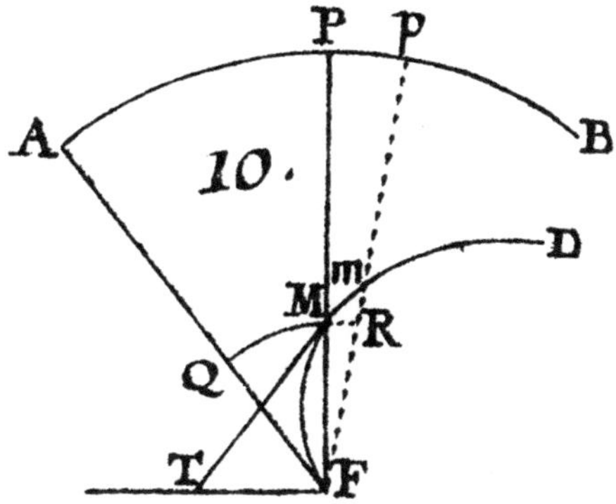

Fig. 10 The Spiral of Archimedes

Spiral of Archimedes

In modern textbooks the Spiral of Archimedes (or Archimedean Spiral) is usually given in polar coordinates as the function $r = \theta_0 + k\theta$, for $r \geq 0$, with constants θ_0 and k. Thus, the distance from the origin to a point on this spiral is proportional to its angle, measured as the offset from the initial angle θ_0. L'Hôpital first mentions this spiral in §23 on page 21, as an example illustrating his Proposition V.

There are no polar coordinates in the *Analyse*, although the variable y, representing the length of the line segment *FM* (see fig. 10) is a polar ordinate, roughly equivalent to the variable r in polar coordinates. Such polar ordinates are used extensively in Chapter 5. In place of an angular coordinate θ, l'Hôpital uses a circular arc *APB* as a reference curve; it is essentially a curved axis, as the semicircle *APB* was in the case of the cycloid. The arc *APB* is a portion of the circle of radius a and center at F, possibly a complete circle. To define the Spiral of Archimedes *FMD*, we let x be the length of the arc *AP* and b the length of the arc *AB*. Then the ordinate y is given by the proportion $b : x :: a : y$. In the *Lectiones* (see page 204), Bernoulli describes this proportional relation in words: "That curve is called the Spiral of Archimedes, which is described from a point, which is moved from the center to the circumference of a circle, the radius rotating uniformly in equal durations of time as the point is moved from the center to the circumference."

If we let θ and α be the radian measures of the arcs *AP* and *AB*, respectively, then we have $x = a\theta$ and $b = a\alpha$, so that

$$y = \frac{a}{b}x = \frac{a}{\alpha}\theta.$$

Because l'Hôpital oriented the arc *APB* clockwise, this is essentially equivalent to $r = \theta_0 + k\theta$, where $k = -\frac{a}{\alpha}$ and θ_0 is the angle between the radius *FA* and the positive x-axis. We note that whereas in the modern treatment, the Spiral of Archimedes is an unbounded curve with arbitrarily large radii, this construction in the *Analyse* describes a bounded portion, which terminates at the point B on the arc *APB*.

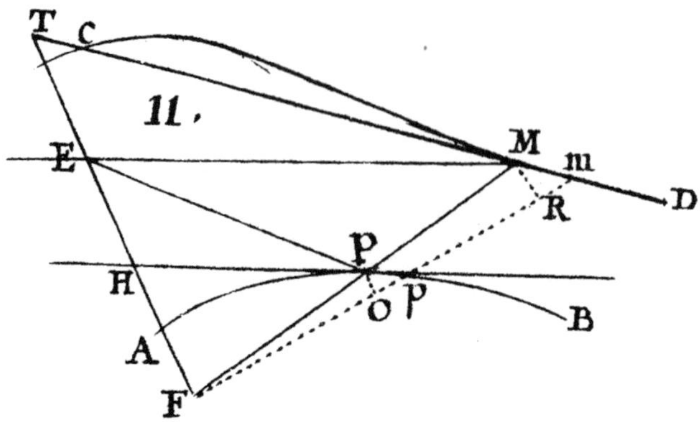

Fig. 11 The Conchoid of Nicomedes

Conchoid of Nicomedes

The word *conchoid* derives from the Ancient Greek word for mussel. The Conchoid of Nicomedes is the curve *CMD* in figure 11 of the *Analyse*. It is defined by means of the straight line *HP*, sometimes called the directrix, and a point F not on that line, called the *pole*. For every point P on the directrix, the line *FP* is drawn and the point M is marked off at a fixed distance $a > 0$ from the point P. The Conchoid of Nicomedes is the locus of all such points M. In the modern definition of this curve, there are two points corresponding to each point P on the directrix, one on the opposite side from F and one on the same side. L'Hôpital considers only the curve that consists of the points M on the opposite side from F. The branch that l'Hôpital does not consider exhibits interesting properties: it has a cusp at the origin when $a = b$ and a double point when $a > b$, but when $a < b$ it has the same general shape as the as curve *CMD*.

A modern equation for the Conchoid of Nicomedes is most conveniently given in polar coordinates. If we choose the coordinate system so that the origin is at the pole F and the directrix is the horizontal line $y = b$, then the branches are given simultaneously by the polar equation

$$r = b \csc \theta \pm a; \quad 0 < \theta < \pi,$$

where l'Hôpital's curve is the branch corresponding to $+a$ and the second branch is given by $-a$. This can be transformed into Cartesian coordinates as $a^2y^2 = (b - y)^2(x^2 + y^2)$. The line $y = b$ is an asymptote and there is a maximum on the illustrated branch at the point $(0, b+a)$. The Conchoid of Nicomedes was of interest to the Ancient Greeks because it could be used to trisect an angle; see Lockwood

(1971, p. 127) for details, although there is no mention of this in the *Analyse*. As an application of higher differentials, l'Hôpital determines the inflection points of this curve in Chapter 4.

Cissoid of Diocles

The word *cissoid* derives from the Ancient Greek word for ivy. In modern sources, the Cissoid of Diocles is defined by considering a circle, a tangent line at some point B on its circumference and the pole F diametrically opposite to B. In figure 14, only the semi-circle *FNB* is given, with the tangent line *Bb*. From the pole F, a straight line *FN* is drawn through the circle at N to the tangent line, and the point M is taken so that $FM = Nb$. The Cissoid of Diocles is then the curve *FMA*, which is the locus of all such points M. Because l'Hôpital only considers the semi-circle with diameter *FB*, he only gets the upper half of the curve that is usually called the Cissoid of Diocles. L'Hôpital gives a different but equivalent construction on page 25, dropping the perpendicular *EN* from the point N to the diameter, considering the abscissa *FL* to be equal to *BE* and determining the ordinate *LM* so that $FL : LM :: FE : EN$. Taking the diameter *FB* as the axis, with the origin at F and the positive direction being from F to B, we denote the length of *FB* as $2a$. With the coordinate system defined in this way, and using the equation $y^2 = 2ax - xx$ for the circle, l'Hôpital derives the equation $y^2(2a - x) = x^3$, which appears in the implicit differentiation section of many modern calculus books.

The Cissoid of Diocles may be used to solve the classical problem of the duplication of the cube. In particular, the curve may be used to find the first of two mean proportionals between a and an arbitrary magnitude b, that is $\sqrt[3]{a^2b}$, which is the cube root of b when a has unit length. For more on this, see Lockwood (1971, p. 131).

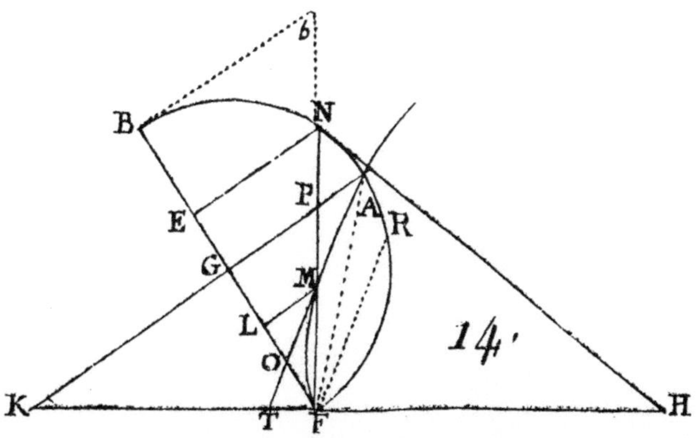

Fig. 14 The Cissoid of Diocles

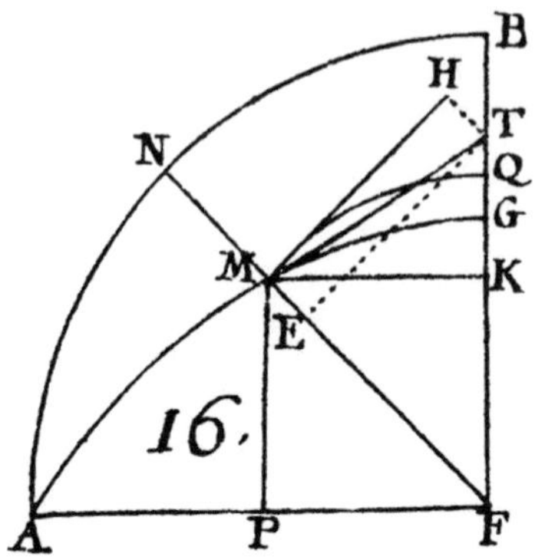

Fig. 16 The Quadratrix or Dinostratus

The Quadratrix of Dinostratus

L'Hôpital calls this curve the Quadratrix of Dinostratus, but it is sometimes called the Quadratrix of Hippias. Both Dinostratus and Hippias were Ancient Greek mathematicians. Apparently Hippias first proposed the curve for the purposes angle trisection, but it was Dinostratus who proved that it could be used to square the circle. For more on Hippias, Dionostratus, and the Quadratrix, see Burton (2007, pp. 132–136).

The Quadratrix of Dinostratus is the curve *AMG*, illustrated in both Figure 16 and Figure 17 of the *Analyse*, in which the curve *ANB* is a quarter circle of radius a. In §30 on page 27, l'Hôpital describes the curve in the following way: We draw any radius *FN* in the quarter-circle and locate the point *P* on the radius *FA* so that as the arc *AN* is to the line segment *AP*, so the quarter-circumference *ANB* (which is denoted b) is to the radius *AF*. We erect a perpendicular at *P* and the point *M* is where this meets the radius *FN*. The Quadratrix of Dinostratus is the locus of all

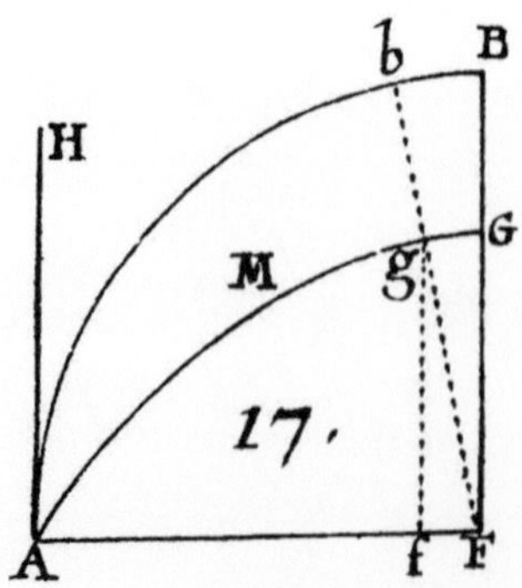

Fig. 17 The Quadratrix or Dinostratus

such points M. If we denote the length of *FP* by u and the measure of the angle *AFN* by ψ, then the proportional relation gives

$$\frac{a\psi}{a-u} = \frac{\pi}{2} \quad \text{so that} \quad \psi = \frac{\pi}{2} - \frac{\pi u}{2a}.$$

If we denote the complimentary angle *BFN* by θ and the length of *PM* by v, then we have

$$\theta = \frac{\pi u}{2a} \quad \text{and} \quad v = u \cot\theta = u \cot\left(\frac{\pi u}{2a}\right).$$

This relationship between u and v explains why modern references use the function

$$f(x) = x \cot\left(\frac{\pi x}{2a}\right)$$

to define the Quadratrix of Dinostratus.

In §31, l'Hôpital determines the length of the segment *FG* to be $\frac{aa}{b}$, which is $\frac{2a}{\pi}$. A modern reader would apply l'Hôpital's rule to $f(x)$, which has the indeterminate form $0 \cdot \infty$ when $x = 0$. Instead, l'Hôpital uses an elementary argument involving the radius *Fb* in figure 17, which is infinitely close to *FB*. L'Hôpital's Rule is not mentioned until §163 in Chapter 9, but even if that rule had been available to him at this point in the text, l'Hôpital still lacked a calculus of trigonometric functions, which only developed over the course of the next century.

To square the quarter circle *ABF* one needs to construct a square of the same area, namely $\frac{\pi a^2}{4}$. To do this, first erect perpendiculars to *FB* at G and B, see figure 17. On the first of these, mark off C so that GC has length a. Now join FC and let D be the point where this line meets the perpendicular at B. Then by similar triangles, *BD* has length $\frac{\pi a}{2}$, so that if we bisect *BD* at L, then the rectangle on *BL* and *BF* has the required area. Finally, the construction of a square with the same area as a rectangle is a standard procedure, described in Proposition 14 of Book II of Euclid's *Elements*.

The Logarithmic Spiral

The Logarithmic Spiral is also called the Equiangular Spiral. It was studied extensively by Johann Bernoulli, who had it engraved on his tombstone. It has many fascinating properties, including the fact that at any point on the curve, the line joining that point to the center always makes the same angle with the tangent at that point. In §42 on page 40, l'Hôpital considers the curve *LM* that is defined with reference to a circle *BN* with center A and radius c, and a hyperbola *FQ*; see Figure 27. When he requires that the circular sector *ANB* have an area equal to half

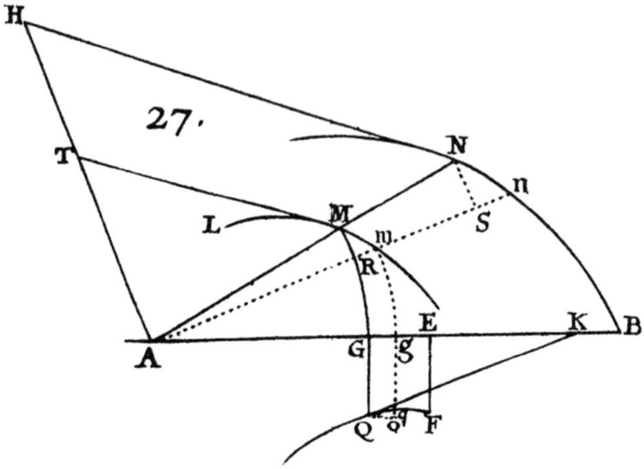

Fig. 27 The Logarithmic Spiral and Related Examples

of the area *EGQF* under the hyperbola, he is able to conclude that the curve has this equiangular property, so that it must be the logarithmic spiral.

In modern texts, the logarithmic spiral is given in polar coordinates by the equation $r = r_0 e^{k\theta}$, where k is the cotangent of the constant angle between the line *AM* and the tangent *MT*. In l'Hôpital's description, the reference circle *BN* plays the role of the coordinate θ, because the area of the sector *ABN*, which is $\frac{c^2}{2}\theta$, is proportional to the angle θ. The radius r (which l'Hôpital denotes by y) is the length of the segment AG and the ordinate GQ is given by $\frac{f^2}{r}$, for a constant f. If we let r_0 denote the length of the initial segment *AE*, then we have $r = r_0 e^{k\theta}$ with $k = \dfrac{c^2}{f^2}$. For more on the logarithmic spiral, see Lockwood (1971, pp. 99–109).

Chapter 3: Maximum and Minimum Ordinates

One of the most important applications of the differential calculus is to finding maximum and minimum values of a function. Because the notion of function did not exist at the time when l'Hôpital wrote the *Analyse*, this problem was stated in terms of finding the maximum or minimum ordinates in the graph of an equation. In Problem XII of the *Lectiones*, Bernoulli taught l'Hôpital that such problems were solved by finding ordinates where the tangent is parallel to the axis, that is where $dy = 0$ with respect to the corresponding dx; see page 205. Modern readers will recognize this as the equivalent of solving $f'(x) = 0$. However, this does not include the case of an extremum being located at a cusp where the tangent is horizontal. In letter 22, written shortly after he had entered into the Contract with l'Hôpital, Bernoulli wrote tell him of this case, where the dy is infinite with respect to dx; see page 251. Thus, when l'Hôpital states his General Proposition in Chapter 3, he includes both of these cases in his discussion, with further elaboration in his Remark in §47 on page 251.

Chapter 3 includes 13 examples in which the least or greatest ordinate is found. Most involve the case where $dy = 0$ with respect to dx, but the second example, in §49 on page 48, is the curve whose equation is $y - a = a^{1/3}(a - x)^{2/3}$, which has a vertical tangent when $x = a$. Other notable examples include the largest cone that can be inscribed in a sphere (§53), the parallelepiped of a given volume with the least surface area (§54 and §55) and three applied problems, all of which appear in some form in the *Lectiones*.

The first of these, Example XI on page 54, is the problem of minimizing the time that it takes a traveler to go from a point in one region, in which he travels at a given speed, to a point in a second region, where he travels with a different speed. L'Hôpital solves the problem in two ways, first using a principle equivalent to Snell's Law, which follows from a more general result derived in Example IX on page 52, and then by a more direct method, using right triangle geometry and the differential calculus. In both cases, the problem reduces to solving a quartic equation.

The second application, Example XII on page 56, is sometimes called the Pulley Problem and has been discussed in a number of modern articles, e.g. (Hahn 1998). The final example in Chapter 3 is an astronomical problem, the problem of the shortest twilight, or crepuscule, on page 57, which presupposes a familiarity with celestial geometry.

Chapter 4: Inflection Points and Cusps

In order to find inflection points and cusps, l'Hôpital first needs to define the second differential, which he also calls *the differential of the differential*. He does so in the first article of Chapter 4 on page 62, and also defines the third and higher order differentials. He denotes these by ddy, $dddy$, and so on, and is careful to distinguish ddy from dy^2 (i.e., $(dy)^2$), $dddy$ from dy^3 and so on. In considering the differential triangle, which has sides dx, dy, and hypotenuse du (the side that is an infinitely small portion of the curve, and which is usually called ds in modern textbooks), all that matters in calculating higher differentials is the ratios among these three quantities. For that reason, l'Hôpital notes that we may consider any one of the increments as being constant; for example, the length of dx may be considered to be the same at every abscissa x, in which case $ddx = 0$, because it is the differential of a constant quantity. L'Hôpital observes that we are free to choose any one of the cases $ddx = 0$, $ddy = 0$, or $ddu = 0$ in order to simplify calculations, but in most cases his choice is $ddx = 0$. This assumption explains the modern use of $\frac{d^2y}{dx^2}$ for the second derivative: if $y = f(x)$, then by the rules of Chapter 1 we clearly have

$dy = f'(x)\,dx$, so by product rule, $ddy = (f''(x)\,dx)dx + f'(x)\,ddx$, which is equal to $f''(x)dx^2$ because ddx is assumed to be zero.

L'Hôpital also defines higher order differentials in the case where the ordinates emanate from a single point, which we sometimes call *polar ordinates* in our commentary. He distinguishes this from the usual case, which we think of as Cartesian coordinates, by referring to that case as being "when the ordinates are parallel to one another." L'Hôpital then defines inflection points and cusps and shows that in the case of parallel ordinates, $ddy = 0$ or ddy is infinite at such points, whereas for polar ordinates, it is the quantity $dx^2 + dy^2 - y\,ddy$ that must be equal to zero or to infinity.

This chapter concludes with 7 examples of curves exhibiting inflection points or cusps, including the prolate cycloid and the Conchoid of Nicomedes. In the case of the Conchoid of Nicomedes, l'Hôpital finds the inflection point using both parallel ordinates and polar ordinates. Example I on page 71 is the curve with equation $axx = xxy + aay$, which was also given in the *Lectiones* on page 218 and is essentially the equation of the curve now called the "Witch of Agnesi." This latter curve is the graph of the function

$$f(x) = \frac{a^3}{x^2 + a^2},$$

but because $y = a - f(x)$ is equivalent to the equation given by Bernoulli and l'Hôpital, their curve is the Witch of Agnes reflected in the line $y = \frac{a}{2}$.

Example VI on page 76, which was given in the *Lectiones* on page 230, is the parabolic spiral, which is closely related to the Spiral of Fermat. The modern equation for this curve is given in polar coordinates as $(a - r)^2 = c^2\theta$. The curve considered by l'Hôpital in Example VI is the branch of this curve corresponding to the positive value of r, with a the radius of the given circle AED and $c = \sqrt{ab}$ for a given magnitude b. The Spiral of Fermat has the equation $r^2 = a^2\theta$; for more on these spirals, see Lockwood (1971, p. 175). L'Hôpital reduces the problem of finding the inflection point in the parabolic spiral to that of solving a quintic equation.

Chapter 5: Evolutes and Involutes

From Chapter 5 onwards, the *Analyse* no longer mirrors the structure of Bernoulli's *Lectiones*. However, Bernoulli's influence is still to be found in many places; Speiss has made a careful cross-reference of places in chapters 5 through 10 where l'Hôpital has drawn on material provided to him in Bernoulli's letters or in his lessons on the integral calculus; see Bernoulli (1955, p. 151).

Suppose that BDF in figure 65 models a thin, rigid wire in the shape of a curve that does not change concavity and that a thread $ABDF$ is laid over the convex side of the curve and fixed at the point F. In this initial position, the portion AB that extends beyond the end of the wire is in a straight line that is tangential to the curve

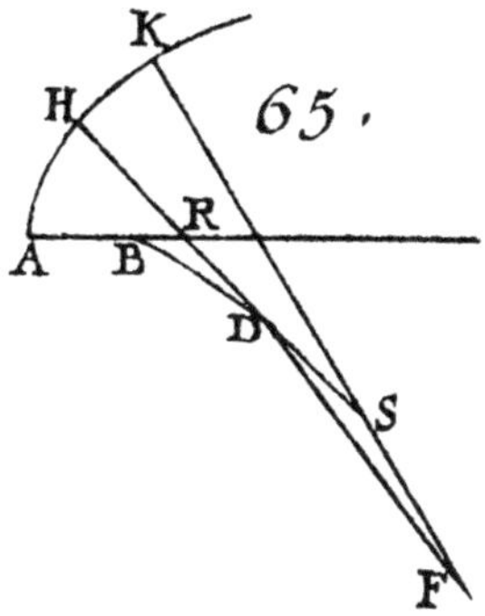

Fig. 65 Definition of Involute and Evolute

at B. If we peel the thread away while holding it taut, the endpoint of the thread initially at A describes the curve AHK by this motion. The curve AHK is called an *involute* of the curve BDF and, reciprocally, the curve BDF is called the *evolute* of AHK. In figure 65, three stages of this process of evolution are illustrated: the initial stage, with the end the thread at A, an intermediate stage, where the thread still lies on the portion FD of the curve and the portion DH is in a straight line tangential to BDF at D, and the line FK is the thread in its final potion.

The central problem of Chapter 5 of the *Analyse* is to determine the evolute of a given curve. It is clear that the line DH is normal to the curve AHK, so the problem reduces to following: given a curve AHK and a normal at any of its points H, to find the length of HD so that the point D lies on the evolute. In modern terminology, the length HD is called the *radius of curvature* of AHK at H, but l'Hôpital uses the term "radius of the evolute" for this length. In §78 on page 84, he derives the expression

$$\frac{\overline{dx^2 + dy^2}\,\sqrt{dx^2 + dy^2}}{-dx\,ddy}, \quad \text{i.e.} \quad \frac{\left(dx^2 + dy^2\right)^{\frac{3}{2}}}{-dx\,d^2y}$$

for the radius of the evolute in the case of rectangular coordinates, which he describes as being when "the ordinates are perpendicular." In the case of polar ordinates, the formula is

$$\frac{\overline{y\,dx^2 + y\,dy^2}\,\sqrt{dx^2 + dy^2}}{dx^3 + dx\,dy^2 - ydx\,ddy}, \quad \text{i.e.} \quad \frac{y\left(dx^2 + dy^2\right)^{\frac{3}{2}}}{dx^3 + dx\,dy^2 - ydx\,d^2y}.$$

Chapter 5, the longest chapter in the *Analyse*, includes 9 examples in which the evolutes of various curves are found, including many of the curves that were encountered in Chapter 2. These include the conics sections, the logarithmic and various spirals. L'Hôpital also shows that the evolute of the cycloid is a congruent cycloid of the opposite orientation, a result originally due to Huygens (1673). The final example is the epicycloid, a curve generated by the motion of one circle rolling on another circle; see Lockwood (1971, p.142) for a modern description of the epicycloid.

This chapter concludes with an extended remark concerning cusps of the second kind, beginning on page 110. The discovery of this type of cusp, which is also called *ramphoid*, seems to have been due to l'Hôpital, although he was assisted by Bernoulli in letters 23 and 24 in clarifying its nature. This cusp became a matter of some interest in the 1740s, when Jean Paul de Gua de Malves (1713–1785) believed he had proved that no such cusp could exist. Leonhard Euler and Jean d'Alembert (1717–1783) independently showed that algebraic curves of low order could in fact be endowed with such a cusp; see Bradley (2006) for more on this curious matter.

Chapters 6–8: Envelopes of Lines and Curves

An *envelope* is a curve that is tangent to every member of some family of lines or curves. Chapters 6 through 8 of the *Analyse* all concern envelopes of various types.

A *caustic by reflection* or *catacaustic* is the envelope of a family of lines that are reflected in a given curve. The study of these curves has its origin in optics. The reflecting curve represents a mirror and the lines that are reflected represent light rays. L'Hôpital studies caustics by reflection in Chapter 6. In most cases the family of rays emanate from a single point B, called the radiant point. However if the radiant point is at a great distance, as it would be in the case of the sun, then the incident rays are considered to be parallel lines. L'Hôpital reduces the problem of finding the envelope of the reflected rays to that of finding the point on each ray where that ray meets the caustic. Having solved this problem, l'Hôpital considers the catacaustics of various conic sections, including the circle, with the radiant point at various locations. He also considers caustics of the cycloid, epicycloid, and logarithmic spiral.

A *caustic by refraction* or *dicaustic* is the envelope of a family of lines that are refracted in a given curve. As with the caustic by reflection, the lines that are refracted represent light rays, but here the curve represents a lens, which refracts the light according to a "given ratio of sines"; that is, the sine of the angle that the incident ray makes with the curve always has a given ratio to the sine of the angle of the refracted ray. L'Hôpital studies caustics by refraction in Chapter 7. Again, the central problem is that of finding the point on each refracted ray where that ray meets the caustic. L'Hôpital finds the caustics by refraction for only a small number of curves: the straight line, the quarter circle, and the logarithmic spiral. This chapter concludes, as does the previous one, with of the solution of the inverse problem: given a caustic, to find the curve that gave rise to it, whether by reflection in Chapter 6 or by refraction in Chapter 7.

Chapter 7 ends with a General Corollary for Chapters 5 through 7: the observation that a given curve has only one evolute, one caustic by reflection and one caustic by refraction, given a ratio of sines. On the converse, however, the same curve may be the evolute of infinitely many lines, and likewise for the caustics, even given the position of the radiant point and the ratio of sines. In the case of involution, we are free to choose the length of the straight line segment (*AB* in figure 4.1) at will. In the case of the caustics, we are free to choose any reflected or refracted ray we may wish, and to choose a point on that ray to be a point of the reflecting or refracting curve. That is, each of these inverse processes involves one arbitrary choice, but becomes fully determined once that choice is made.

Chapter 8 of the *Analyse* concerns the problem of finding envelopes of various families of lines and curves. There are 6 propositions, each dealing with a different kind of problem. Most of the problems deal with envelopes of various families of straight lines, but the problem in Proposition I on page 140 is that of finding the envelope of a family of parabolas, all of which pass through the origin. The vertices of the parabolas in this family are the points on a given curve. L'Hôpital gives the general solution for this problem and then finds the envelope explicitly in the case where the indexing curve is also a parabola.

On the surface, Problem II does not seem to be about envelopes. In this problem, a curve and an axis are given. One seeks a second curve, whose normal from any point on the axis is equal to the ordinate of the given curve from the same point. What makes this a problem about envelopes is that l'Hôpital solves it by considering the family of circles that have the normals of the curve that we seek as their radii. The envelopes of these circles is then the curve that we wish to find.

Taken together, these three chapters on envelopes are only slightly longer than Chapter 5 or Chapter 2. However, these topics seem to have been of particular interest to l'Hôpital and it is worth noting that he published an article 1693, before he had entered into The Contract with Bernoulli, describing an "easy method" for finding the points on the caustic by refraction (L'Hôpital 1693). Bernoulli had published his method for finding these points earlier the same year in the *Acta Eruditorum*, and l'Hôpital wrote this article to clarify that method. Already in 1693, we see the partnership between the researcher and the expositor beginning to take shape.

Chapter 9: L'Hôpital's Rule and Other Problems

Chapter 9 contains "the solution of various problems that depend upon the previous Methods." The first of these is the celebrated rule that we now call L'Hôpital's Rule. The rule is Bernoulli's discovery and, as we have already seen, l'Hôpital became acquainted with it because of Bernoulli's $\frac{0}{0}$ Challenge Problem, see page xiv. L'Hôpital's treatment of this Proposition closely follows the solution that Bernoulli gave him in Letter 28 on p. 267. In Figure 130, which is probably the most famous illustration in the *Analyse*, the curves *ANB* and *COB* represent functions that both take the value zero when x corresponds to the abscissa *AB*, while *AMD* is the curve that represents the quotient of these two quantities. Modern readers will understandably think that *COB* represents negative values of y, because the graph is below the x-axis, but this is not necessarily the case. Here, as well as many other places as in the *Analyse*, graphs that contain more than one curve often have the curves represented above and below the axis, simply in order to make the graph easier to understand, rather than to represent opposite signs of the ordinates in question.

There are 5 propositions in Chapter 9. Although Proposition II is a general proposition concerning drawing tangents to curves that are defined by means of evolutes and involutes, it is in fact applied to the case of the epicycloid. The remaining three propositions also concern the epicycloid. Of particular note is the final proposition, on page 162, in which l'Hôpital finds the quadrature of regions bounded by the epicycloid. Strictly speaking, this is a problem of the integral calculus, but l'Hôpital is able to present what little he needs of the integral calculus in a self-contained manner. Spiess' table (Bernoulli 1955, p. 151) cross-references the portion of Bernoulli's lessons on integral calculus that l'Hôpital needed for this proposition.

Chapter 10: The Method of Descartes and Hudde

When l'Hôpital left military service and took up the academic life, he studied the work of Descartes as a member of Malebranche's intellectual circle. Evidently, he had mastered much of the mathematics of Descartes, and of those who followed him, by the time he met Bernoulli in November 1691. Descartes had shown how to use analytic geometry to find the osculating circle of an algebraic curve, although his method is cumbersome and generally leads to solving an equation that has double the order of the original equation. Because this gives the normal to the curve at that point, the method can be used to find tangents. In the years following the publication of Descartes' *Géometrie* (Descartes 1954), various other

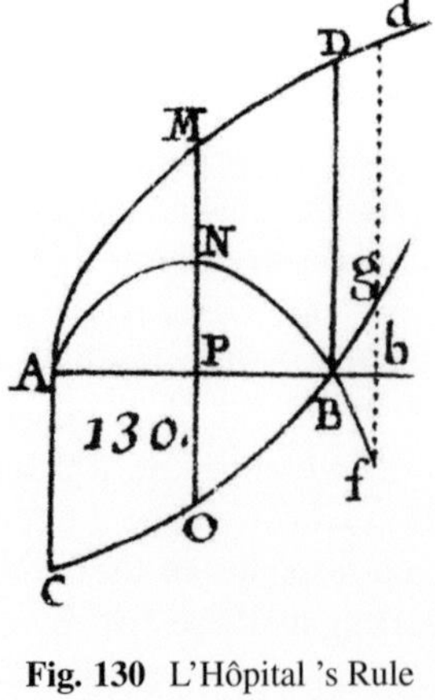

Fig. 130 L'Hôpital 's Rule

mathematicians, particularly Johannes Hudde (1628–1704), elaborated on and simplified Descartes' methods.

In his final chapter of the *Analyse*, l'Hôpital demonstrates how all of the methods of Descartes and Hudde may be easily derived and justified using Leibniz' differential calculus. Because Leibniz' calculus can handle transcendental curves as well as algebraic ones (which l'Hôpital calls "geometric," following Descartes), and does not require removing roots in the case of algebraic curves, he concludes on the final page of the *Analyse* that the new calculus is vastly superior to the older methods.

There is no need here to describe the methods of Descartes and Hudde, because l'Hôpital's exposition of them is very clear and lucid. For a modern exposition of the work of Hudde, see Suzuki (2005).

General Remarks

The core of this book consists of the translated text of the *Analyse* and the *Lectiones*, both made available in English for the first time and cross-referenced so that modern readers may judge for themselves the extent to which l'Hôpital's work is based on Bernoulli's and the extent to which l'Hôpital adds original content. To add further context to this mathematical issue, as well as to give insight into the personalities of these two men and the nature of their relationship, we have included translations of a significant portion of their correspondence in Chapter 12. It must be noted that, although the material here fills almost sixty pages, the full correspondence included in Bernoulli (1955) is 225 pages long. We have been selective, concentrating on places in the correspondence where Bernoulli provides l'Hôpital with material that will appear in the *Analyse*, as well as on anything of a personal or professional nature that illuminates the complex friendship between these two mathematicians.

This volume concludes with a translation of Fontenelle's Eulogy of l'Hôpital, which he wrote in 1704. Much of what we know about l'Hôpital's life comes from this document. Indeed, it was virtually the only source of biographical material on the Marquis until Spiess' research appeared (Bernoulli 1955).

We have attempted to provide accurate translations of all of these primary sources, but that does not always mean literal translations. In the *Analyse*, we have taken a few liberties that we believe make the text more readable, without in any way compromising either the mathematical content or l'Hôpital's expository prowess. Specifically, we have used the present tense in his mathematical exposition, whereas l'Hôpital generally used the future tense, as was traditional at the time and continued to be so well into the 19th century. Also, we have tried to simplify l'Hôpital's complex sentence structure in places, including breaking long sentences into smaller pieces.

In translating the *Lectiones*, the correspondence, and Fontenelle's eulogy, we have generally been much more literal.

The *Analyse* was richly illustrated with 156 figures, which are now in the public domain. These were printed on 11 plates, which were printed with extremely wide left margins, so that when they were bound into the book, they could be folded out. In this way, they extended to the right of the text pages, and could be followed along with the text. We have instead included the illustrations within the body of the pages, as is the custom in modern textbooks. At the time of this writing, many copies of the French editions of the *Analyse* are available as free downloads on the internet. Readers may wish to get copies of the original plates from one of these sources to duplicate the experience of 17th century readers with regard to the illustrations. We have added captions to the figures in this edition, but these were composed by us – there are no captions at all in any of the French editions. In the French editions, the figures were numbered consecutively from 1 to 156. We have used the same numbering in this preface. In our translation of the *Analyse*, we have adopted the following convention: Figure 3.7, for example, is the seventh figure in Chapter 3. The original figure numbers can still be seen within the figures themselves. Our figures have been reproduced from a copy of the 1768 edition that belongs to our friend and colleague V. Frederick Rickey. He obtained this book many years ago as a gift from Philip S. Jones (1912–2002), one of the founding members of the International Study Group on the Relations Between the History and Pedagogy of Mathematics.

Other illustrations from the *Analyse*, including the frontispiece, are from a copy of the 1696 edition that has generously been made available by Google Books.

The illustrations in the *Lectiones* originally appeared in Schafheitlin (1922). The numbering of these figures is due to Schafheitlin, not Bernoulli. They are also in the public domain. We retrieved our figures from a copy of the journal that was made available on the internet by the Biodiversity Heritage Library (BHL). The original volume of the journal was in the collection of the Marine Biological Library of the Woods Hole Oceanographic Institute. We are grateful to the BHL and to Diane M. Rielinger, Director of Library Services at the Marine Biological Library.

The illustrations we have included in the correspondence chapter originally appeared in Bernoulli (1955). They are reproductions of drawings included by l'Hôpital and Bernoulli in their letters. We have assigned consecutive figure numbers to them, but these were not numbered in the originals. We are grateful to Springer Science+Business Media for their kind permission to reproduce these line drawings from the letters of l'Hôpital and Bernoulli.

A number in square brackets in our translation of the *Analyse* denotes the place in L'Hôpital (1696) where the page with that number began. We have not cross-referenced pages with any other edition, although we note that the second edition of 1715/1716 had identical pagination to the 1696 edition. In the French editions, all articles were numbered consecutively from 1 to 209 and these were used in internal citations. We have reproduced the original article numbers in parentheses, such as (§22). Numbers in square brackets in the *Lectiones* are page numbers of the original manuscript of ca. 1705; these numbers were provided in Schafheitlin (1922).

Garden City, NY, USA Robert E. Bradley

Acknowledgements

We are very grateful to the following who offered assistance with translation questions: Richard Garner, Stacy Langton, Kate McDonnell, Kim Plofker, Lyne St. Jacques, and Pria Wadhera.

We thank Larry D'Antonio, Doug Furman, Ross Gingrich, Jim Kiernan, Toke Knudsen, Andy Perry, Kim Plofker, Fred Rickey, Chuck Rocca, Joel Silverberg, and Jeff Suzuki who, as participants in the ARITHMOS reading group, read early drafts of portions of this translation.

We thank Fred Rickey for lending us his copy of the 1768 edition of the *Analyse*. The mathematical figures from that book are reproduced here. Other illustrations in l'Hôpital's Preface and in Chapters 1–10 of the *Analyse* are from the 1696 edition, courtesy of Google Books. Illustrations in Bernoulli's *Lectiones* are courtesy of the Biodiversity Heritage Library. We are grateful to Springer Science+Business Media for their kind permission to reproduce the illustrations in our chapter of excerpts from the l'Hôpital-Bernoulli correspondence.

Garden City, NY, USA — Robert E. Bradley
Garden City, NY, USA — Salvatore J. Petrilli
Newtown, CT, USA — C. Edward Sandifer
February 2015

Contents

Analysis

of the

Infinitely Small

For the Understanding of Curved Lines[1]

Guillaume François Antoine, Marquis de l'Hôpital

---

[1] *Analyse des infiniment petits, pour l'intelligence des lignes courbes*, Imprimerie Royale, Paris, 1696.

L'Hôpital's Preface

The analysis that we explain in this work assumes common analysis, but it is very different. Ordinary analysis considers only finite magnitudes. This analysis reaches all the way to infinity itself. It compares infinitely small differences of finite magnitudes; it uncovers the ratios among these differences; and in this way it makes known the ratios among finite magnitudes, which are as though infinite when compared to the infinitely small. We could even say that this analysis extends beyond the infinite, because it is not limited only to infinitely small differences, but it uncovers the ratios among the differences of these [ii] differences, and also the third and fourth differences, and so on, without ever finding a term that could end it. In this way, it not only includes the infinite, but the infinity of the infinite, or an infinity of infinities.

Only an analysis of this kind could lead us to the true principles of curved lines. This is because curves are nothing but polygons with infinitely many sides and differ from each other only in the differences in the angles that these infinitely small sides make among themselves. It is only the analysis of the infinitely small that can determine the positions of these sides to find the curvatures that they form, that is to say, the tangents to these curves, their perpendiculars, their inflection points or cusps, the rays that they reflect or refract, etc.

Polygons inscribed in or circumscribed on curves, which by the infinite multiplication of their sides finally conform with them, have always been taken to be the curves themselves. However, we had to remain there; it is only since the discovery of the analysis with which we are concerned here that we have [iii] well understood the extent and the fertility of this idea.

What we have from the Ancients on these matters, principally from *Archimedes*,[1] is surely worthy of admiration. However, in addition to the fact that they touched upon very few curves, which they touched only lightly upon, almost all of what we have from them are particular propositions, without order, in which we cannot see any regular and coherent method. However, it would not be fair to reproach them for this; they needed the strength of great genius[2] to penetrate such darkness and to be the first to enter entirely unknown lands. Even if they did not go far, and even if they took circuitous paths, at least, according to Viète,[3] they were not just wandering, and the more difficult and thorny their paths were, the more admirable it is that they did not get lost. In a word, it does not seem that the Ancients could have done more in their times. They did what our best minds would have done in their place, and if they were in our place, we should believe that they would have the same insights as we do. All this [iv] follows from the natural equality of spirits and from the necessary succession of discoveries.

Thus, it is not surprising that the Ancients did not go any further. However, we cannot be so surprised that great men, no doubt men as great as the Ancients, should have remained there for so long, and that by an almost superstitious admiration of their works, satisfied themselves by reading and commenting on those works, without allowing themselves any other use of their insights except what was necessary to follow them, and without daring to commit the crimes of sometimes thinking for themselves and taking their ideas beyond what the Ancients had discovered. Many people worked in this way. They wrote, the books multiplied, and nevertheless nothing advanced. All the works of several centuries accomplished nothing but to fill the world with respectful commentaries and repetitious translations of the originals, themselves so frequently contemptible.

Such was the state of mathematics, and above all of philosophy, until Mr. *Descartes*.[4] Propelled by his genius and by the superiority that he felt, this great

---

[1] Archimedes of Syracuse (ca. 287 BCE-212 BCE).

[2] In (L'Hôpital 1696), the following Latin quote was given: Though I have two or three times read Archimedes' *Treatise of Spirals* with the utmost Attention, to comprehend the Art employed in his subtle Demonstrations relating to the Tangents of Spirals, yet I could never rise from him without some suspicion that I had not taken the whole Force of the Demonstration, etc. I. Bulliadlus, from the Preface of the *de Lioneis Spiralibus*.

[3] Françoos Viète (1540–1603). In (L'Hôpital 1696), the following Latin quote was given: If Archimedes is truth, Euclid infers falsely. *Vietae Supplementum geometriae*.

[4] René Descartes (1596–1650).

man abandoned the Ancients to follow none other than the same [v] reason that the Ancients had followed, and this fortunate audacity, which was treated as a revolt, that brought to us an infinity of new ideas that were so useful in Physics and Geometry. So we opened our eyes and dared to think.

Speaking only of the mathematics, which is all that we consider here, Mr. *Descartes* started where the Ancients had stopped, and he began with the solution of a problem that *Pappus*[5] said[6] had entirely confounded everyone. We know to what point he brought Analysis and Geometry, and how the alliance he made of them simplifies the solution of infinitely many problems which seemed previously impenetrable before him. However, because he applied himself mostly to the solution of equations, he did not pay much attention to curves, other than those that he could use for finding roots. These being sufficient for ordinary analysis, it never occurred to him to seek others. Nevertheless, he was able to make good use of these in his research on Tangents, and the Method that he discovered for this seemed so beautiful to him that he had no difficulty in saying[7] that *this Problem is the most useful* [vi] *and most general not only that he knew, but even that he could ever have desired to know in Geometry*.[8]

Because the Geometry of Mr. *Descartes* put the construction of Problems by solution of equations strongly into fashion, and because he made some great first steps in this, most Geometers applied themselves to it and also made new discoveries which increase and are perfected with every passing day.

Mr. *Pascal*[9] set his sights in an entirely different direction. He examined the curves both in and of themselves, and in the form of polygons. He researched the lengths of several of these, the spaces that they enclose, the solids that these spaces describe, the centers of gravity of these, etc. Solely by the consideration of their elements, that is to say of the infinitely small, he discovered Methods so general and especially so surprising, that it seemed that he could only reach them by strength of mind alone and without analysis.

Shortly after the publication of Mr. *Descartes'* Method of tangents, Mr. *de Fermat*[10] also discovered a method, which Mr. *Descartes* himself finally admitted[11] to be [vii] simpler in many cases than his own. It is furthermore true that this was not as simple as what Mr. *Barrow*[12] gave, after a closer consideration of the nature of polygons, which naturally present themselves to the mind as a little triangle made up of a particle of the curve, contained between two infinitely close ordinates, of

---

[5]Pappus of Alexandria (ca. 290 CE-ca. 350 CE).

[6]Reference given in L'Hôpital (1696): *Collect. Mathem. Lib. 7. initio.*

[7]Reference given in L'Hôpital (1696): *Gèom. liv. 2.*

[8]The italicized passage is a paraphrase of a quotation in Descartes (1954, p. 95).

[9]Blaise Pascal (1623–1662). In L'Hôpital (1696) the spelling was given as "Paschal."

[10]Pierre de Fermat (1601–1665).

[11]Reference given in L'Hôpital (1696): *Lett. 71. Tom. 3.*

[12]Isaac Barrow (1630–1677).

the difference of these two ordinates, and of the difference of the corresponding abscissas. This triangle is similar to the one that is formed by the tangent, the ordinate and the subtangent, so that by a simple analogy, this last method saves all the calculation previously required by the method of Mr. *Descartes* and by the method of Mr. *de Fermat*.

Mr. *Barrow*[13] did not stop there. He also invented a type of calculus suitable for this method but, like the method of Mr. Descartes, his method required him to remove fractions and to make radical signs vanish, in order to make use of it.

The defects of this calculus were overcome by that of the celebrated Mr. *Leibniz*,[14] and this Learned Geometer began where Mr. *Barrow* and the others left off. His calculus brought him to [viii] previously unknown lands, and he made discoveries that astonished the most able Mathematicians of Europe. Messrs. *Bernoulli*[15] were the first to have perceived the beauty of this calculus. They brought it to a point that put them in a state to overcome difficulties that one would never previously have dared to challenge.

The scope of this calculus is immense. It applies to mechanical curves as well as to geometric curves;[16] it handles radical signs and often even conveniently so; it extends to as many indeterminates as we wish; and it easily compares infinitely small quantities of all orders equally well. Furthermore, this gives rise to an infinity of surprising discoveries with respect to Tangents, whether curved or straight, as well as to questions *De maximis & minimis*,[17] to inflection points and cusps of curves, to Evolutes, to Caustics by reflection and by refraction, etc., as we will see in this work.

I divide it into ten chapters.[18] The first contains the principles of the Calculus of differentials. The second reveals the way in which we should make use of this to find Tangents [ix] to all kinds of curves, however many indeterminates there may be in the equation that expresses it, although Mr. *Craig*[19] did not believe that it could be extended to mechanical or transcendental curves. The third chapter shows how it is used to resolve all questions *De maximis & minimis*. The fourth shows how it

---

[13]Reference given in L'Hôpital (1696): *Lect. geomet.* p. 80.

[14]Gottfried Wilhelm von Leibniz (1646–1716). In L'Hôpital (1696), the spelling was given as "Leibnis." Additionally, the following reference was given in L'Hôpital (1696): *Acta Erud. Lips.* an. 1684, p. 467.

[15]Jakob (Jacques) Bernoulli (1654–1705) and Johann Bernoulli (1667–1748).

[16] This distinction is essentially due to Descartes, who called a curve "geometric" if it could be described by a polynomial equation in two variables (Descartes 1954, p. 48). Thus, a geometric curve is what would be called "algebraic" in modern terminology. A curve that is not geometrical was called "mechanical" by Descartes and is called "transcendental" in modern usage.

[17]This Latin term was used in L'Hôpital (1696).

[18]We translate (L'Hôpital 1696) word *Séctions* as "chapters."

[19]John Craig (1663–1731). In (L'Hôpital 1696), the spelling was given as "Craige." Additionally, the following reference was given in (L'Hôpital 1696): *De figurarum curvilinearum quadraturis, part. 2.*

gives inflection points and cusps of curves. The fifth uncovers its use for finding the Evolutes of Mr. *Huygens*[20] in all kinds of curves. The sixth and seventh show how they give the Caustics, both by reflection and by refraction, of which the illustrious Mr. *Tschirnhaus*[21] is the inventor, for all kinds of curves. The eighth shows the further use for finding the points of curved lines that touch an infinity of lines given in position, whether straight or curved. The ninth contains the solution to several Problems that depend on the preceding discoveries. Finally, the tenth consists of a new method for using the Calculus of differentials on geometric curves, from which we deduce the Method of Messrs. *Descartes* and *Hudde*,[22] which applies only to these kinds of curves.

[x] We note that in chapters 2, 3, 4, 5, 6, 7, and 8, there are only a very few Propositions, but they are entirely general and because with such Methods, it is easy to apply them to as many particular propositions as one might wish, I have done it for only a few well-chosen examples. I am persuaded that with regard to mathematics the great benefit is in the methods, and that books that consist of nothing but detail or particular propositions are good for nothing but to waste the time of those who write them and those who read them. Also, I have only added the Problems of the ninth chapter because they appear quite curious and are very universal. Likewise, the tenth Chapter contains only what the Methods of the Calculus of differentials gives to the procedures of Messrs. *Descartes* and *Hudde*. If they are so limited, we see by all that precedes it that this is not a defect of the calculus but of the Cartesian Method to which it is subject. On the contrary, nothing better proves the immense usefulness of this calculus than this great variety of methods, and for the little attention that we pay to it, we see that it draws all that can be drawn from [xi] the method of Messrs. *Descartes* and *Hudde*. The universal proof that it gives to the use that we have made of arithmetic progressions leaves nothing to be wished for, concerning the infallibility of this latter Method.

I had originally intended to add yet another chapter to this, to also make known the marvellous use of this calculus in Physics, to the degree of precision that it can bring, and how much utility Mechanics may draw from it. However, an illness prevented me from doing this; yet, the public will lose nothing because of this, as they will eventually have it someday.

In all of this there is only the first part of Mr. *Leibniz'* calculus, the one that consists in descending from whole magnitudes to their infinitely small differences, and of comparing these infinitely small quantities, of whatever order they may be, to each other – this is what we call *differential Calculus*. The other part, which we call *integral Calculus*, consists in going back from these infinitely small quantities to the magnitudes or the wholes of which they are the differences, that is to say in finding their sums. I had also intended to present this. However, Mr. *Leibniz*, having written me that he is working on a Treatise titled *De Scientiâ infiniti*, [xii] I

[20] Christiaan Huygens (1629–1695). In (L'Hôpital 1696), the spelling was given as "Hugens."

[21] Ehrenfried Walter von Tschirnhaus (1651–1708).

[22] Johann van Waveren Hudde (1628–1704).

took care not to deprive the public of such a beautiful Work, which must include all that is most interesting about the inverse Method of Tangents, the Rectification of curves, the Quadrature of the spaces they enclose, that of the surfaces of the bodies they describe, the size of these bodies, the discovery of the centers of gravity, etc. I would not even have published this work, except that he asked me to do so in his letters and because I believe it to be necessary to prepare the minds to understand all that we may yet discover on those matters.

Furthermore, I acknowledge having received much from the illuminations of Messrs. *Bernoulli*, particularly those of the younger, presently Professor at Groningen. I have made plain use of their discoveries and those of Mr. *Leibniz*. This is why I grant that they may claim as much of this as they may wish, being content with that which they are willing to leave for me.

In fairness, there is also credit due to the learned Mr. *Newton*,[23] which Mr. *Leibniz* himself rendered[24] him: that he also discovered something similar to the differential Calculus, [xiii] which appeared in the excellent book titled *Philosophia naturalis principia Mathematica*, which he gave us in 1687, which is almost all of this calculus. However, the Notation of Mr. *Leibniz* makes his calculus much easier and more expeditious, as well as providing marvelous assistance in many situations.

As we were printing the last page of this Treatise, the Book of Mr. *Nieuwentijt*[25] fell into my hands. Its title, *Analysis infinitorum*, gave me the curiosity to look it over. However, I found that it is quite different from this one, because in addition to the fact that this author does not use the Notation of Mr. *Leibniz*, he absolutely rejects second differentials, third differentials, etc. Because I have based the best part of this Work on this foundation I would have believed myself to be obliged to respond to his objections and to show how fragile they are, had Mr. *Leibniz* not already done so to complete satisfaction in the *Acts of Leipzig*.[26] What's more, the two postulates that I have made at the beginning of this Treatise and upon which it solely rests, seem to me so clear, that I do not believe that they can leave any [xiv] doubt in the minds of attentive Readers. I could even easily have proven them in the manner of the Ancients had I not intended to be brief on things that are already known, and to concentrate principally on those that are new.

[23] Sir Isaac Newton (1643–1727).

[24] Reference given in (L'Hôpital 1696): *Journal des Sçavans du 30 Aoust 1694*.

[25] Bernard Nieuwentijt (1654–1718). In L'Hôpital (1696), the spelling was given as Nieuwentiit.

[26] Reference given in (L'Hôpital 1696): *Acta Erud. an. 1695, pp. 310 & 369*. In L'Hôpital (1696), the spelling was given as Leypsick.

Chapter 1
In Which We Give the Rules of This Calculus

Analysis
of the
Infinitely Small

First Part
On the Differential Calculus

Definition I. Those quantities are called *variable* which increase or decrease continually, as opposed to *constant* quantities that remain the same while others change. Thus in the parabola the ordinate and the abscissa are variable quantities, as opposed to the parameter, which is a constant quantity.[1]

---

[1]In L'Hôpital (1696) the terms *appliquée* for ordinate and *coupée* for abscissa are used, literally meaning "applied" and "cut."

R.E. Bradley et al., *L'Hopital's Analyse des infiniments petits*, Science Networks. Historical Studies 50, DOI 10.1007/978-3-319-17115-9_1

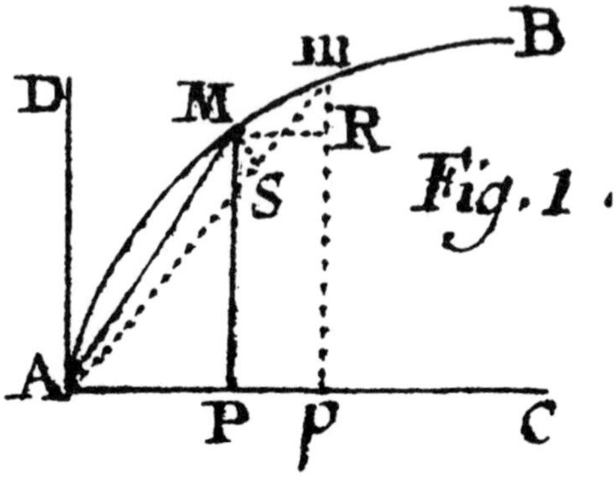

Fig. 1.1 Definition of the Differential

Definition II. [2] The infinitely small portion by which a variable quantity continually increases or decreases is called the *Differential*.[2] For example, let *AMB* be an arbitrary curved line (see Fig. 1.1) which has the line *AC* as its axis or diameter,[3] and has *PM* as one of its ordinates. Let *pm* be another ordinate, infinitely close to the first one. Given this, if we also draw *MR* parallel to *AC*, and the chords *AM Am*, and describe the little circular arc *MS* of the circle with center *A* and radius *AM*, then *Pp* is the differential of *AP*, *Rm* the differential of *PM*, *Sm* the differential of *AM*, and *Mm* the differential of the arc *AM*. Furthermore, the little triangle *MAm*, which has the arc *Mm* as its base is the differential of the segment *AM*, and the little region *MPpm* is the differential of the region contained by the straight lines *AP* and *PM*, and by the arc *AM*.

Corollary. (§1) *It is evident that the differential of a constant quantity is null or zero, or (what amounts to the same thing) that constant quantities do not have a differential.*

Note. In what follows, we will make use of the symbol d to denote the differential of a variable quantity that is expressed by a single letter and, in order to avoid confusion, the letter d will not be used in any other way in the following calculations. If, for example, we denote *AP* by x, *PM* by y, *AM* by z, the arc *AM* by u, the curvilinear region *APM* by s, and the segment *AM* by t, then dx denotes the value of *Pp*, dy that of *Rm*, dz that of *Sm*, du that of the little arc *Mm*, ds that of the little region *MPpm*, and dt that of the little curvilinear triangle *MAm*.

[2]In L'Hôpital (1696), the same word *différence* is used for both the differential and the difference of ordinary subtraction. In this translation, it is consistently translated as "differential" when the difference is infinitely small.

[3]In L'Hôpital (1696), these terms are used interchangeably for the x-axis. Technically, the term *diameter* should only be used in the sense for a curve that is symmetric about the axis.

Postulate I.[4] (§2) We suppose that two quantities that differ by an infinitely small quantity may be used interchangeably, or (what amounts to the same [3] thing) that a quantity which is increased or decreased by another quantity that is infinitely smaller than it is, may be considered as remaining the same. We suppose, for example, that we may take Ap for AP, pm for PM, the region Apm for the region APM, the little region $MPpm$ for the little rectangle $MPpR$, the little sector AMm for the little triangle AMS, the angle pAm for the angle PAM, and so forth.

Postulate II.[5] (§3) We suppose that a curved line may be considered as an assemblage of infinitely many straight lines, each one being infinitely small, or (what amounts to the same thing) as a polygon with an infinite number of sides, each being infinitely small, which determine the curvature of the line by the angles formed amongst themselves. We suppose, for example, that the portion Mm of the curve and the arc MS of the circle may be considered to be straight lines on account of their infinite smallness, so that the little triangle mSM may be considered to be rectilinear.

Note. In what follows, we will normally suppose that the final letters of the alphabet, z, y, x, etc., denote variable quantities, and conversely, that the first letters a, b, c, etc., denote constant quantities, so that as x becomes $x + dx$, y, z, etc., become $y + dy$, $z + dz$, etc., and a, b, c, etc. remain as a, b, c, etc. (see §1).

Proposition I.

Problem. (§4) *To take the differential of several quantities added together or subtracted from one another.*[6]

Let $a+x+y-z$ be the expression whose differential we are to take. If we suppose that x is increased by an infinitely small quantity, i.e. that it becomes $x + dx$, then y [4] becomes $y+dy$, and z becomes $z+dz$. As for the constant a, it remains a (see §1). Thus, the given quantity $a+x+y-z$ becomes $a+x+dx+y+dy-z-dz$, and the differential, which is found by subtracting the former from the latter, is $dx+dy-dz$. It is similar for other expressions, giving rise to the following rule.

Rule I. For quantities added or subtracted.

We take the differential of each term in the given quantity and, keeping the signs the same, we add them together in a new quantity which is the differential that we wish to find.

Proposition II.

Problem. (§5) *To take the differential of a product composed of several quantities multiplied together.*

---

[4]Compare to Postulate 1 on p. 187.

[5]Compare to Postulate 2 on p. 187.

[6]Compare to the section "On the Addition and Subtraction of Differentials" on p. 187.

1. The differential of xy is $y\,dx + x\,dy$. This is because y becomes $y + dy$ while x becomes $x + dx$, and consequently xy becomes $xy + y\,dx + x\,dy + dx\,dy$, which is the product of $x + dx$ and $y + dy$. The differential is $y\,dx + x\,dy + dx\,dy$, that is to say (see §2) $y\,dx + x\,dy$. This is because the quantity $dx\,dy$ is infinitely small with respect to the other terms $y\,dx$ and $x\,dy$: If we were to divide $y\,dx$, for example, and $dx\,dy$ by dx we would find, on the one hand y, and on the other dy, which is the differential of the former, and is consequently infinitely smaller than it. From this it follows that the differential of the product of two quantities is equal to the product of the differential of the first of these two quantities with the second, plus the product of the differential of the second quantity with the first.
2. The differential of xyz is $yz\,dx + xz\,dy + xy\,dz$. This is because in considering the product xy as a single quantity, it is necessary, as we have just proven, to take the product of its differential $y\,dx + x\,dy$ with the second quantity z (which gives $yz\,dx + xz\,dy$), plus the product of the differential dz [5] of the second quantity z by the first xy (which gives $xy\,dz$). Consequently the differential of xyz is seen to be $yz\,dx + xz\,dy + xy\,dz$.
3. The differential of $xyzu$ is $uyz\,dx + uxz\,dy + uxy\,dz + xyz\,du$. This is proved as in the previous case, by considering xyz as a single quantity. It is similar for other cases to infinity,[7] from which we form the following rule.

Rule II. For multiplied quantities.

The differential of a product of several quantities multiplied together is equal to the sum of the products of the differential of each of the quantities multiplied by the product of the others.[8]

Thus, the differential of ax is $x\,0 + a\,dx$, that is $a\,dx$. The differential of[9] $\overline{a + x} \times \overline{b - y}$ is $b\,dx - y\,dx - a\,dy - x\,dy$.

Proposition III.

Problem. (§6) *To take the differential of any fraction.*[10]

The differential of $\frac{x}{y}$ is $\frac{y\,dx - x\,dy}{yy}$. This is because, if suppose that $\frac{x}{y} = z$, we have $x = yz$. Because these two variable quantities x and yz must always be equal to one another, whether increasing or decreasing, it follows that their differentials, that is to say their increments or decrements, are equal to each other. It then follows (see §5) that $dx = y\,dz + z\,dy$, and so

$$dz = \frac{dx - z\,dy}{y} = \frac{y\,dx - x\,dy}{yy},$$

[7]In L'Hôpital (1696) the expression "to infinity" is frequently employed meaning roughly "of all orders."

[8]Compare this with the product rule as discussed on p. 189.

[9]In L'Hôpital (1696), the overline is used for grouping terms. We note, however, that a negative sign does not seem to distribute through terms grouped under an overline; see §13 ff.

[10]Compare to the section "On the Differentials of Divided Quantities" on p. 189.

substituting the value $\frac{x}{y}$ for z. This is what we wished to find, and we form the following rule.

Rule III. For divided quantities, or fractions.

The differential of any fraction is equal to [6] the product of the differential of the numerator by the denominator minus the product of the differential of the denominator by the numerator all divided by the square of the denominator.

Thus, the differential of $\frac{a}{x}$ is $\frac{-a\,dx}{xx}$ and the differential of $\frac{x}{a+x}$ is $\frac{a\,dx}{aa+2ax+xx}$.

Proposition IV.

Problem. (§7) *To take the differential of any power, whether perfect or imperfect,*[11] *of a variable quantity.*[12]

In order to give a general rule that applies to perfect and imperfect powers, it is necessary to explain the analogy that we find among their exponents.

If we consider a geometric progression[13] whose first term is unity and whose second term is any quantity x, and if under each term we write its exponent, it is clear that these exponents form an arithmetic progression.

Geom. prog.:	1,	x,	xx,	x^3,	x^4,	x^5,	x^6,	x^7,	etc.
Arith. prog.:	0,	1,	2,	3,	4,	5,	6,	7,	etc.

If we continue the geometric progression below unity and the arithmetic progression below zero, the terms of the one are the exponents of the corresponding terms in the other. Thus, -1 is the exponent of $\frac{1}{x}$, -2 that of $\frac{1}{xx}$, etc.

Geom. prog.:	x,	1,	$\frac{1}{x}$,	$\frac{1}{xx}$,	$\frac{1}{x^3}$,	$\frac{1}{x^4}$,	etc.
Arith. prog.:	1,	0,	-1,	-2,	-3,	-4,	etc.

If we introduce a new term in the geometric progression, we must introduce a similar term in the arithmetic progression to be its exponent.

Thus, $\sqrt{x}$ has $\frac{1}{2}$ as its exponent, $\sqrt[3]{x}$ has $\frac{1}{3}$, $\sqrt[5]{x^4}$ has $\frac{5}{4}$, $\frac{1}{\sqrt{x^3}}$ has $-\frac{3}{2}$, $\frac{1}{\sqrt[3]{x^5}}$ has $-\frac{5}{3}$, $\frac{1}{\sqrt{x^7}}$ has $-\frac{7}{2}$, etc., so that the expressions [7] $\sqrt{x}$ and $x^{\frac{1}{2}}$, $\sqrt[3]{x}$ and $x^{\frac{1}{3}}$, $\sqrt[5]{x^4}$ and $x^{\frac{4}{5}}$, $\frac{1}{\sqrt{x^3}}$ and $x^{-\frac{3}{2}}$, etc., signify the same thing.

Geom. prog.:	1,	$\sqrt{x}$,	x.		
Arith. prog.:	0,	$\frac{1}{2}$,	1.		
Geom. prog.:	1,	$\sqrt[3]{x}$,	$\sqrt[3]{xx}$,	x.	
Arith. prog.:	0,	$\frac{1}{3}$,	$\frac{2}{3}$,	1.	

Geom. prog.:	1,	$\sqrt[5]{x}$,	$\sqrt[5]{xx}$,	$\sqrt[5]{x^3}$,	$\sqrt[5]{x^4}$,	x.
Arith. prog.:	0,	$\frac{1}{5}$,	$\frac{2}{5}$,	$\frac{3}{5}$,	$\frac{4}{5}$,	1.

[11] As we will see on p. 6 (L'Hôpital 1696, p. 8), a "perfect" power is an integer and an "imperfect" power is a fraction.

[12] Compare to the section "On the Differentials of Surd Quantities" on p. 190.

[13] Compare the discussion that follows with Bernoulli's discussion of geometric and arithmetic progressions on p. 190.

Geom. prog.: $\frac{1}{x}$, $\frac{1}{\sqrt{x^3}}$, $\frac{1}{xx}$.
Arith. prog.: -1, $-\frac{3}{2}$, -2.
Geom. prog.: $\frac{1}{x}$, $\frac{1}{\sqrt[3]{x^4}}$, $\frac{1}{\sqrt[3]{x^5}}$, $\frac{1}{xx}$.
Arith. prog.: -1, $-\frac{4}{3}$, $-\frac{5}{3}$, -2.
Geom. prog.: $\frac{1}{x^3}$, $\frac{1}{\sqrt{x^7}}$, $\frac{1}{x^4}$.
Arith. prog.: -3, $-\frac{7}{2}$, -4.

From this we see that just as $\sqrt{x}$ is the geometric mean between 1 and x, so also is $\frac{1}{2}$ the arithmetic mean between their exponents zero and 1, and just as $\sqrt[3]{x}$ is the first of two means that are geometrically proportional between 1 and x, so also is $\frac{1}{3}$ the first of two means that are arithmetically proportional between their exponents zero and 1. It is the same for the others. Now, two things follow from the nature of these two progressions:

1. That the sum of the exponents of any two terms of the geometric progression is the exponent of the term that is their product. Thus, x^{4+3} or x^7 is the product of x^3 by x^4, $x^{\frac{1}{2}+\frac{1}{3}}$ or $x^{\frac{5}{6}}$ is the product of $x^{\frac{1}{2}}$ by $x^{\frac{1}{3}}$, $x^{-\frac{1}{3}+\frac{1}{5}}$ or $x^{-\frac{2}{15}}$ is the product of $x^{-\frac{1}{3}}$ by $x^{\frac{1}{5}}$, etc. Similarly, $x^{\frac{1}{3}+\frac{1}{3}}$ or $x^{\frac{2}{3}}$ is the product of $x^{\frac{1}{3}}$ by itself, that is to say its square, x^{+2+2+2} or x^6 is the product of x^2 by x^2 by x^2, that is to say its cube, $x^{-\frac{1}{3}-\frac{1}{3}-\frac{1}{3}-\frac{1}{3}}$ or $x^{-\frac{4}{3}}$ is the fourth power of $x^{-\frac{1}{3}}$, and so on for other powers. From this it is clear that the double, triple, etc., of the exponent of any term in the geometric progression is the exponent of the square, the cube, etc., of this term, and consequently that the half, the third, etc., of the exponent of any term of the geometric progression is the exponent of the square root, the cube root, etc., of this term;
2. That the difference of the exponents of any two terms of the geometric progression is the exponent of [8] the quotient of the division of these terms. Thus, $x^{\frac{1}{2}-\frac{1}{3}} = x^{\frac{1}{6}}$ is the exponent of the quotient of the division of $x^{\frac{1}{2}}$ by $x^{\frac{1}{3}}$, and $x^{-\frac{1}{3}-\frac{1}{4}} = x^{-\frac{7}{12}}$ is the exponent of the quotient of the division of $x^{-\frac{1}{3}}$ by $x^{\frac{1}{4}}$, from which we see that multiplying $x^{-\frac{1}{3}}$ by $x^{-\frac{1}{4}}$ is the same thing as dividing $x^{-\frac{1}{3}}$ by $x^{\frac{1}{4}}$. It is the same for the others.

This being well understood, there are two cases that may arise.

The first case is that of a perfect power, that is to say when the exponent is a whole number.[14] The differential of xx is $2x\,dx$, of x^3 is $3xx\,dx$, of x^4 is $4x^3\,dx$, etc. Because the square of x is nothing but the product of x by x, its differential is $x\,dx + x\,dx$ (see §5), that is to say $2x\,dx$. Similarly, the cube of x being nothing but the product of x by x by x, its differential is $xx\,dx + xx\,dx + xx\,dx$ (see §5), that is to say $3xx\,dx$. However, because it is thus for other powers up to infinity, it follows that if we suppose that m indicates whatever whole number we might wish, the differential of x^m is $mx^{m-1}\,dx$.

[14]This case is proved by Bernoulli on p. 188, without using the Product Rule.

If the exponent is negative, we find that the differential of x^{-m} or of $\frac{1}{x^m}$ is

$$\frac{-mx^{m-1}\,dx}{x^{2m}} = -mx^{-m-1}\,dx.$$

The second case is that of an imperfect power, that is to say when the exponent is a fractional number. Suppose we are to take the differential of $\sqrt[n]{x^m}$, or $x^{\frac{m}{n}}$, where $\frac{m}{n}$ denotes any fractional number. We let $x^{\frac{m}{n}} = z$ and, raising both sides to the power n, we have $x^m = z^n$. Taking differentials as we have explained in the previous case, we find that $mx^{m-1}\,dx = nz^{n-1}\,dz$, and so

$$dz = \frac{mx^{m-1}\,dx}{nz^{n-1}} = \frac{m}{n}x^{\frac{m}{n}-1}\,dx,$$

or $\frac{m}{n}\,dx\,\sqrt[n]{x^{m-n}}$, by substituting the value $nx^{m-\frac{m}{n}}$ in place of nz^{n-1}. If the exponent is negative, we find that the differential of $x^{-\frac{m}{n}}$ or $\frac{1}{x^{\frac{m}{n}}}$ is

$$\frac{-\frac{m}{n}x^{\frac{m}{n}-1}\,dx}{x^{\frac{2m}{n}}} = -\frac{m}{n}x^{-\frac{m}{n}-1}\,dx.$$

[9] This gives us the following general rule.

Rule IV. For powers, both perfect and imperfect.

The differential of any power, whether perfect or imperfect, of a variable quantity is equal to the product of the exponent of this power by the same quantity raised to a power diminished by unity, and multiplied by its differential.

Thus, if we suppose that m denotes either a whole or fractional number, whether positive or negative, and x denotes any variable quantity, the differential of x^m is always $mx^{m-1}\,dx$.

Examples. The differential of the cube of $ay - xx$, that is to say of $\overline{ay - xx}^3$, is $3 \times \overline{ay - xx}^2 \times \overline{a\,dy - 2x\,dx} = 3a^3yy\,dy - 6aaxxy\,dy + 3ax^4\,dy - 6aayyx\,dx + 12ayx^3\,dx - 6x^5\,dx$.

The differential of $\sqrt{xy + yy}$, or of $\overline{xy + yy}^{\frac{1}{2}}$, is

$$\frac{1}{2} \times \overline{xy + yy}^{-\frac{1}{2}} \times \overline{y\,dx + x\,dy + 2y\,dy},$$

or

$$\frac{y\,dx + x\,dy + 2y\,dy}{2\sqrt{xy + yy}}.$$

The differential of $\sqrt{a^4 + axyy}$, or of $\overline{a^4 + axyy}^{\frac{1}{2}}$, is

$$\frac{1}{2} \times \overline{a^4 + axyy}^{-\frac{1}{2}} \times \overline{ayy\,dx + 2axy\,dy},$$

or

$$\frac{ayy\,dx + 2axy\,dy}{2\sqrt{a^4 + axyy}}.$$

The differential of[15] $\sqrt[3]{ax + xx}$, or of $\overline{ax + xx}^{\frac{1}{3}}$, is

$$\frac{1}{3} \times \overline{ax + xx}^{-\frac{2}{3}} \times \overline{a\,dx + 2x\,dx},$$

or

$$\frac{a\,dx + 2x\,dx}{3\sqrt[3]{ax + xx}^{\,2}}.$$

The differential of[16] $\sqrt{ax + xx + \sqrt{a^4 + axyy}}$, or of

$$\overline{ax + xx + \sqrt{a^4 + axyy}}^{\frac{1}{2}},$$

is

$$\frac{1}{2} \times \overline{ax + xx + \sqrt{a^4 + axyy}}^{-\frac{1}{2}} \times \overline{a\,dx + 2x\,dx + \frac{ayy\,dx + 2axy\,dy}{2\sqrt{a^4 + axyy}}},$$

or

$$\frac{a\,dx + 2x\,dx}{2\sqrt{ax + xx + \sqrt{a^4 + axyy}}} + \frac{ayy\,dx + 2axy\,dy}{2\sqrt{a^4 + axyy} \times 2\sqrt{ax + xx + \sqrt{a^4 + axyy}}}.$$

[10] According to this rule and the rule of fractions (see §7 and 6), the differential of[17]

$$\frac{\sqrt[3]{ax + xx}}{\sqrt{xy + yy}}$$

---

[15] Bernoulli gave this example on p. 192.

[16] Bernoulli gave a similar example on p. 192.

[17] Bernoulli gave this example on p. 192.

is[18]

$$\frac{\frac{a\,dx+2x\,dx}{3\sqrt[3]{ax+xx}^{\,2}} \times \sqrt{xy+yy} + \frac{-y\,dx-x\,dy-2y\,dy}{2\sqrt{xy+yy}} \times \sqrt[3]{ax+xx}}{xy+yy}.$$

Remark. (§8) It is appropriate to note carefully that in taking differentials, we always suppose that, as one of the variables x increases, the others, y, z, etc., also increase. That is to say, as x becomes $x+dx$, then y, z, etc., become $y+dy$, $z+dz$, etc. This is why if it happens that some of them increase while others decrease, it is necessary to regard the differentials as negative quantities with respect to those others which we suppose are increasing, and consequently to change the signs of the terms containing the differentials of those which are decreasing. Thus, if we suppose that x increases and y and z decrease, that is to say that x becomes $x+dx$ and y and z become $y-dy$ and $z-dz$, and if we wish to take the differential of the product xyz, it is necessary in the differential $xy\,dz+xz\,dy+yz\,dx$ already found (see §5), to change the signs of the terms which contain dy and dz. This gives the differential $yz\,dx-xy\,dz-xz\,dy$ that we wished to find.

[18]In L'Hôpital (1696), the + sign separating the terms of the numerator was omitted.

Chapter 2
Use of the Differential Calculus for Finding the Tangents of All Kinds of Curved Lines

Definition. [11] If we prolong one of the little sides *Mm* (see Fig. 2.1) of the polygon that makes up a curved line (see §3), this little side, thus prolonged, is called the *Tangent* to the curve at the point *M* or *m*.

Proposition I.

Problem. (§9) *Let AM be a curved line* (see Fig. 2.2), *where the relationship between the abscissa AP and the ordinate PM is expressed by any equation. At a given point M on this curve, we wish to draw the tangent MT.*

We draw the ordinate *MP* and suppose that the straight line *MT* that meets the diameter at the point *T* is the tangent we wish to find. We imagine another ordinate *mp* infinitely close to the first one, with a little straight line *MR* parallel to *AP*. Now denoting the given quantities *AP* by x and *PM* by y (so that *Pp* or $MR = dx$ and $Rm = dy$), the similar triangles *mRM* and *MPT* give[1] $mR(dy) : RM(dx) :: MP(y) : PT = \frac{y\,dx}{dy}$. Now, by means of the differential of the given equation, we find a value for dx in terms that are multiplied by dy. This (being multiplied by y and divided by dy) will give the value of the subtangent *PT* in terms that are entirely known and free of differentials, which can be used to draw the tangent that we wish to find.

Remark. (§10) When the point T falls on the opposite side of the point A, the origin of the x's, it is clear that as x increases, y decreases (see Fig. 2.3). [12] Consequently (see §8) we must change the signs of all the terms containing dy in the differential of the given equation, otherwise the value of dx over dy would be negative and therefore also the sign of PT $\left(\frac{y\,dx}{dy}\right)$. However, in order not to get muddled, it is better always to take the differential of the given equation by the prescribed rules

[1] In L'Hôpital (1696) the notation $a\ .\ b :: c\ .\ d$ was used to express equal proportions; we write this instead as $a : b :: c : d$. We note further that in L'Hôpital (1696) the right parenthesis following dx was omitted.

R.E. Bradley et al., *L'Hopital's Analyse des infiniments petits*, Science Networks. Historical Studies 50, DOI 10.1007/978-3-319-17115-9_2

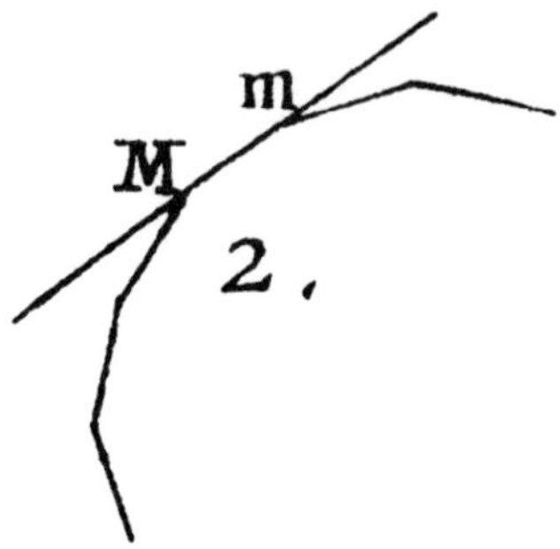

Fig. 2.1 Definition of the Tangent

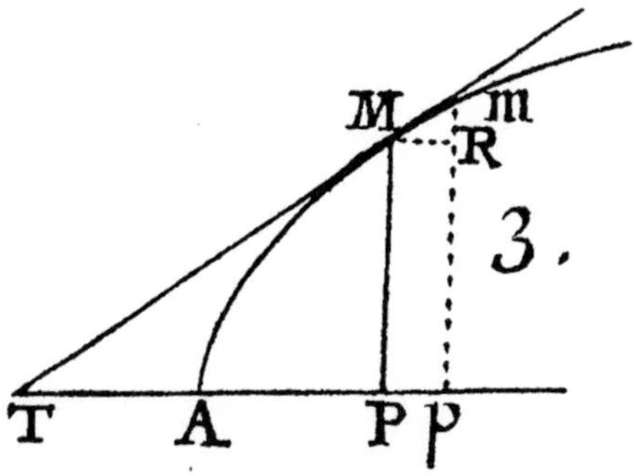

Fig. 2.2 Tangent to a Curve: First Case

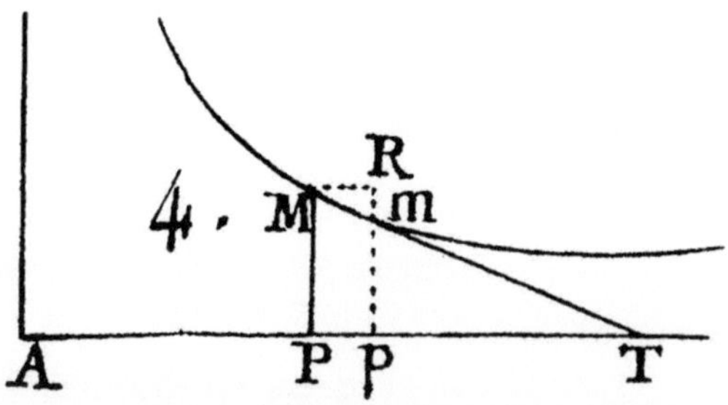

Fig. 2.3 Tangent to a Curve: Second Case

(see Ch. 1) without any changes. Then if it turns out at the end of the operation that the sign of PT is positive, it follows that we must take the point T on the same side as A, the origin of the x's, as we had supposed in making the calculation. On the other hand, if it is negative, we must take it on the opposite side. This will be clarified by the following examples.

Example I. (§11) Examples:

1. If we let $ax = yy$ express the relationship of *AP* to *PM*, the curve *AM* (see Fig. 2.2) is a parabola[2] that has the given straight line a as parameter. If we take differentials of both sides, we have $a\,dx = 2y\,dy$, $dx = \frac{2y\,dy}{a}$ and $PT\left(\frac{y\,dx}{dy}\right) = \frac{2yy}{a} = 2x$, substituting the value ax for yy. From this it follows that if we take *PT* to be twice *AP* and if we draw the straight line *MT*, it will be tangent at the point M. This is what we wish to find.
2. Let the equation be $aa = xy$, which expresses the nature of a hyperbola[3] between the asymptotes (see Fig. 2.3). Taking differentials, we have $x\,dy + y\,dx = 0$ and thus $PT\left(\frac{y\,dx}{dy}\right) = -x$. From this it follows that if we take $PT = PA$ on the opposite side from the point A and we draw the straight line *MT*, it will be tangent to M.
3. Let the general equation[4] be $y^m = x$, which expresses the nature of all parabolas to infinity, when the exponent m denotes a whole or fractional positive number, and of all hyperbolas when it denotes a negative number. Taking differentials, we have $my^{m-1}\,dy = dx$ and thus $PT\left(\frac{y\,dx}{dy}\right) = my^m = mx$, substituting the value x for y^m.

[13] If $m = \frac{3}{2}$, the equation is $y^3 = axx$, which expresses the nature of one of the cubic parabolas, and the subtangent $PT = \frac{3}{2}x$. If $m = -2$, the equation is $a^3 = xyy$, which expresses the nature of one of the cubic hyperbolas and the subtangent $PT = -2x$. It is similar for the others.

To draw the tangent to parabolas at the point A, the origin of the x's, it is necessary to find what the ratio of dx to dy should be at that point, for it is evident that if this ratio were known, the angle that the tangent makes with the axis or diameter would also be determined. In this example, we have $dx : dy :: my^{m-1} : 1$. From this we see that because y is zero at A, the ratio of dy to dx should be infinitely large if m is greater than 1 and infinitely small if it is less than 1. That is to say, the tangent at A should be parallel to the ordinates in the first case and it coincides with the diameter in the second case.

Example II. (§12) Let *AMB* be a curved line[5] (see Fig. 2.4) such that $AP \times PB(x \times \overline{a-x}) : \overline{PM}^2(yy) :: AB(a) : AD(b)$. Thus $\frac{ayy}{b} = ax - xx$ and taking differentials, we have $\frac{2ay\,dy}{b} = a\,dx - 2x\,dx$. From this we get $PT\left(\frac{y\,dx}{dy}\right) = \frac{2ayy}{ab-2bx} = \frac{2ax-2xx}{a-2x}$, substituting the value $ax - xx$ for $\frac{ayy}{b}$. Thus $PT - AP$ or $AT = \frac{ax}{a-2x}$.

---

[2]Compare this to Bernoulli's Problem I on p. 192.

[3]Compare this to Bernoulli's Problem III on p. 195.

[4]Compare this to the treatment of higher order parabolas beginning on p. 193.

[5]*AMB* is an ellipse with center $\left(\frac{a}{2}, 0\right)$, horizontal axis a, and vertical axis $\sqrt{ab}$. Compare this to Bernoulli's discussion on p. 194.

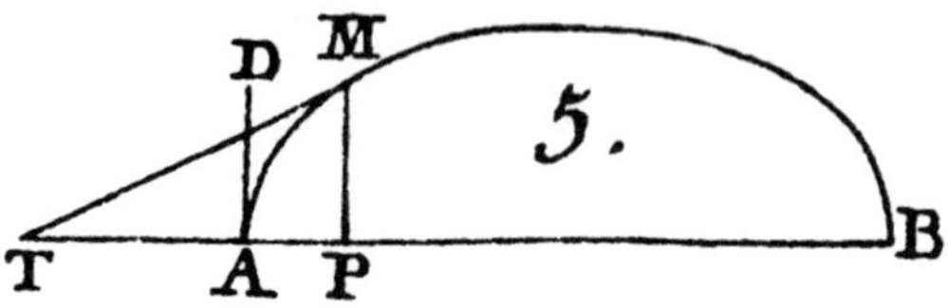

Fig. 2.4 Tangent to an Ellipse

Supposing for the moment that $\overline{AP}^3 \times \overline{PB}^2$ $(x^3 \times \overline{a-x}^2) : \overline{PM}^5$ $(y^5) :: AB(a) : AD(b)$, we have $\frac{ay^5}{b} = x^3 \times \overline{a-x}^2$. Taking differentials,[6]

$$\frac{5ay^4 dy}{b} = 3xx\,dx \times \overline{a-x}^2 - \overline{2a\,dx + 2x\,dx} \times x^3,$$

from which we conclude[7]

$$\begin{aligned} \frac{y\,dx}{dy} &= \frac{5x^3 \times \overline{a-x}^2}{3xx \times \overline{a-x}^2 - \overline{2a+2x} \times x^3} \\ &= \frac{5x \times a - x}{3a - 3x - 2x} \\ &= \frac{5ax - 5xx}{3a - 5x} \\ \text{and} \quad AT &= \frac{2ax}{3a-5x}. \end{aligned}$$

[14] Generally,[8] if we let m denote the exponent of the power of AP and n the exponent of the power of PB, we have $\frac{ay^{m+n}}{b} = x^m \times \overline{a-x}^n$, which is the general equation of all ellipses up to infinity, the differential of which is

$$\frac{\overline{m+n}\,ay^{m+n-1}\,dy}{b} = mx^{m-1}\,dx \times \overline{a-x}^n - n\,\overline{a-x}^{n-1}\,dx \times x^m,$$

[6]The term $-\overline{2a\,dx + 2x\,dx} \times x^3$ in what follows would be written today as $(-2a\,dx + 2x\,dx)x^3$. Thus, the negative sign before the grouping of expressions under the overline symbol is not meant to obey the distributive law. This does not affect the validity of the computed value for AT.

[7]In the numerator of the second line below, the expression $5x \times a - x$ seems to mean $5x(a - x)$. Thus, the use of the symbol $\times$ appears to imply a grouping of expressions.

[8]See p. xxi for a discussion of generalized ellipses.

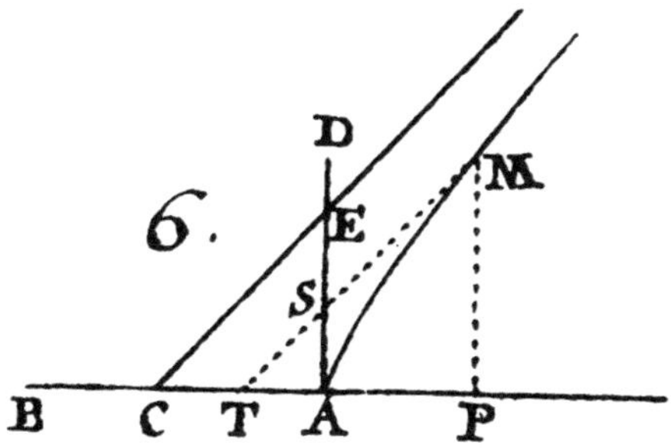

Fig. 2.5 Generalized Hyperbolas

from which we conclude (substituting the value $x^m \times \overline{a-x}^n$ for $\frac{ay^{m+n}}{b}$)

$$PT\left(\frac{y\,dx}{dy}\right) = \frac{\overline{m+n}\,x^m \times \overline{a-x}^n}{mx^{m-1} \times \overline{a-x}^n - n\,\overline{a-x}^{n-1} \times x^m} = \frac{\overline{m+n}\,x \times \overline{a-x}}{m\,\overline{a-x} - nx}$$

or

$$PT = \frac{\overline{m+n} \times \overline{ax-xx}}{ma - \overline{m-n}\,x} \quad \text{and} \quad AT = \frac{nax}{ma - \overline{m-n}\,x}.$$

Example III. (§13) With the same assumptions as in the previous example, except that here we suppose that the point B falls on the other side of the point A with respect to P (see Fig. 2.5), we have the equation $\frac{ay^{m+n}}{b} = x^m \times \overline{a+x}^n$, which expresses the nature of all hyperbolas considered with respect to their diameters. From this we conclude, as above, that[9]

$$PT = \frac{\overline{m+n} \times \overline{ax+xx}}{ma + \overline{m+n}\,x} \quad \text{and} \quad AT = \frac{nax}{ma + \overline{m+n}\,x}.$$

If we now suppose that AP is infinitely large, the tangent TM will only meet the curve at an infinite distance, that is to say it will become the asymptote CE and in this case we have

$$AT\left(\frac{nax}{ma + \overline{m+n}\,x}\right) = \frac{n}{m+n}a = AC,$$

because since a is infinitely smaller than x, the term ma is zero with respect to $\overline{m+n}\,x$. For the same reason, in this case the equation of the curve becomes

[9]In L'Hôpital (1696) the equal sign following AT was omitted.

$ay^{m+n} = bx^{m+n}$. Thus, abbreviating $m + n = p$ and extracting the pth root from both sides, we have $y\sqrt[p]{a} = x\sqrt[p]{b}$, the differential of which is $dy\sqrt[p]{a} = dx\sqrt[p]{b}$, so that by drawing AE parallel to the ordinates and imagining a little triangle at the point where the asymptote CE meets the curve, we form the following proportion:[10] $dx : dy$ or $\sqrt[p]{a} : \sqrt[p]{b} :: AC\left(\frac{n}{p}a\right) : AE = \frac{n}{p}\sqrt[p]{ba^{p-1}}$. Now the [15] values of CA and of AE being thusly determined, we may draw the indefinite straight line CE, which is the asymptote that we wish to find.

If $m = 1$ and $n = 1$, the curve is the ordinary hyperbola and we have $AC = \frac{1}{2}a$ and $AE = \frac{1}{2}\sqrt{ab}$, that is to say half of the conjugate diameter,[11] as we know from elsewhere to be true.

Example IV. (§14) Let the nature of the curve AM be given by the equation[12] $y^3 - x^3 = axy$, where $AP = x$, $PM = y$ and a is a given straight line (see Fig. 2.5). The differential is $3yy\,dy - 3xx\,dx = ax\,dy + ay\,dx$. Thus

$$\frac{y\,dx}{dy} = \frac{3y^3 - axy}{3xx + ay}$$

and

$$AT\left(\frac{ydx}{dy} - x\right) = \frac{3y^3 - 3x^3 - 2axy}{3xx + ay} = \frac{axy}{3xx + ay},$$

substituting the value $3axy$ for $3y^3 - 3x^3$.

If we now suppose that AP and PM are both infinitely large, the tangent TM becomes the asymptote CE and the straight lines AT and AS become AC and AE, which determine the position of the asymptote. Now we denote AT by $t = \frac{axy}{3xx+ay}$, from which we conclude that $y = \frac{3txx}{ax-at} = \frac{3tx}{a}$ when AT becomes AC, because then at is null with respect to ax. Thus, substituting this value $\frac{3tx}{a}$ for y in $y^3 - x^3 = axy$, we have $27t^3x^3 - a^3x^3 = 3a^3txx$, from which we conclude that AC $(t) = \frac{1}{3}a$ by deleting the term $3a^3txx$, which is zero with respect to the two others $27t^3x^3$ and a^3x^3, because x is infinite. Furthermore, we denote AS $\left(y - \frac{x\,dy}{dx}\right)$ by $s = \frac{axy}{3yy-ax}$, from which we conclude that $x = \frac{3syy}{ay+as} = \frac{3sy}{a}$, because y is infinite with respect to s, so the term as is zero with respect to the term ay. Substituting this value into the equation of the curve, we find that AE $(s) = \frac{1}{3}a$. From this it follows that if we take the lines AC and AE both equal to $\frac{1}{3}a$ and we draw the indefinite straight line CE, it is the asymptote to the curve AM.

[16] We use these two last examples to guide ourselves in finding asymptotes to other curved lines.

---

[10] In L'Hôpital (1696) the last term was written as $\frac{n}{p}\sqrt[p]{ba^{p-1}}$.

[11] Also called the conjugate axis. See p. xxi for a discussion of generalized hyperbolas.

[12] This is the Folium of Descartes reflected in the x-axis.

Proposition II.

Problem. (§15) *If we suppose in the previous proposition that the abscissas AP are the portions of a curved line for which we know how to draw the tangents PT* (see Fig. 2.6), *then from a given point M on the curve AM, we wish to draw the tangent MT.*[13]

Having drawn the ordinate MP with tangent PT, suppose that the straight line MT that meets it in T is the desired tangent. We imagine another ordinate mp infinitely close to the first one and a little straight line MR parallel to PT. Denoting the given quantities AP by x and PM by y we have, as before, Pp or $MR = dx$ and $Rm = dy$. The similar triangles mRM and MPT give $mR\ (dy) : RM\ (dx) :: MP\ (y) : PT = \frac{y\,dx}{dy}$. We complete the solution by means of the equation that expresses the relationship of the abscissas $AP\ (x)$ to the ordinates $PM\ (y)$, as we have seen in the previous examples, and as we will further see in the ones that follow.

Example I. (§16) Let $\frac{yy}{x} = \frac{x\sqrt{aa+yy}}{a}$, the differential of which is

$$\frac{2xy\,dy - yy\,dx}{xx} = \frac{dx\,\sqrt{aa+yy}}{a} + \frac{xy\,dy}{a\sqrt{aa+yy}}.$$

Reducing this to a single proportion, we have

$$dy : dx\ \ (MP : PT) :: \frac{\sqrt{aa+yy}}{a} + \frac{yy}{xx} : \frac{2xy}{xx} - \frac{xy}{a\sqrt{aa+yy}}.$$

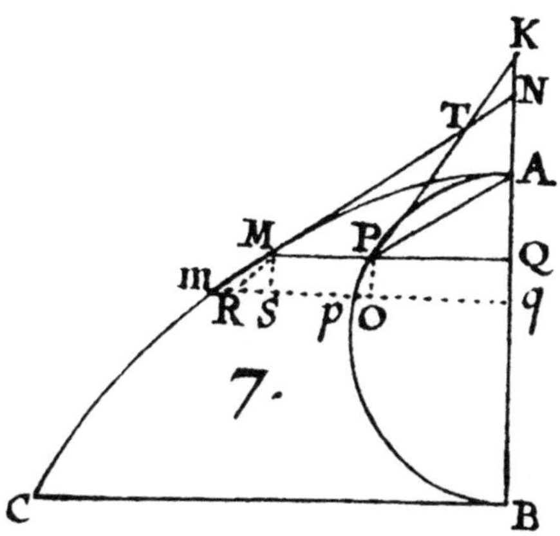

Fig. 2.6 The Cycloid and Related Examples

[13]See p. xxvii for a discussion of this construction.

From this, the ratio of the given MP to the subtangent PT that we wish to find is expressed entirely by known quantities, without differentials. This is what was proposed.

Example II. [17] (§17) Let $x = \frac{ay}{b}$, the differential of which is $dx = \frac{a\,dy}{b}$. We have

$$PT\left(\frac{y\,dx}{dy}\right) = \frac{ay}{b} = x.$$

If we suppose that the curved line APB is a semi-circle and that the ordinates MP, being prolonged to Q, are perpendicular to the diameter AB, the curve AMC is the *half-roulette*,[14] or *cycloid*[15] – *simple* when $b = a$, *curtate*[16] when b is greater and *prolate*[17] when it is smaller.

Corollary. (§18) *If the roulette is simple*[18] *and we draw the chord AP, then I say it is parallel to the tangent MT. This is because the triangle MPT is therefore isosceles, so the external angle TPQ is twice the internal opposite angle TMQ. Now the angle APQ is equal to the angle APT, because both have as measure half the arc AP, and thus it is half of the angle TPQ. The angles TMQ and APQ are therefore equal to each other and consequently the lines MT and AP are parallel.*

Proposition III.

Problem. (§19) *Let AP be any curved line that has the straight line KNAQ as diameter* (see Fig. 2.6) *and for which we know how to draw tangents PK. Furthermore, let AM be another curve such that, if we draw at will the ordinate MQ which cuts the first curve at the point P, the relationship of the arc AP to the ordinate MQ is expressed by any equation. We wish to draw the tangent MN from any given point M.*

If we denote the given quantities PK by t, KQ by s, the arc AP by x and MQ by y, we have (imagining another ordinate mq infinitely close to MQ and drawing PO and MS parallel to AQ) $Pp = dx$ and $mS = dy$. Because of the similar triangles KPQ and PpO we have [18] $PK(t) : KQ(s) :: Pp(dx) : PO$ or $MS = \frac{s\,dx}{t}$. Also, because of the similar triangles mSM and MQN, $ms(dy) : SM\left(\frac{s\,dx}{t}\right) :: MQ(y) : QN = \frac{sy\,dx}{t\,dy}$. Now, by means of the differential of the given equation we find a value of dx in terms that are all multiplied by dy. Consequently, if we substitute this value in place of dx in $\frac{sy\,dx}{t\,dy}$, the dy cancels and the value of the subtangent QN that we wish to find is expressed entirely in known terms. This is what we were required to find.

---

[14]The five terms defined here were not italicized in L'Hôpital (1696).

[15]See p.xxvii for a discussion of this construction of the cycloid.

[16]In L'Hôpital (1696) the term *allongée* (elongated) was used; "curtate" is the standard modern term.

[17]In L'Hôpital (1696) the term *accourcie* (shortened) was used; "prolate" is the standard modern term.

[18]Compare this to Bernoulli's Problem VI on p. 198.

Proposition IV.

Problem. (§20) *Let AQC and BCN be two curved lines* (see Fig. 2.7) *that have the straight line TEABF as diameter and such that we know how to draw the tangents QE and NF. Furthermore, let MC be another curved line such that the relationship of the ordinates MP, QP, and NP is expressed by any equation. From a given point M on this last curve we wish to draw the tangent MT.*

Imagine the little triangles QOq, MRm, and NSn at the points Q, M, and N. Denote the given quantities PE by s, PF by t, PQ by x, PM by y, and PN by z, so that $Oq = dx$, $Rm = dy$, and $Sn = -dz$ (see §8), because as x and y increase, z decreases. Because the triangles QPE and qOQ are similar, we have $QP\ (x) : PE\ (s) :: qO\ (dx) : OQ$ or MR or $SN = \frac{s\,dx}{x}$. Because the triangles NPF and nSN are similar, we have $NP\ (z) : PF\ (t) :: nS\ (-dz) : SN = \frac{-t\,dz}{z} = \frac{s\,dx}{x}$ (from which we conclude that $dz = \frac{-szdx}{tx}$). Because the triangles MPT and mRM are similar, we have $mR\ (dy) : RM\ \left(\frac{s\,dx}{x}\right) :: MP\ (y) : PT = \frac{sy\,dx}{x\,dy}$. If in the given equation, we substitute the value $-\frac{sz\,dx}{tx}$ in place of the differential dz, then we find the value of dx in terms of dy. When this is substituted in $\frac{sy\,dx}{x\,dy}$, the dy cancels and the value of the subtangent PT is expressed entirely in known terms.

Example. [19] (§21) Let $yy = xz$, the differential of which is $2y\,dy = z\,dx + x\,dz = \frac{tz\,dx - sz\,dx}{t}$, substituting the negative value $-\frac{sz\,dx}{tx}$ for dz. From this we conclude that $dx = \frac{2ty\,dy}{tz - sz}$, and consequently $PT \left(\frac{sy\,dx}{x\,dy}\right) = \frac{2styy}{txz - sxz} = \frac{2st}{t-s}$, substituting the value xz for yy.

Now let the general equation be $y^{m+n} = x^m z^n$, the differential of which is

$$\overline{m+n}\,y^{m+n-1}\,dy = mz^n x^{m-1}\,dx + nx^m z^{n-1}\,dz = \frac{m\,tz^n x^{m-1}\,dx - nsz^n x^{m-1}\,dx}{t},$$

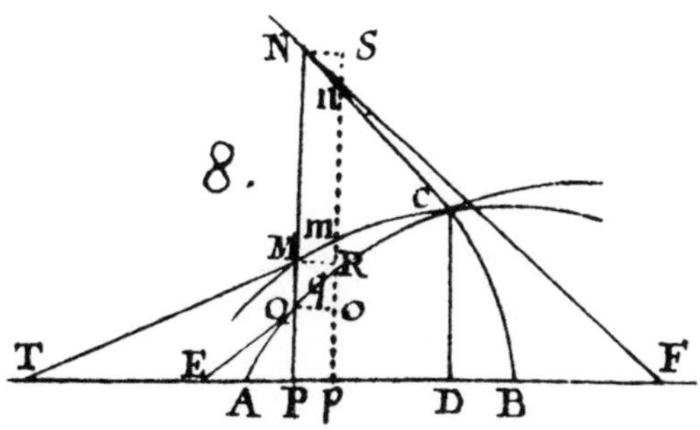

Fig. 2.7 Conic Sections and Related Examples

substituting the value $-\frac{sz\,dx}{tx}$ for dz. From this we conclude that

$$PT\left(\frac{sy\,dx}{x\,dy}\right)=\frac{\overline{mst+nst}\,y^{m+n}}{mtz^nx^m-nsz^nx^m}=\frac{mst+nst}{mt-ns},$$

substituting the value x^mz^n for y^{m+n}.

We may note that if the curves *AQC* and *BCN* become straight lines, the curve *MC* would then be one of the conic sections[19] to infinity: namely, an Ellipse if the ordinate *CD*, which emanates from the intersection point *C*, falls between the extremities *A* and *B*, a Hyperbola if it falls on one side or the other, and finally a Parabola if either of the extremities *A* or *B* is infinitely distant from the other, that is to say when one of the straight lines *CA* or *CB* is parallel to the diameter *AB*.

Proposition V.

Problem. (§22) *Let APB be a curved line* (see Fig. 2.8) *which has a fixed and invariable beginning at the point A, such that we know how to draw the tangents PH. Now let F be another fixed point not on this line and let CMD be another curved line such that when we draw any straight line FMP, the relationship of the part FM to the portion of the curve AP is expressed by whatever equation we wish. We wish to draw the tangent MT from the given point M.*

We erect the perpendicular *FH* on *FP*, which meets [20] the given tangent *PH* at the point *H* and the *MT* that we wish to find at the point *T*. We imagine the straight line *FRmOp*, which makes an infinitely small angle with *FP* and describe the little circular arcs *PO* and *MR* with center *F*. The little triangle *pOP* is similar to the right triangle *PFH*, because the angles *HPF* and *HpF* are equal (see §2), since they differ from each other only by the angle *PFp*, which we suppose to be infinitely small. Furthermore, the angle *pOP* is right, because the tangent at *O* (which is none other than the continuation of the little arc *PO* which is taken to be a straight line) is perpendicular to the ray *FO*. For the same reasons the triangles *mRM* and *MFT* are similar. Now it is clear that the little triangles or sectors *FPO* and *FMR* are similar.

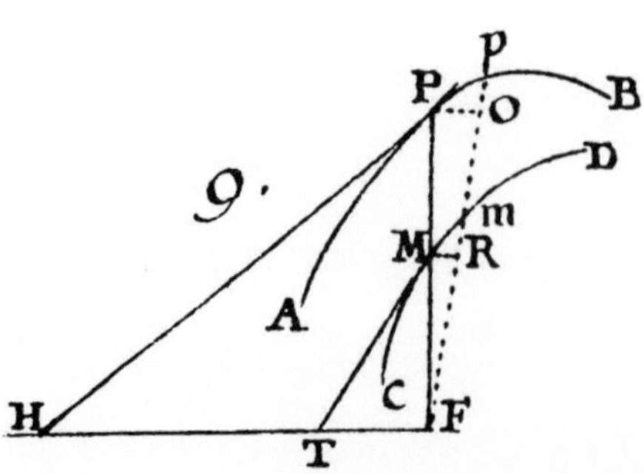

Fig. 2.8 The Spiral of Archimedes and Related Examples

[19] See p. xxiii for a discussion of this construction of the conic sections.

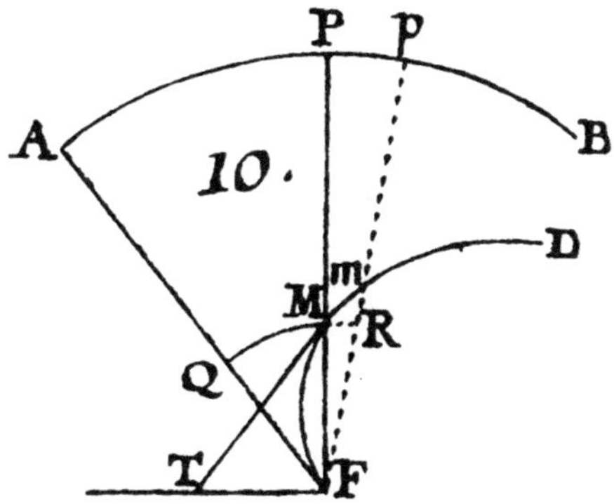

Fig. 2.9 The Spiral of Archimedes

Therefore, if we denote the known quantities *PH* by t, *HF* by s, *FM* by y, *FP* by z, and the arc *AP* by x, then we will have that

$$PH\ (t) : HF\ (s) :: Pp\ (dx) : PO = \tfrac{s\,dx}{t},$$
$$FP\ (z) : FM\ (y) :: PO\ \left(\tfrac{s\,dx}{t}\right) : MR = \tfrac{ys\,dx}{tz}, \quad \text{and}$$
$$mR\ (dy) : RM\ \left(\frac{sy\,dx}{tz}\right) :: FM\ (y) : FT = \frac{syy\,dx}{tz\,dy}.$$

We complete the solution by means of the differential of the given equation.

Example. (§23)[20] If we wish the curve *APB* (see Fig. 2.9) to be a circle that has its center at the fixed point F, then it is clear that the tangent *PH* becomes parallel and equal to the subtangent *FH*, because *HP* is also perpendicular to *PF*. Thus, in this case we have $FT = \frac{yy\,dx}{z\,dy} = \frac{yy\,dx}{a\,dy}$, denoting the straight line $FP(z)$ by a, because it is constant and no longer variable. Given this, if we denote the entire circumference by b, or some determined portion of it, and we make $b : x :: a : y$, then the curve *CMD*, or in this case *FMD*, is the *Spiral of Archimedes*[21] and we have $y = \frac{ax}{b}$, the differential of which is $dy = \frac{a\,dx}{b}$. From this we conclude that $y\,dx = \frac{by\,dy}{a} = x\,dy$, substituting the value $\frac{ax}{b}$ for y, and consequently $FT \left(\frac{yy\,dx}{a\,dy}\right) = \frac{xy}{a}$. This gives us the following construction.

[21] Let the circular arc *MQ* be described with center F and radius *FM*, which is terminated at Q by the radius *FA* that joins the fixed points A and F. Let *FT* be

[20] Compare this to Problem XI on p. 204.

[21] See p. xxix for a discussion of this construction of the Spiral of Archimedes.

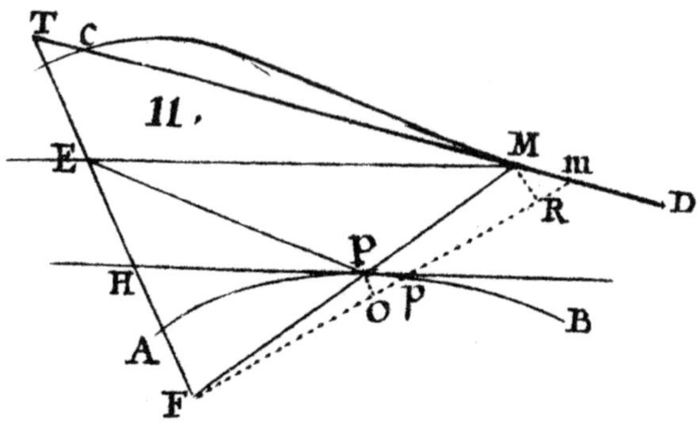

Fig. 2.10 The Conchoid of Nicomedes and Related Examples

taken equal to the arc MQ. I say that the straight line MT is tangent at M. This is because the sectors FPA and FMQ are similar, so we have $FP(a) : FM(y) :: AP(x) : MQ = \frac{yx}{a} = FT$.

More generally, if we make $b : x :: a^m : y^m$ (where the exponent m denotes whatever whole or fractional number we wish), then the curve FMD is one of the spirals to infinity and we have $y^m = \frac{a^m x}{b}$, the differential of which is $my^{m-1}\,dy = \frac{a^m\,dx}{b}$, from which we conclude that $y\,dx = \frac{mby^m\,dy}{a^m} = mx\,dy$, substituting the value $\frac{a^m x}{b}$ for y^m. Consequently, $FT\left(\frac{yy\,dx}{a\,dy}\right) = m \times MQ$.

Proposition VI.

Problem. (§24) *Let APB be a curved line* (see Fig. 2.10) *such that we know how to draw the tangents PH and let F be a fixed point not on this line. Let CMD be another curved line such that when we draw any straight line FPM we wish, the relationship of FP to FM is expressed by any equation. We wish to draw the tangent MT from the given point M.*

If we erect a perpendicular FHT on FM and imagine, as in the previous proposition, the little triangles POp and MRm similar to the triangles HFP and TFM, we denote the known quantities FH by s, FP by x, and FM by y, then we have

$$PF(x) : FH(s) :: pO(dx) : OP = \tfrac{s\,dx}{x},$$
$$FP(x) : FM(y) :: OP\left(\tfrac{s\,dx}{x}\right) : RM = \tfrac{sy\,dx}{xx}, \quad \text{and}$$
$$mR(dy) : RM\left(\tfrac{sy\,dx}{xx}\right) :: FM(y) : FT = \tfrac{syy\,dx}{xx\,dy}.$$

We then complete the solution by means of the differential of the given equation.

Example. [22] (§25) If we wish that the curve APB be the straight line PH and the relationship between FP and FM be expressed by the equation $y - x = a$, that is to say, that PM is always equal to the same given straight line a, then the differential is $dy = dx$, and hence $FT\left(\frac{syy\,dx}{xx\,dy}\right) = \frac{syy}{xx}$. This gives the following construction.

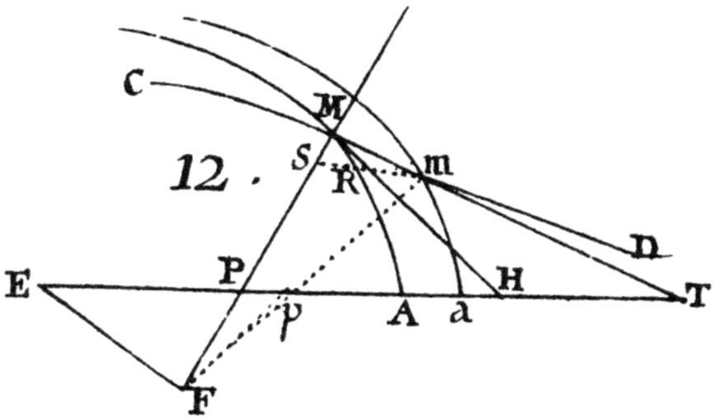

Fig. 2.11 Hyperbola, Conchoid, and Related Examples

Let *ME* be drawn parallel to *PH* and *MT* parallel to *PE*; I say that *MT* is tangent at M.

This is because $FP(x) : FH(s) :: FM(y) : FE = \frac{sy}{x}$ and $FP(x) : FE\left(\frac{sy}{x}\right) :: FM(y) : FT = \frac{syy}{xx}$. It is clear that the curve *CMD* is the *Conchoid*[22] *of Nicomedes*,[23] for which the asymptote is the straight line *PH* and the pole is the fixed point F.

Proposition VII.

Problem. (§26) *Let ARM be a curved line* (see Fig. 2.11) *such that we know how to draw the tangents MH, and which has the straight line $EPAHT$ as diameter. Let F be a point not on this diameter, from which emanates an indefinite straight line $FPSM$ that cuts the diameter at P and the curve at M. If we now imagine that the straight line FPM revolves around the point F, making the plane PAM move in a manner always parallel to itself along the immobile and indefinite straight line ET so that the distance PA always remains the same. It is clear that the continual intersection M of the lines FM and AM describes the motion of a curved line CMD.*[24] *We wish to draw the tangent MT to this curve from the given point M.*

If we imagine that the plane *PAM* is brought to the infinitely close position *pam* and we draw the line *mRS* parallel to *AP*, it is clear by the generating process that $Pp = Aa = Rm$ and consequently $RS = Sm - Pp$. Now denote the known quantities *FP* or *Fp* by x, *FM* or *Fm* by y, *PH* by s, *MH* by t, and [23] the differential *Pp* by dz. The similar triangles *FPp* and *FSm* give $Fp(x) : Fm(y) :: Pp(dz) : Sm = \frac{y\,dz}{x}$ $\left(\text{therefore } SR = \frac{y\,dz - x\,dz}{x}\right)$. Also, the similar triangles *MPH* and *MSR*

[22] See p. xxx for a discussion of this construction of the Conchoid of Nichomedes. Compare this to Bernoulli's Problem VII on p. 199.

[23] Nicomedes (ca. 280 BCE-ca. 210 BCE).

[24] That is, in this construction, the curve *ARM* moves rigidly along the axis *ET* as the line *FPM* revolves around F and that each point on *CMD* is the intersection of the rotated line *FPM* and the translated curve *ARM*.

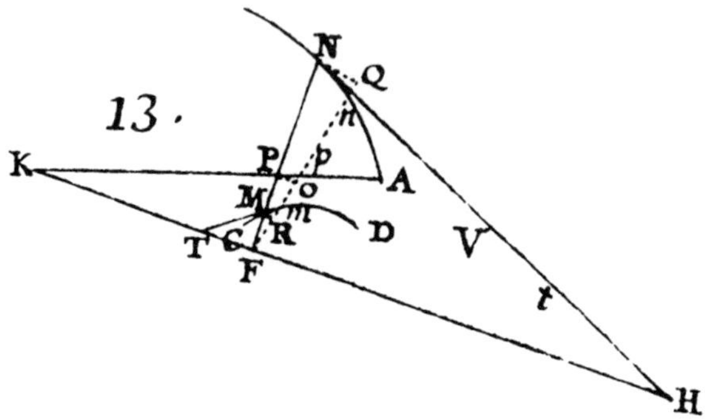

Fig. 2.12 The Cissoid and Related Examples

give $PH(s) : HM(t) :: SR\left(\frac{y\,dz - x\,dz}{x}\right) : RM = \frac{ty\,dz - tx\,dz}{sx}$. Finally, the similar triangles *MHT* and *MRm* give $MR\left(\frac{ty\,dz - tx\,dz}{sx}\right) : Rm(dz) :: MH(t) : HT = \frac{sx}{y-x}$. Therefore, if we draw *FE* to be parallel to *MH* and we take $HT = PE$, then the line *MT* is the tangent we wish to find.

If the line *AM* were a straight line, then the curve *CMD* would be a hyperbola that has the line *ET* as one of its asymptotes.[25] If it were a circle that has its center at the point P, then the curve *CMD* would be the *Conchoid of Nicomedes*, which has the line *ET* as its asymptote and the point F as its pole. However, if it were a parabola, then the curve *CMD* is the companion of the *Paraboloid of Descartes*,[26] which is also described below the straight line *ET* by the intersection of *FP* with the other half of the parabola.

Proposition VIII.

Problem. (§27) *Let AN be a curved line* (see Fig. 2.12) *that has the straight line AP as its diameter, with a fixed point F not on these lines. Let CMD be another curved line such that, if we draw the straight line FMPN as we may wish, the relationship among its parts FN, FP, and FM is expressed by any equation. We wish to draw the tangent MT at the given point M.*

Let *HK* be the perpendicular erected to the line *FN* at the point F, which meets the diameter *AP* at K and the given tangent *NH* at H. Let the little circular arcs *NQ*, *PO*, and *MR* be described with center F and radii *FN*, *FP*, and *FM*, terminated by the straight line *Fn*, which we imagine making an infinitely small angle with *FN*.

Given this, if we denote the known quantities *FK* by s, *FH* by t, *FP* by x, *FM* by y, and *FN* by z, then the similar triangles *PFK* and *pOP* [24] give $PF(x) : FK(s) :: pO(dx) : OP = \frac{s\,dx}{x}$. Also, the similar triangles *FMR*, *FPO*, and *FNQ* give $FP(x)$:

[25] See p. xxiv for a discussion of this construction of the hyperbola.

[26] In L'Hôpital (1696) this is footnoted as "*Géométrie*, Book 3." The "Paraboloid of Descartes" is usually called the *Trident of Descartes* or the *Parabola of Descartes*. It is a cubic curve with two branches, one of which resembles a parabola. The "companion" is the other branch.

$FM(y) :: PO\left(\frac{s\,dx}{x}\right) : MR = \frac{sy\,dx}{xx}$ and $FP(x) : FN(z) :: PO\left(\frac{s\,dx}{x}\right) : NQ = \frac{sz\,dx}{xx}$. Also, the similar triangles HFN and NQn give $HF(t) : FN(z) :: NQ\left(\frac{sz\,dx}{xx}\right) : Qn(-dz) = \frac{szz\,dx}{txx}$. Finally, the similar triangles mRM and MFT give $mR(dy) : RM\left(\frac{sy\,dx}{xx}\right) :: FM(y) : FT = \frac{syy\,dx}{xx\,dy}$. Now by means of the differential of the given equation we find the value of dy in terms of dx and dz, in which we substitute the negative value $\frac{-szz\,dx}{txx}$ in place of dz, because as x increases, z decreases. All the terms are multiplied by dx, so when this value is substituted into $\frac{syy\,dx}{xx\,dy}$, the dx will cancel. Consequently, the value of FT is expressed in known terms and free of differentials.

If we supposed that the straight line AP were a curved line and that we drew the tangent PK, we would still find the same value for FT, and the reasoning would remain the same.

Example. (§28) Suppose that the curved line AN (see Fig. 2.13) is a circle that passes through the point F (situated with respect to the diameter AP so that the line FB, perpendicular to this diameter, passes through the center G of this circle) and that PM is always equal to PN. It is clear that the curve CMD, which is in this case becomes FMA, is the *Cissoid*[27] *of Diocles*[28] and that we have the equation $z + y = 2x$, the differential of which is

$$dy = 2\,dx - dz = \frac{2txx\,dx + szz\,dx}{txx},$$

substituting the value $-\frac{szz\,dx}{txx}$, which we found above (see §27), for dz. Consequently, $FT\left(\frac{syy\,dx}{xx\,dy}\right) = \frac{styy}{2txx+szz}$.

If the given point M falls on the point A, then the lines FM, FN, and FP are all equal to FA, as are also [25] the straight lines FK and FH. Consequently we would have in this case $FT = \frac{x^4}{3x^3} = \frac{1}{3}x$, that is to say that if we take $FT = \frac{1}{3}AF$, and draw the line AT it will be tangent at A.

We may also find the tangents of the Cissoid by means of the first Proposition, by drawing the perpendiculars NE and ML on the diameter FB, and seeking the equation that expresses the ratio of the abscissa FL to the ordinate LM, which is done as follows. Denoting the known quantities FB by $2a$, FL or BE by x, and LM by y, the similar triangles FEN and FLM, and the property of the circle, give

[27] See p. xxxi for a discussion of this construction of the Cissoid of Diocles. Compare this to Bernoulli's Problem VIII on p. 201.

[28] Diocles of Carystus (ca. 240 BCE-ca. 180 BCE).

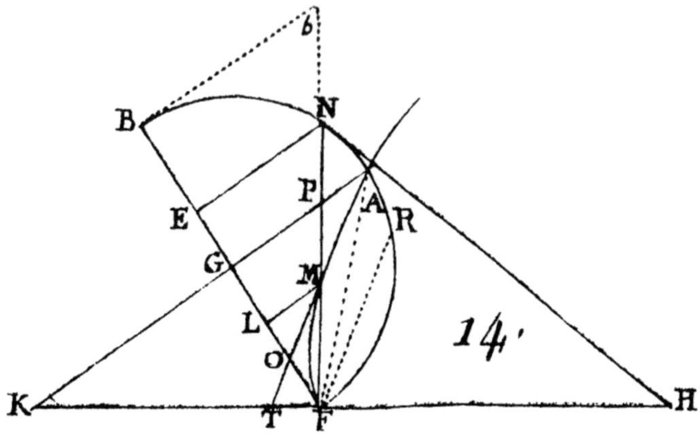

Fig. 2.13 The Cissoid of Diocles

$FL(x) : LM(y) :: FE : EN :: EN\left(\sqrt{2ax - xx}\right) : EB(x)$. From this we conclude $yy = \frac{x^3}{2a-x}$, the differential of which is

$$2y\,dy = \frac{6axx\,dx - 2x^3\,dx}{\overline{2a - x}^2}.$$

Consequently[29] (see §9)

$$LO\left(\frac{y\,dx}{dy}\right) = \frac{yy \times \overline{2a - x}^2}{3axx - x^3} = \frac{2ax - xx}{3a - x},$$

substituting the value $\frac{x^3}{2a-x}$ for yy.

Proposition IX.

Problem. (§29) *Let ANB and CPD be two curved lines* (see Fig. 2.14) *and FKT be a straight line, on these lines the fixed points A, C, and F are marked. Furthermore, let EMG be another curved line, such that if we draw the straight line FMN from any of its points M and draw MP parallel to FK, then the relationship of the arc AN to the arc CP is expressed by any equation. We wish to draw the tangent MT at a given point M on the curve EG.*

[29]In L'Hôpital (1696) the numerator of the second term in the equation that follows was given as $yy \times 2\overline{a - x^2}$.

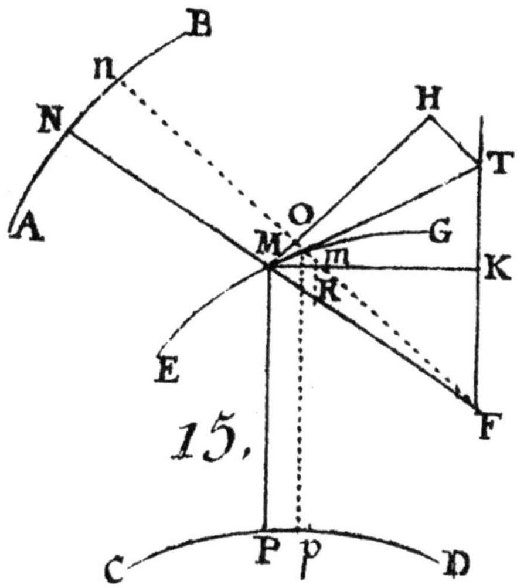

Fig. 2.14 The Quadratrix and Related Examples

We draw the line *TH* through the point T that we wish to find, parallel to *FM*, and the straight lines *MRK* and *MOH* through the given point M, parallel to the tangents at P and N. We now draw $FmOn$ infinitely close to *FMN* and *mRp* parallel to *MP*.

Given this, we denote the known quantities *FM* by s, *FN* by t, *MK* by u, *CP* by x, and *AN* by y (thus *Pp* or $MR = dx, Nn = dy$). The similar triangles *FNn* and *FMO* give $FN(t) : FM(s) :: Nn(dy) : MO = \frac{s\,dy}{t}$. [26] Also, the similar triangles *MOm* and *MHT*, and the similar triangles *MRm* and *MKT* give $MR(dx) : MO\left(\frac{s\,dy}{t}\right) :: MK(u) : MH = \frac{su\,dy}{t\,dx}$. Now, by using the differential of the given equation, we will have a value for *dy* in terms that are all multiplied by *dx*. When this is substituted into $\frac{su\,dy}{t\,dx}$, the *dx* will cancel. Consequently the value of *MH* is expressed in entirely known quantities. This gives the following construction.

Let *MH* be drawn parallel to the tangent at N and equal to the value we have just found. Let *HT* be drawn parallel to *FM*, which meets the straight line *FK* at T, from which and through the given point M the tangent *MT* that we wish to find is drawn.

Example. (§30) If we wish that the curve *ANB* be the quarter circle (see Fig. 2.15) that has the fixed point F as its center, and that the curve *CPD* be the radius *APF*, which is perpendicular to the straight line $FKGQTB$, and that the arc $AN(y)$ is always to the straight line $AP(x)$ as the quarter circle *ANB* (b) is to the radius $AF(a)$,

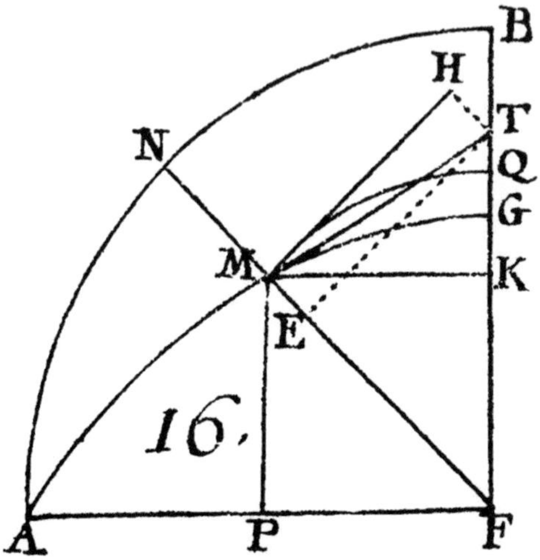

Fig. 2.15 The Quadratrix of Dinostratus

then the curve *EMG* is the *Quadratrix*[30] *of Dinostratus*[31] *AMG*, and we will have $MH\left(\frac{su\,dy}{t\,dx}\right) = \frac{as\,dy - sx\,dy}{a\,dx}$, because *FP* or $MK(u) = a - x$ and $FN(t) = a$. However, the proportional relationship that we supposed gives $ay = bx$ and so $a\,dy = b\,dx$. Thus, substituting the values $\frac{ay}{b}$ and $\frac{b\,dx}{a}$ in place of x and dy in *MH*, we find that $MH = \frac{bs - ys}{a}$. This gives the following construction.

Let *MH* be drawn perpendicular to *FM* and equal to the arc *MQ* described with center F, and let *HT* be drawn parallel to *FM*. I say that the line *MT* is tangent at M. This is because the similar sectors *FNB* and *FMQ* give $FN(a) : FM(s) :: NB(b - y) : MQ = \frac{bs - sy}{a}$.

Corollary. (§31) *If we wish to determine the point G where the quadratrix AMG meets the radius*[32] *FB (see Fig. 2.16), then we imagine another radius Fgb infinitely close to FGB. Drawing gf parallel to FB, the property of the quadratrix [27] and the similar triangles FBb and gfF, with right angles at B and f, give AB : AF :: Bb : Ff :: FB or AF : gf or FG. From this we see that if we take a third proportional*

[30]See p. xxxii for a discussion of this construction of the Quadratrix of Dinostratus. Compare this to Bernoulli's Problem IX on p. 202.

[31]Dinostratus (ca. 390 BCE-ca. 320 BCE). Dinostratus was a Greek mathematician who discovered the quadratrix and supposedly used it to solve the problem of squaring the circle.

[32]Compare this to Bernoulli's Problem X on p. 204.

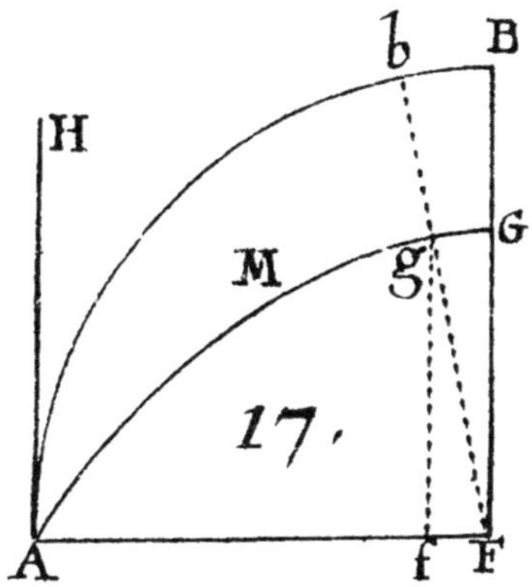

Fig. 2.16 The Quadratrix of Dinostratus

to the quarter circle AB and the radius AF,[33] *it is equal to FG, that is to say that* $FG = \frac{aa}{b}$*. This gives rise to the shortening of the construction of tangents.*[34]

If we draw TE parallel to MH (see Fig. 2.15), then the similar triangles FMK and FTE give $MK(a - x) : MF(s) :: ET \text{ or } MH \left(\frac{bs-sy}{a}\right) : FT = \frac{bss-yss}{aa-ax} = \frac{bss}{aa}$, *by substituting the value* $\frac{ay}{b}$ *for x and then dividing numerator and denominator by* $b - y$*. From this it is clear that the line FT is the third proportional to FG and FM.*

Proposition X.

Problem. (§32) *Let AMB be a curved line* (see Fig. 2.17) *such that if we draw the straight lines MF, MG, MH, etc., from any one of its points M to the foci*[35] *F, G, H, etc., their relationship is expressed by any equation. We wish to draw MP, which is perpendicular to the tangent at the given point M.*

Let *Mm* be an infinitely small arc taken on the curve *AB* and let the straight lines *FRm*, *GmS*, and *HmO* be drawn. Describe the little circular arcs *MR*, *MS*, and *MO* with centers *F*, *G*, and *H*. Next, with center *M* and any interval we also describe the circle *CDE*, which cuts the straight lines *MF*, *MG*, and *MH* at the points *C*, *D*, and *E*, from which we drop the perpendiculars *CL*, *DK*, and *EI* to *MP*. With this preparation done, I remark that:

1. The right triangles *MRm* and *MLC* are similar, because if we remove the common angle *LMR* from the right angles *LMm* and *RMC*, the remainders *RMm* and *LMC* are equal, and furthermore the angles at *R* and *L* are right. Similarly, we prove

[33]The third proportional to a and b is the value of x such that $a : b :: b : x$.

[34]See p. xxxiii for a discussion of how the Quadratrix may be used to construct π and to solve the problem of squaring the circle.

[35]In L'Hôpital (1696), the term *foyer* is used in optics to mean focus. The curves discussed in §32 are in some sense a generalization of the conic sections.

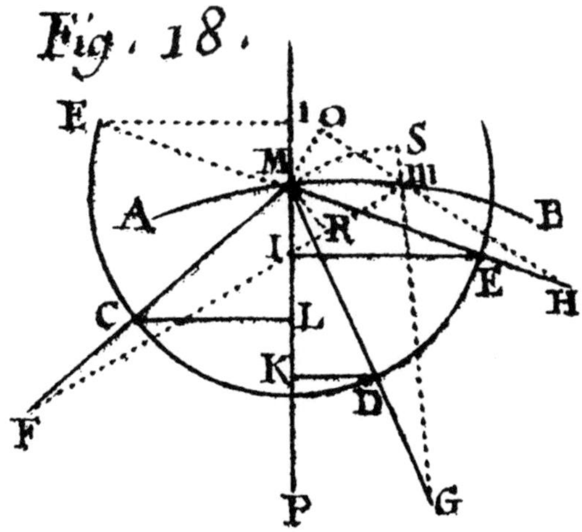

Fig. 2.17 Generalized Conic Sections, Defined via Multiple Foci

that the right triangles *MSm* and *MKD* are similar, and that the right triangles *MOm* and *MIE* are similar. Hence, because the hypotenuse *Mm* is common to the little triangles *MRm*, *MSm*, and *MOm* and the [28] hypotenuses *MC*, *MD*, and *ME* of the triangles *MLC*, *MKD*, and *MIE* are equal to one another, it follows that the perpendiculars *CL*, *DK*, and *EI* have the same ratios among themselves as the differentials *Rm*, *Sm*, and *Om*.

2. The lines that emanate from the foci situated on the same side of the perpendicular *MP* increase while the others decrease, or vice versa. For example, in figure 2.17, *FM* increases by its differential *Rm*, while the others, *GM* and *HM*, decrease by their differentials *Sm* and *Om*.

We suppose for the moment, to illustrate these ideas, that the equation that expresses the relationship of the straight lines *FM* (x), *GM* (y) and *HM* (z) is $ax + xy - zz = 0$, whose differential is $a\,dx + y\,dx + x\,dy - 2z\,dz = 0$. It is clear that the tangent at M (which is nothing more than the continuation of the little side *Mm* of the polygon, that we conceive of (see §3) as composing the curve *AMB*) must be placed so that if we draw from any one of its points m the lines *mR*, *mS*, and *mO* parallel to the straight lines *FM*, *GM*, and *HM*, terminated at R, S, and O by *MR*, *MS*, and *MO*, perpendiculars to the same straight lines, then we still have the equation $\overline{a+y} \times Rm + x \times Sm - 2z \times Om = 0$. By substituting for *Rm*, *Sm*, and *Om* their proportional values *CL*, *DK*, and *EI*, this amounts to the same thing as that the perpendicular *MP* to the curve must be placed so that $\overline{a+y} \times CL + x \times DK - 2z \times EI = 0$. This gives the following construction.

We imagine that the point C (see Figs. 2.17, 2.18) is loaded with a weight $a + y$ that multiplies the differential dx of the straight line *FM* on which it is situated. Similarly, the point D is loaded with a weight x and the point E, taken on the other side of M with respect to the focus H (because the term $-2z\,dz$ is negative)

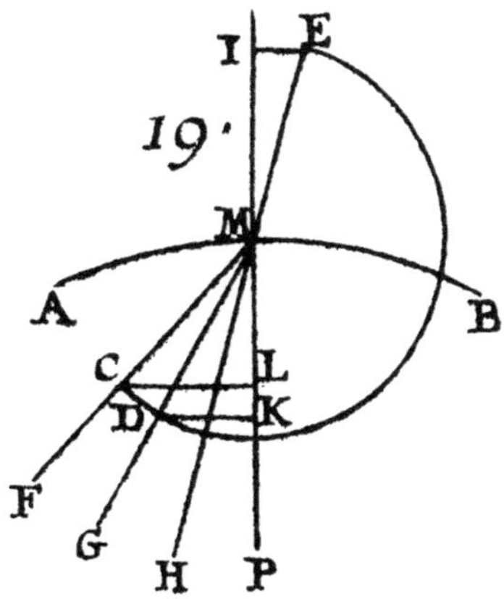

Fig. 2.18 Generalized Conic Sections, Case II

is loaded with the weight $2z$. I say that the straight line *MP* that passes through the common center of gravity of the weights that we suppose to be at C, D, and E is the required perpendicular. For it is clear from the principles of Mechanics that any straight line which passes through the center of gravity of several weights separates them so that the weights on one side multiplied by their distances from the line are precisely equal to the weights [29] on the other side, each one also multiplied by its distance to the same straight line. Thus, considering the case that as x increases, y and z also increase, (see Fig. 2.18) that is to say that the foci F, G, and H lie on the same side of *MP*, as we always suppose in taking the differential of the given equation according to the prescribed rules, it follows from this that the line *MP* has the weights at C and D on one side and the weight at E on the other. Therefore, we have[36] $\overline{a+y} \times CL + x \times DK - 2z \times EI = 0$, which is the equation to be constructed.

I say now that because the construction is correct in this case it will also be correct in all the others. For if we suppose, for example, that the point M changes position on the curve so that as x increases, y and z decrease (see Fig. 2.17), that is to say that the foci G and H move to the other side of *MP*, it follows that:

1. We must change (see §8) the signs of the terms multiplied by dy and dz, or by their proportionals DK and EI, in the differential of the given equation, so that the equation to be constructed in this new case is $\overline{a+y} \times CL - x \times DK + 2z \times EI = 0$.
2. The weights D and E change sides with respect to *MP* and therefore we have, by the property of the center of gravity, $\overline{a+y} \times CL - x \times DK + 2z \times EI = 0$, which is the equation to be constructed.

Because this always happens in every possible case it follows, etc.

[36] In L'Hôpital (1696) the overline grouping of $a + y$ was omitted here and in the next occurrence, but not in the third occurrence. They have been added for consistency.

It is clear that the same arguments always remain valid for any number of foci and for any given equation, so that we may therefore state the general construction.

Let the differential of the equation be taken so that I suppose that one side is equal to zero and let a circle *CDE* be described as we may wish with center M, which cuts the straight lines *MF*, *MG*, and *MH* at the points C, D, and E, at which we imagine weights that have the same ratios among themselves as the quantities that multiply the differentials of the lines on which they are situated. I say that the line *MP* that passes through their common center of gravity is the perpendicular that we wish to find. We should remark that if one of these [30] weights is negative in the differential of the given equation, we must then imagine it on the other side of the point M with respect to its focus.

If we wish that the foci F, G, and H be straight or curved lines, which the straight lines *MF*, *MG*, and *MH* (see Fig. 2.19) meet at right angles, the same construction still holds. For let the point m be taken infinitely close to M and drop the perpendiculars *mf*, *mg*, and *mh* to the foci and drop the little perpendiculars *MR*, *MS*, and *MO* from the point M to these lines. It is then clear that *Rm* is the differential of *MF* because the straight lines *MF* and *Rf* are perpendicular between the parallel lines *Ff* and *MR*, and so they are equal. Similarly, *Sm* is the differential of *MG* and *Om* is the differential of *MH*, and we then prove the rest as above.

We may also imagine that some or all the foci F, G, and H (see Fig. 2.20) are curved lines that have fixed and invariable origins at the points F, G, and H. Furthermore, suppose that the curved line *AMB* is such that, when we draw, for example, the tangents *MV* and *MX* and the straight line *MG* from any of its points M, the relationship among the curvilinear lines *FVM* and *HXM* and the straight line *GM* is expressed by any equation. If we take the tangent *mu* from a point m that is infinitely close to M, it is clear that it will meet the other tangent at the point

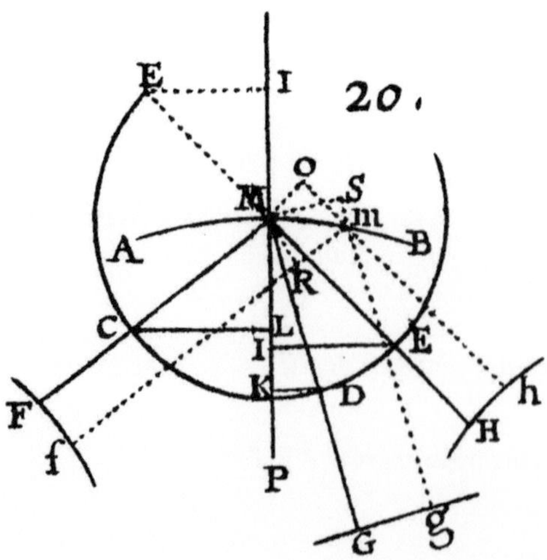

Fig. 2.19 Generalized Conic Sections – Foci as Straight or Curved Lines

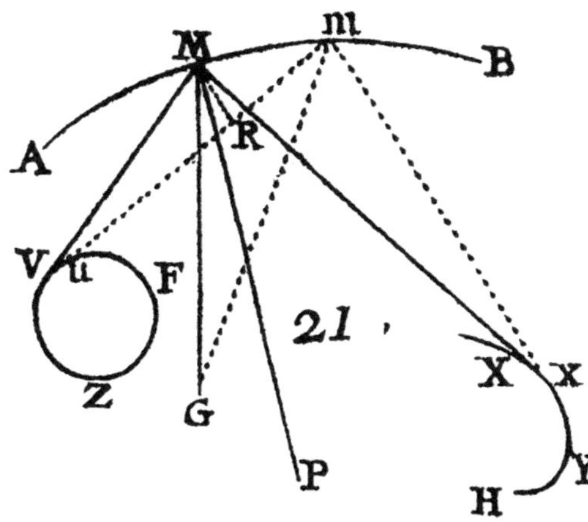

Fig. 2.20 Generalized Conic Sections – Curved Foci Joined by Tangents

V (because it is nothing other than the continuation of the little arc Vu considered to be a little straight line). Consequently, if we describe the little circular arc MR with center V, Rm is the differential of the curvilinear line FVM, which becomes $FVuRm$. All the rest is demonstrated as before.

Mr. Tschirnhaus *was the first to have given the idea for this problem in his book on* la Medecine de l'esprit.[37] *Mr.* Fatio[38] *then discovered a very ingenious solution which he had inserted in the* Journaux D'Hollande.[39] *However, in the way they conceived it, it is only a particular case of the general construction that I have just given.*

Example I. [31] (§33) Let $axx + byy + czz - f^3 = 0$ (the lines a, b, c, and f are given), the differential of which is $ax\,dx + by\,dy + cz\,dz = 0$. For this reason we imagine the weight ax at C (see Fig. 2.21), the weight by at D, and the weight cz at E, that is to say that these weights are to one another as these rectangles.[40] The line MP that passes through their common center of gravity is perpendicular to the curve at the point M.

However, if we draw FO parallel to CL and we take the radius MC to be the unit, then the similar triangles MCL and MFO give $FO = x \times CL$. Similarly, if we draw GR parallel to DK and HS parallel to EI, we find that $GR = y \times DK$ and $HS = z \times EI$, so that if we imagine the weights a, b, and c at the foci F, G, and H, then the line MP that passes through the center of gravity of the weights ax, by,

[37] *Medicina mentis* [1687].

[38] Nicolas Fatio de Duillier (1664–1753).

[39] *Journal des sçavans*, March 1687.

[40] The quantities ax, by, and cz are products of two lines and are therefore referred to as "rectangles," following the Euclidean tradition.

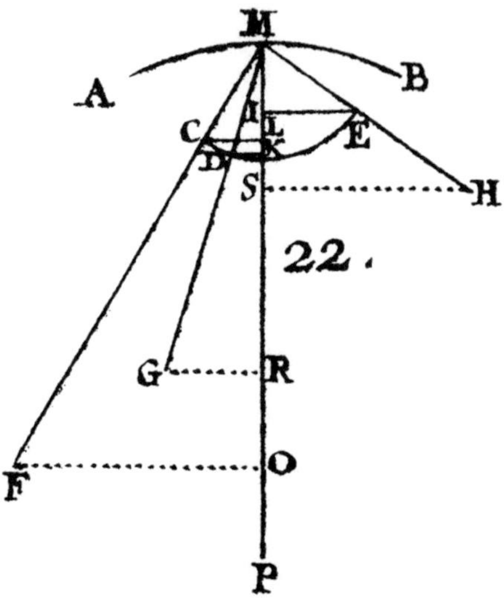

Fig. 2.21 Curve Defined by Three Focal Points

and cz, which we suppose to be at C, D, and E, also passes through the center of gravity of these new weights. Now, this center is a fixed point, because the weights at F, G, and H, namely a, b, and c, are constant straight lines that always remain the same in whatever place the point M is found. From this it follows that the curve *AMB* must be such that all its perpendiculars meet at the same point, that is to say, it is a circle that has this point as its center. Here, therefore is a very remarkable property of the circle that can be stated as follows.[41]

Suppose we have as many weights a, b, c, etc., as we may wish, situated on the same plane at F, G, H, etc., and we describe from their common center of gravity a circle *AMB*. I say that if we draw the straight lines *MF*, *MG*, *MH*, etc., from any of its points M, then the sum of their squares, each multiplied by the weight that corresponds to it, will always be equal to the same amount.

Example II. (§34) Let *AMB* be a curve (see Fig. 2.22) and let M be any of its points. [32] If we draw the straight line *MF* to the focus F, which is a fixed point, and drop a perpendicular *MG* to the focus G, which is a straight line, then the ratio of *MF* to *MG* is always the same as the ratio of the given a to the given b.

If we denote the quantities *FM* by x and *MG* by y, then we have[42] $x : y :: a : b$ and consequently $ay = bx$, the differential of which is $a\,dy - b\,dx = 0$. This is why, if we imagine the weight b at C taken on the other side of M with respect to F and

[41]This was first proved by Huygens and is given in Part IV, Proposition 12 of *Horologium Oscillatorium sive de motu pendulorum* (Huygens 1673, p. 124).

[42]In L'Hôpital (1696) there was a comma between a and b in the proportional relation that follows.

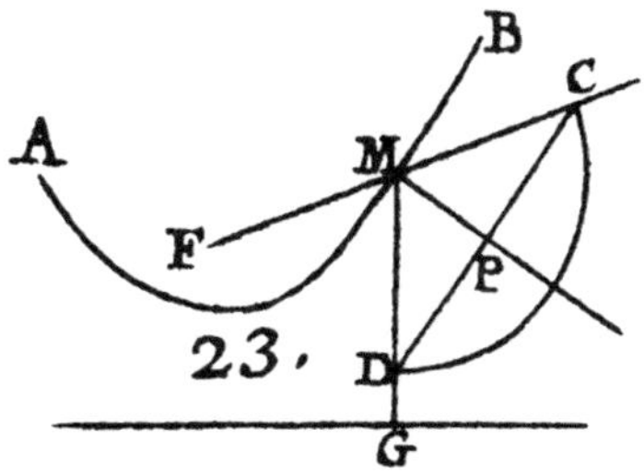

Fig. 2.22 Conic Sections Defined by Focus and Directrix

the weight a at D (the same distance from M), and draw the line *MP* through their common center of gravity, *MP* is the desired perpendicular.

Clearly, by the law of the lever, if we divide the chord *CD* at P so that $CP : DP :: a : b$, then the point P is the common center of gravity of the weights that we supposed to be at C and D.

The curve *AMB* is a conic section,[43] namely a Parabola if $a = b$, a Hyperbola if a is greater b, and finally an Ellipse when a is less than b.

Example III. (§35) Attach the extremities of a thread $FZVMGMXYH$ at F and H (see Fig. 2.23) and affix a small pin at G. We stretch this thread evenly by the means of a pen placed at M, so that the parts *FZV* and *HYX* are wound around curves that have their origins at F and H and the part *MG* is doubled, that is to say that it is folded back at G. With these things remaining in this state, we move the pen M; it is clear that it describes a curve *AMB*. We wish to erect the perpendicular *MP* on this curve from a given point M, where the position of the thread which is used to describe it is given at that point.

I note that the straight parts *MV* and *MX* of the thread are always tangent at V and X. If we denote the curvilinear lines $FZVM$ by x and $HTXM$ by z, the straight line *MG* by y, and a straight line taken equal to the length of the thread by a, then we always have $x + 2y + z = a$. From this I know that the curve *AMB* is included in the general construction. For this reason, we take the differential $dx + 2\,dy + dz = 0$ and imagining the weight 1 at C, the weight 2 at D, and the weight 1 [33] at E, I say that the line *MP*, which passes through the common center of gravity of these weights, is the perpendicular that we wish to find.

Proposition XI.

Problem. (§36)[44] *Let APB and EQF be any two lines* (see Fig. 2.24), *on which we know how to draw the tangents PG and QH and let PQ be a straight line on which a point M is marked. If we imagine the extremities P and Q of this straight line sliding along the lines AB and EF, then it is clear that the point M describes a*

[43] See p. xxvi for a discussion of this construction of the conic sections.

[44] Compare this to Bernoulli's Letter 28 p. 267.

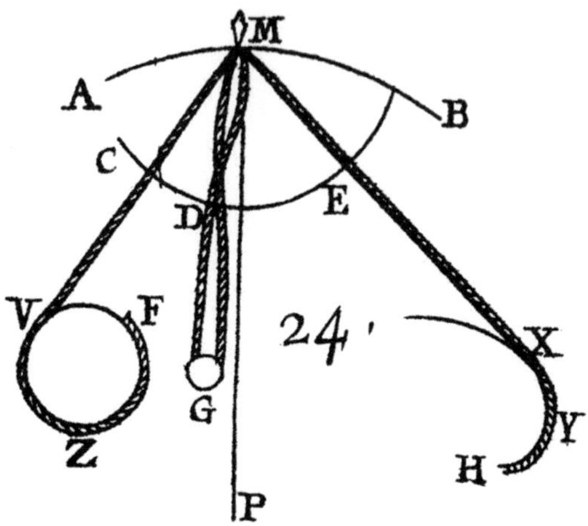

Fig. 2.23 Thread and Pen Example

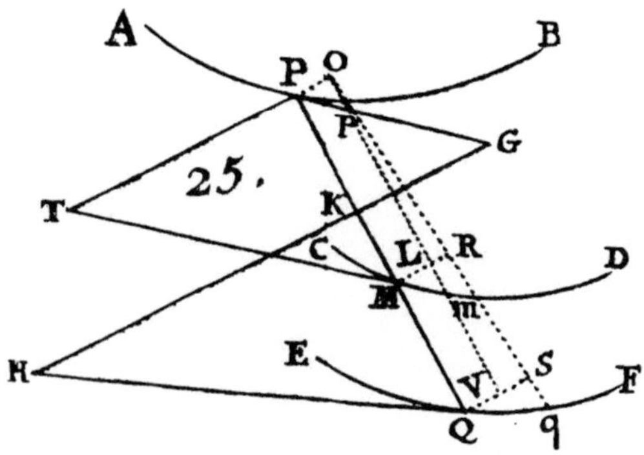

Fig. 2.24 Moving Straight Line Segment

curved line CD by this motion. We wish to draw the tangent MT to this curve at the given point M.

If we imagine that the mobile straight line *PMQ* is brought to the infinitely close position *pmq*, we draw the little straight lines *PO*, *MR*, and *QS* perpendicular to *PQ*, which form the little right triangles *pOP*, *mRM*, and *qSQ*. Taking *PK* equal to *MQ*, we draw the straight line *HKG* perpendicular to *PQ*, and we extend *OP* to T, where I suppose that it meets the tangent *MT* that we wish to find. Given this, it is clear that the little straight lines *Op*, *Rm*, and *Sq* are equal to one another, because by the construction, *PM* and *MQ* are everywhere the same.

We denote the known quantities *PM* or *KQ* by a, *MQ* or *PK* by b, *KG* by f, and *KH* by g and the little straight line *Op*, *Rm*, or *Sq* by dy. The similar triangles *PKG* and *pOP* give $PK(b) : KG(f) :: pO(dy) : OP = \frac{f\,dy}{b}$. Additionally, the similar triangles *QKH* and *qSQ* give $QK(a) : KH(g) :: qS(dy) : SQ = \frac{g\,dy}{a}$. Now, we know by common Geometry[45] that

$$MR = \frac{OP \times MQ + QS \times PM}{PQ} = \frac{f\,dy + g\,dy}{a+b}.$$

Thus, the similar triangles *mRM* and *MPT* give[46]

$$mR(dy) : RM\left(\frac{f\,dy + g\,dy}{a+b}\right) :: MP(a) : PT = \frac{af+ag}{a+b}.$$

This is what we were required to find.

Proposition XII.

Problem. [34] (§37) *Let BN and FQ be any two lines* (see Fig. 2.25) *having the straight lines BC and ED as axes, which intersect each other at right angles at the point A. Let LM be a curved line and draw the straight lines MGQ and MPN parallel to AB and AE from any of its point M. If the point E is a given fixed point on the straight line AE and the line EF is parallel to AC, then the relationship among the spaces EGQF and APND, and the straight lines AP, PM, PN, and GQ is expressed by any equation. We wish to draw the tangent MT from a given point M on the curve LM.*

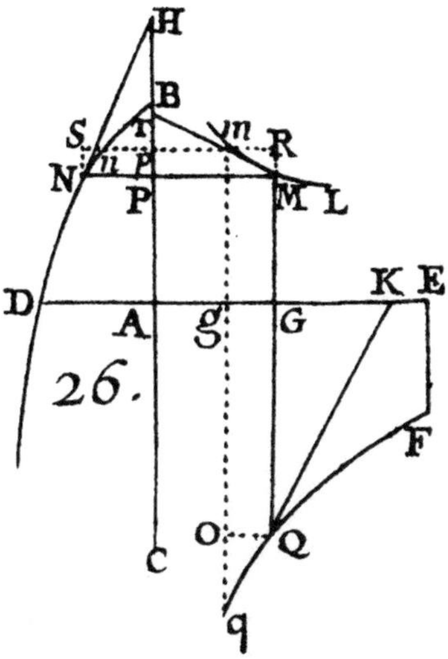

Fig. 2.25 The Logarithmic and Related Examples

[45] This follows by similar triangles that $LR : OL :: VS : DV$.

[46] In L'Hôpital (1696) the expression for *RM* did not have parentheses. They were added for consistency.

We denote the given and the variable quantities *AP* or *GM* by x, *PM* or *AG* by y, *PN* by u, *GQ* by z, the space $EGQF$ by s, the space $APND$ by t, and the given subtangents *PH* by a and *GK* by b. We have *Pp* or *NS* or $MR = dx$, *Gg* or *Rm* or $OQ = -dy$, and $Sn = -du = \frac{u\,dx}{a}$, because of the similar triangles *HPN* and *NSn*. Also, $Oq = dz = -\frac{z\,dy}{b}$, $NPpn = dt = u\,dx$, and $QGgq = ds = -z\,dy$, where we should note that the values of *Rm* and *Sn* are negative because as $AP(x)$ increases, $PM(y)$ and $PN(u)$ decrease. Given this, we take the differential of the given equation, in which we substitute the values $u\,dx$, $-z\,dy$, $-\frac{u\,dx}{a}$ and $-\frac{z\,dy}{b}$ for dt, ds, du, and dz, which gives a new equation that expresses the ratio of dy to dx or of *MP* to *PT*, that we wish to find.

Example I. (§38) Let $s + zz = t + ux$; taking the differential we have $ds + 2z\,dz = dt + u\,dx + x\,du$. Substituting the values for ds, dt, dz, and du, we find $-z\,dy - \frac{2zz\,dy}{b} = 2u\,dx - \frac{ux\,dx}{a}$, from which we conclude

$$PT\left(\frac{y\,dx}{dy}\right) = \frac{2ayzz + aybz}{bux - 2abu}.$$

Example II. [35] (§39) Let $s = t$; therefore, $ds = dt$. That is to say, $-z\,dy = u\,dx$ and hence $PT\left(\frac{y\,dx}{dy}\right) = -\frac{yz}{u}$. Now, because this quantity is negative, it follows (see §10) that we must take the point T on the opposite side of the point A, the origin of x's. If we suppose that the line *FQ* is a hyperbola that has that straight lines *AC* and *AE* as its asymptotes, so that $GQ(z) = \frac{cc}{y}$, and that the straight line *BND* is parallel to *AB*, so that $PN(u)$ is always equal to the given straight line c, then it is clear that the curve *LM* has the straight line *AB* as an asymptote and that its subtangent $PT\left(-\frac{yz}{u}\right) = -c$, that is to say, that it always remains the same.

In this case, the curve *LM* is called *Logarithmic*.[47]

Proposition XIII.

Problem. (§40) *Let BN and FQ be any two lines* (see Fig. 2.26) *with the same axis BA on which two fixed points A and E are marked. Let LM be a third curved line such that if we draw the straight line AN through any of its points M, describe the circular arc MG with center A, and draw GQ parallel to EF, which is perpendicular to AB, then the relationship among the spaces $EGQF(s)$ and $ANB(t)$ and the straight lines AM or $AG(y)$, $AN(z)$, and $GQ(u)$ is expressed by any equation. We wish to draw the tangent MT to the given point M on the curve LM.*

---

[47]In Problem V on p. 198, Bernoulli shows that the curve with constant subtangent is the curve "whose ordinates make a Geometric progression and abscissas an Arithmetic"; that is, this curve is an exponential curve. L'Hôpital uses that fact here, without proof, to identify this curve. The result had been published in Leibniz (1684). An exponential curve was called *logarithmic* at this time, because the logarithm of an ordinate is the corresponding abscissa. In Figure 2.25, the x-axis is vertical and the y-axis is horizontal, and curve *LM* has the equation $y = y_0 e^{-x/c}$, if we take the length of the segment *AE* to be y_0.

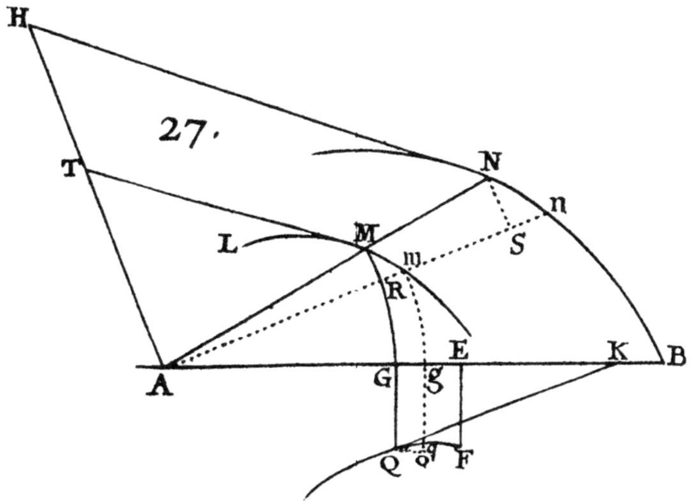

Fig. 2.26 The Logarithmic Spiral and Related Examples

After erecting the perpendicular ATH on AMN, let another straight line Amn be imagined, which is infinitely close to AMN, and another arc mg, another perpendicular gq, and the little circular arc NS with center A. We denote the given subtangents AH by a and GK by b, and we have Rm or $Gg = dy$ and $Sn = dz$. The similar triangles HAN and NSn, and KGQ [36] and QOq also give $SN = \frac{a\,dz}{z}$, $Oq = -du = \frac{u\,dy}{b}$, $GQqg = -ds = u\,dy$, and ANn or $AN \times \frac{1}{2}NS = -dt = \frac{1}{2}a\,dz$. We substitute all these values in the differential of the given equation, and form a new equation, from which we get a value for dz in terms of dy. Now, because of the similar sectors ANS and ARM, we find $AN(z) : AM(y) :: NS\left(\frac{a\,dz}{z}\right) : MR = \frac{ay\,dz}{zz}$. Also, because of the similar triangles mRM and MAT, we find $mR(dy) : RM\left(\frac{ay\,dz}{zz}\right) :: AM(y) : AT = \frac{ayy\,dz}{zz\,dy}$. Therefore, if we substitute for dz in this formula its value in terms of dy, then the differentials will cancel, and the value of the subtangent AT that we wish to find will be expressed in completely known terms. This is what we were required to find.

Example I. (§41) Let $uy - s = zz - t$, the differential of which is $u\,dy + y\,du - ds = 2z\,dz - dt$, which gives (after making the substitution)

$$dz = \frac{4bu\,dy - 2uy\,dy}{4bz + ab}$$

and substituting this value in $\frac{ayy\,dz}{zz\,dy}$ we find that

$$AT = \frac{4abuyy - 2auy^3}{4bz^3 + abzz}.$$

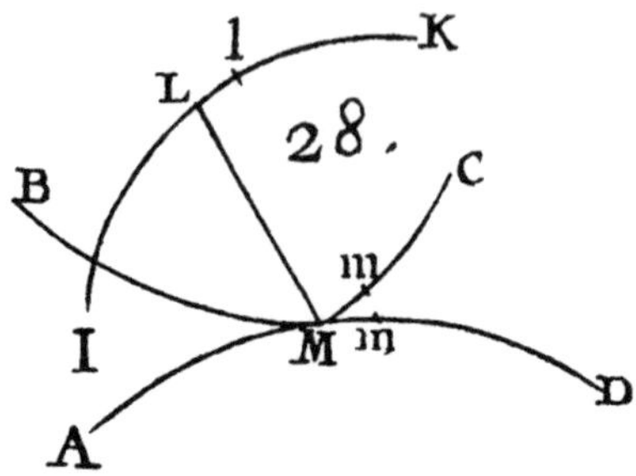

Fig. 2.27 Generalized Roulette

Example II. (§42) Let $s = 2t$; therefore, $ds = 2\,dt$, that is to say, $-u\,dy = -a\,dz$ or $dz = \frac{u\,dy}{a}$, and consequently $AT\left(\frac{ayy\,dz}{zz\,dy}\right) = \frac{uyy}{zz}$.

If the line *BN* is a circle that has the point *A* as its center, and radius $AB = AN = c$ and if *FQ* is a hyperbola such that $GQ(u) = \frac{ff}{y}$, then it is clear that curve *ML* makes an infinity of revolutions around the center *A* before it arrives there (because the space $FEGQ$ becomes infinite when the point *G* falls on *A*) and that $AT = \frac{ffy}{cc}$. From this, we see that the ratio of *AM* to *AT* is constant, and consequently that the angle *AMT* is everywhere the same.

In this case, the curve *LM* is called the *Logarithmic Spiral.*[48]

Proposition XIV.

Theorem. [37] (§43) *Let AMB and BMC* (see Fig. 2.27) *be any two curves on the same plane that touch each other at a point M and let L be a fixed point in the region*[49] *of the curve BMC. If we imagine for the moment that the curve BMC rolls on the curve AMD continually applying itself so that the revolved parts AM and BM are always equal to each other, it is clear that as the region BMC carries the point L, this point describes a kind of roulette ILK by this motion. Given this, I say that if we draw the straight line LM from each different positions on the curve BMC (from the describing point L to the point of contact M) it is perpendicular to the curve ILK.*

If we imagine two infinitely little and equal parts *Mm* and *Mm* on the two curves *AMD* and *BMC*, then we may consider them (see §3) as two little straight lines that makes an infinitely small angle at the point *M*. Now, as the little side *Mm* of the curve or polygon *BMC* falls on the little side *Mm* of the polygon *AMD*, it must be the case that the point *L* describes a little arc *Ll* with the point of contact *M* as the center. Therefore, it is evident that this little arc is a part of the curve *ILK* and consequently that the straight line *ML*, which is perpendicular to it, is also perpendicular to the curve *ILK* at the point *L*. This is what we were required to prove.

[48] See p. xxxiii for a discussion of the Logarithmic Spiral.

[49] In L'Hôpital (1696) the same word (*plan*) is used for both the plane that contains the two curves and the region of the plane bounded on the convex side of the curve *BMC*. Here and in the next sentence we translate this latter meaning by "region."

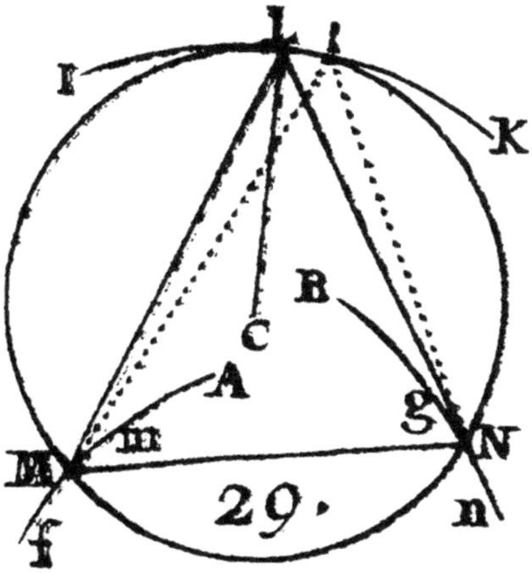

Fig. 2.28 Moving Rectilinear Angle

Proposition XV.

Problem. (§44) *Let MLN be any rectilinear angle* (see Fig. 2.28), *whose sides LM and LN touch any two curves AM and BN. If we make these sides slide around these curves, so that they continually touch them, then it is clear that the vertex L describes a curve ILK by this motion. We wish to draw a perpendicular LC to this curve, given the position of the angle MLN.*

[38] Let a circle be described that passes through the vertex L and through the points of the contact M and N. Let the straight line *CL* be drawn from the center of this circle. I say that it is perpendicular to the curve *ILK*.

Consider the curves *AM* and *BN* as polygons with an infinity of sides, such as *Mm* and *Nn*. If we suppose that the sides *LM* and *LN* of the rectilinear angle *MLN* always remain the same, then it is evident that if we make them slide around the points M and N (we consider the tangents *LM* and *LN* as the continuation of the little sides *Mf* and *Ng*) until the side *LM* of the angle falls on the little side *Mm* of the polygon *AM* and the other side *LN* falls on the little side *Nn* of the polygon *BN*, then the vertex L describes a little part *Ll* of arc of the circle *MLN*, because by the construction, this arc contains the given angle *MLN*. This little part *Ll* therefore coincides with the curve *ILK* and consequently the straight line *CL* which is perpendicular to it is also perpendicular to this curve at the point L. This is what we were required to show.

Proposition XVI.

Problem. (§45)[50] *Let* $ABCD$ *be a perfectly flexible rope* (see Fig. 2.29), *to which different weights* A, B, C, *etc. are attached, having whatever intervals AB, BC, etc.,*

[50]Compare this to Bernoulli's Letter 28 p. 265.

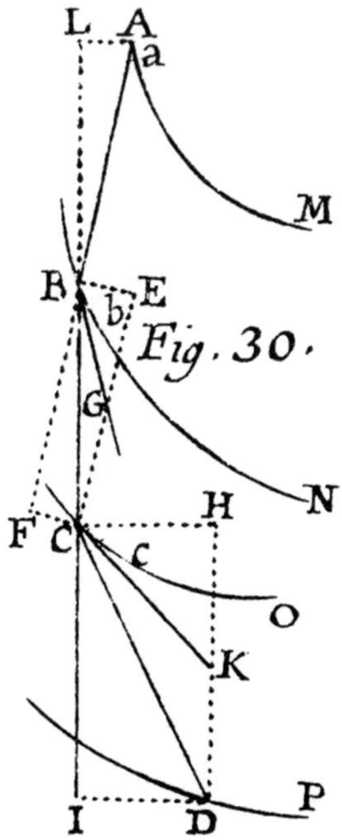

Fig. 2.29 The Tractrix

between them that we might wish. If we drag this rope in a horizontal plane by the extremity D, along a given curve DP, then it is clear that these weights will position themselves so that they stretch the rope and that they thereby describe the curves AM, BN, CO, etc. We wish to find the method for drawing the tangents given the position ABCD of the rope and the magnitude of the weights.

In the first instant that the extremity D moves towards P, the weights A, B, and C describe or tend to describe the same number of little sides *Aa*, *Bb*, and *Cc* of the polygons that compose the curves *AM*, *BN*, and *CO*. Consequently, in order to draw the tangents *AB*, *BG*, and *CK* we need to only determine the [39] directions of the weights A, B, and C in this first instant, that is to say the position of the straight lines that they tend to describe. In order to find these I remark that:

1. In this first instant, the weight A is drawn in the direction *AB*, and because there is no obstacle that opposes this direction, since it does not drag any weights behind it, it must follow this direction. Consequently, the straight line *AB* is tangent to the curve *AM* at A.

2. The weight B is drawn in the direction BC, however, because it drags the weight A, which is not in this direction, behind it, and this must consequently bring some change to it, the weight B does not go in the direction BC, but follows another straight line BG whose position we must find. I do that as follows.

   I describe the rectangle EF with BC as its diagonal, the side BF of which is prolonged from AB. Supposing that the force with which the weight B is drawn towards BC is expressed by BC, it is clear by the rules of Mechanics that this force BC may be divided up into two other forces, BE and BF, that is that the weight B, being pulled in the direction of BC by the force of BC is the same thing as if it were drawn by the force of BE in the direction of BE and by the force BF in the direction BF, both at the same time. Now, the weight A does not oppose the direction BE, because it is perpendicular to it, and consequently, the force BE in this direction remains entirely whole, but A opposes the direction BF with all of its weight. Therefore, insofar as the weight B with its force BF overcomes the resistance of the weight A, it is necessary that this force is distributed between these weights in proportion to the their masses or magnitudes. This is why, if we divide EC at the point G, so that CG is to GE as the weight A is to B, it is clear that EG expresses the remaining force with which the weight B tends to move in the direction of BF after overcoming the resistance of the weight A. Therefore, it is clear that the weight B is drawn at the same time by the force BE in the direction BE and by the force EG in the direction [40] BF or EC. Thus, it will tend to go along BG with the force BG, that is to say that BG will be its direction and consequently the tangent to the curve BN at B.
3. To find the tangent CK, I form the rectangle HI with CD as its diagonal, the side CI of which is prolonged from BC. I see that the weight B does not resist the force CH for which the weight C is drawn in the direction CH, but rather it resists the force CI with which it is drawn in the direction CI and furthermore that the weight A also resists this force. To know by how much, I draw the perpendicular AL on CB, extended on the side of B, and I remark that if AB expresses the force with which the weight A is drawn in the direction of AB, then BL expresses the force with which the same weight A is drawn in the direction BC, so that the weight C, with force CI, must overcome the entire weight B and, furthermore, a part of the weight A that is to the weight A as BL is to BA, or as BF is to BC. Therefore, if we make $B + \frac{A \times BF}{BC} : C :: DK : KH$, then it is clear that CK is the direction of the weight C, and consequently the tangent to the third curve CO at C.

If the number of curves were greater, then we would find the tangent of the fourth, fifth, etc., in the same manner. Finally, if we wished to have the tangents to the curves described at the points between the weights, then we would find them by means of §36.

Chapter 3
Use of the Differential Calculus for Finding the Greatest And the Least Ordinates, to Which Are Reduced Questions *De Maximis & Minimis*

Definition I. [41] Let *MDM* be a curved line, for which the ordinates *PM*, *ED*, and *PM* are parallel to one another, and are such that as the abscissa *AP* increases continually the ordinate *PM* also increases up to a certain point E, after which it decreases (see Figs. 3.1, 3.4), or, on the contrary, that it decreases up to a certain point E, after which it increases (see Figs. 3.2, 3.3). Given this, the line *ED* is called *the greatest* or *the least* ordinate.

Definition II. If we consider a quantity such as *PM*, which is composed of one or several indeterminates such as *AP*, so that as *AP* continually increases, this quantity *PM* also increases up to a certain point E, after which it decreases, or on the contrary. Suppose also that we wish to find a value *AE* for *AP* such that the quantity *ED*, of which it is composed, is greater or less than all other quantities *PM* similarly formed from *AP*. This is called a question *De maximis & minimis.*[1]

General Proposition. (§46) *Given the nature of the curved line MDM, we wish to find a value AE of AP such that the ordinate ED is the greatest or the least of its similar ordinates PM.*

If, while *AP* increases, *PM* also increases, then it is evident (see §8, 10) that the differential *Rm* is positive with respect to the differential of *AP*. On the contrary, when *PM* decreases while the abscissa *AP* still increases [42], the differential is negative. Now any quantity that continually increases or decreases may not change from positive to negative without passing through infinity or through zero,[2] namely through zero when it proceeds by decreasing, and through infinity when it proceeds by increasing. From this it follows that the differential of a quantity that expresses a

[1] In L'Hôpital (1696) this italicized Latin expression meaning "of maximums and minimums" was used within the French text. See Problem XII on p. 205 for the treatment of these types of problems.

[2] This is an implicit use of the Intermediate Value Theorem; the author clearly considers this to be self-evident.

R.E. Bradley et al., *L'Hopital's Analyse des infiniments petits*, Science Networks. Historical Studies 50, DOI 10.1007/978-3-319-17115-9_3

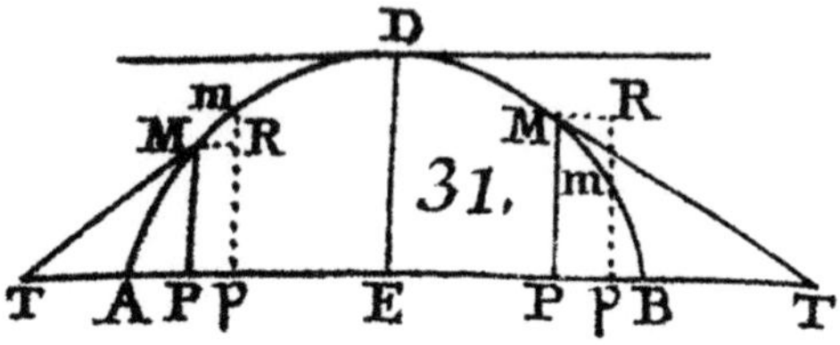

Fig. 3.1 Definition of Greatest Ordinate

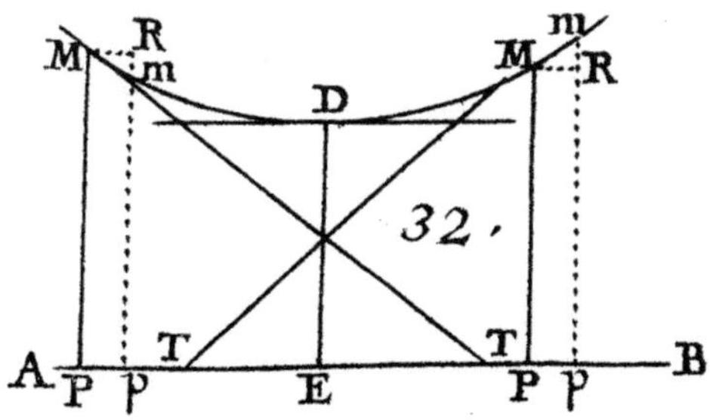

Fig. 3.2 Definition of Least Ordinate

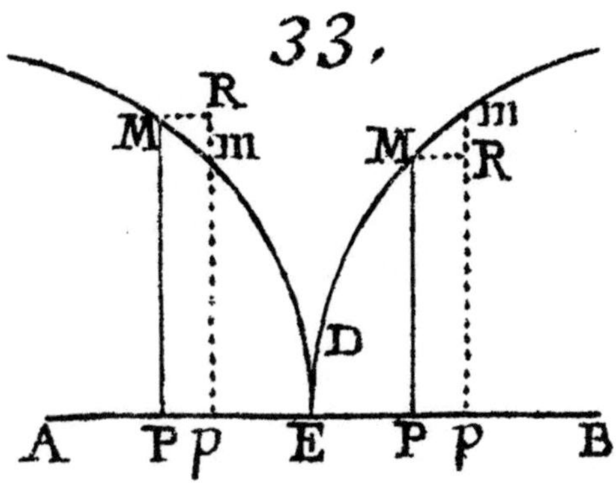

Fig. 3.3 Least Ordinate

greatest or a *least* must be equal to zero or to infinity. Now, given the nature of the curve *MDM*, we find (see Ch. 1 or 2) a value for *Rm*, first equal to zero, and then equal to infinity, which we will use to find the values of *AE* that we wish to find, under one or the other of these assumptions.

Remark. (§47) The tangent at *D* is parallel to the axis *AB* when the differential *Rm* becomes zero at this point (see Figs. 3.1, 3.2). However, when it becomes infinite, then the tangent coincides with the ordinate *ED* (see Figs. 3.3, 3.4). From this we see that the ratio of *mR* to *RM*, which expresses the ratio of the ordinate to the subtangent, is null or infinite at the point *D*.[3]

---

[3]The case of an extremum at a cusp was omitted from the *Lectiones*. Bernoulli alerted L'Hôpital to this case in Letter 22 (p. 251).

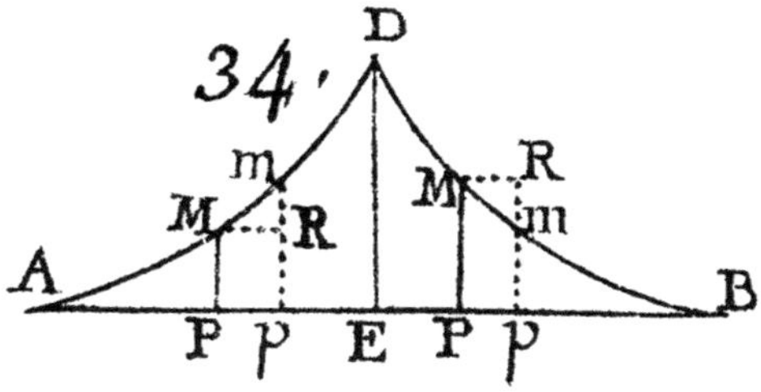

Fig. 3.4 Greatest Ordinate

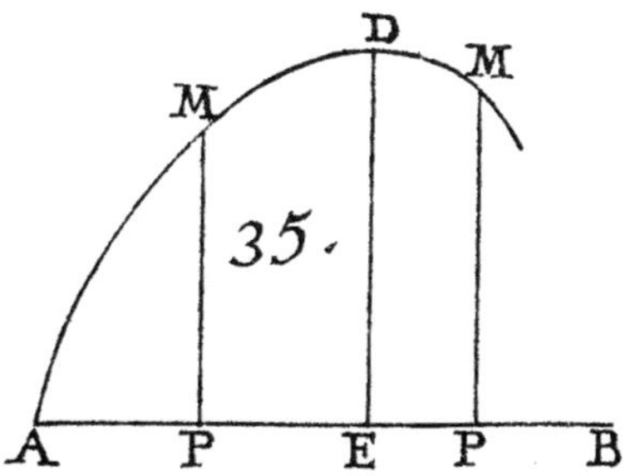

Fig. 3.5 The Folium of Descartes

We easily understand that a quantity that continually decreases cannot go from positive to negative without passing through zero, but we do not see with the same clarity that when it increases, it must pass through infinity. This is why, to aid the imagination, we consider the tangents at the points M, D, and M (see Figs. 3.1, 3.2). It is clear that in curves for which the tangent at D is parallel to the axis AB, the subtangent PT increases continually as long as the points M and P approach the points D and E and that when the point M falls on D, it becomes infinite. Therefore, when AP becomes greater than AE, the subtangent PT becomes (see §10) negative, from the positive that it had been, or vice versa.

Example I. (§48) Suppose that $x^3 + y^3 = axy$ ($AP = x$, $PM = y$ and $AB = a$) expresses the nature of the curve MDM (see Fig. 3.5). Taking differentials, we have $3xx\,dx + 3yy\,dy = ax\,dy + ay\,dx$ [43] and

$$dy = \frac{ay\,dx - 3xx\,dx}{3yy - ax} = 0$$

when the point P coincides with the point E that we wish to find, from which we conclude that $y = \frac{3xx}{a}$. Substituting this value in place of y in the equation

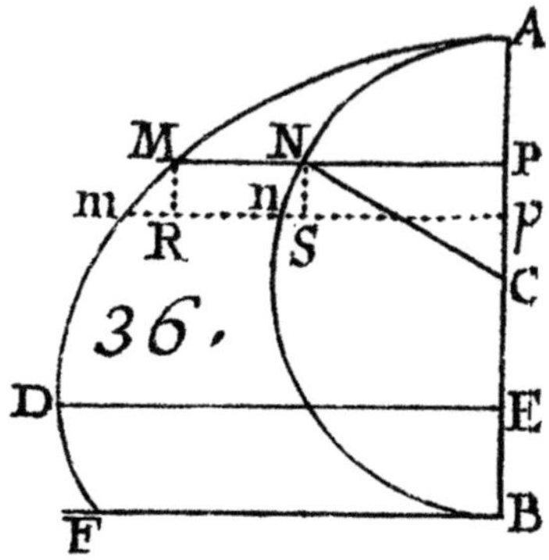

Fig. 3.6 Prolate Cycloid

$x^3 + y^3 = axy$, we find a value of $x = \frac{1}{3}a\sqrt[3]{2}$ for AE so that the ordinate ED is greater than all similar ordinates PM.[4]

Example II. (§49) Let $y - a = a^{1/3} \times \overline{a-x}^{2/3}$ be the equation that expresses the nature of the curve MDM (see Fig. 3.3). Taking differentials, we have $dy = -\frac{2\,dx\sqrt[3]{a}}{3\sqrt[3]{a-x}}$, which I first equate to zero. However, because this assumption gives me $-2\,dx\sqrt[3]{a} = 0$, which cannot give the value of AE, I then equate $\frac{-2\,dx\sqrt[3]{a}}{3\sqrt[3]{a-x}}$ to infinity, which gives me $3\sqrt[3]{a-x} = 0$, from which we conclude that $x = a$, which is the value of AE that we wish to find.

Example III. (§50) Let AMF be a prolate half-roulette (see Fig. 3.6), for which the base BF is less than the semi-circumference ANB of the generating circle that has the point C as its center. We wish to determine the point E on the diameter AB, so that the ordinate ED is the greatest possible.

We draw the ordinate PM at will, which cuts the semi-circle at N, and imagine as usual the little triangles MRm and NSn at the points M and N. Denoting the indeterminates AP by x, PN by z, and the arc AN by u, and the given quantities ANB by a, BF by b and CA or CN by c, then by the property of the roulette, $ANB\ (a) : BF\ (b) :: AN\ (u) : NM = \frac{bu}{a}$. Thus, $PM = z + \frac{bu}{a}$ and its differential $Rm = \frac{a\,dz + b\,du}{a} = 0$ when the point P falls on the point E that we wish to find. Now the right triangles NSn and NPC are similar, for if we remove the common angle CNS from the right angles CNn and PNS, the remainders SNn and PNC are equal. Consequently, [44] $CN\ (c) : CP\ (c - x) :: Nn\ (du) : Sn\ (dz) = \frac{c\,du - x\,du}{c}$. Thus, substituting this value in place of dz in $a\,dz + b\,du = 0$, we find that

[4]We note that the case where dy is infinite is not considered, presumably because this does not correspond to a greatest or least ordinate.

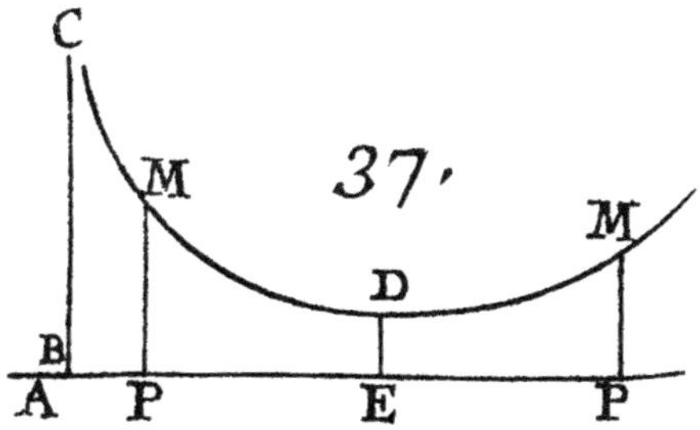

Fig. 3.7 General Problem, $m = 2$ and $n = -1$

$\frac{ac\,du - ax\,du + bc\,du}{c} = 0$, from which we conclude that x (which is AE in this case) $= c + \frac{bc}{a}$.

It is therefore clear that if we take CE on the side of B to be the fourth porportional[5] of the semi-circumference ANB to the base BF and to the radius CB, the point E is the one we wish to find.

Example IV. (§51) To cut the given line AB at a point E so that the product of the square of one of the parts AE by the other part EB is the greatest among all the other products formed in the same way (see Fig. 3.5).

If we denote the unknown AE by x and the given quantity AB by a, then we have $\overline{AE}^2 \times EB = axx - x^3$, which must be a *maximum*.[6] This is why we imagine a curved line MDM, such that the relationship of the ordinate MP (y) to the abscissa AP (x) is expressed by the equation $y = \frac{axx - x^3}{aa}$, and we wish to find a point E such that the ordinate ED is the greatest of all similar ordinates PM. This gives $dy = \dfrac{2ax\,dx - 3xx\,dx}{aa} = 0$, from which we conclude that AE $(x) = \frac{2}{3}a$.

If we wish, in general, that $x^m \times \overline{a-x}^n$ is a *maximum* (where m and n may denote whatever numbers we might wish), it is necessary that the differential of this product be equal to zero or to infinity. This gives $mx^{m-1}\,dx \times \overline{a-x}^n - n\,\overline{a-x}^{n-1}\,dx \times x^m = 0$, from which, dividing by $x^{m-1} \times \overline{a-x}^{n-1}\,dx$, we conclude that $am - mx - nx = 0$ and AE $(x) = \frac{m}{m+n}a$.

If $m = 2$ and $n = -1$, we have $AE = 2a$ and it is therefore necessary to express the problem as follows:

Extend the given line AB on the side of B to a point E (see Fig. 3.7), so that the quantity $\frac{\overline{AE}^2}{BE}$ is a *minimum*[7] and not a *maximum*, because the equation of the curve

---

[5]In other words, we wish that $ANB : BF :: CB : CE$.

[6]This is the first of many places in L'Hôpital (1696) where the expression *un plus grand* (a greatest) is used, with the last two words italicized. We translate this as "a *maximum*," preserving the emphasis. The words "maximum" or "minimum" are not used in this chapter after Definition II.

[7]The is the first of many places in L'Hôpital (1696) where *un moindre* (a least) is used, with both words italicized. We translate this as "a *minimum*."

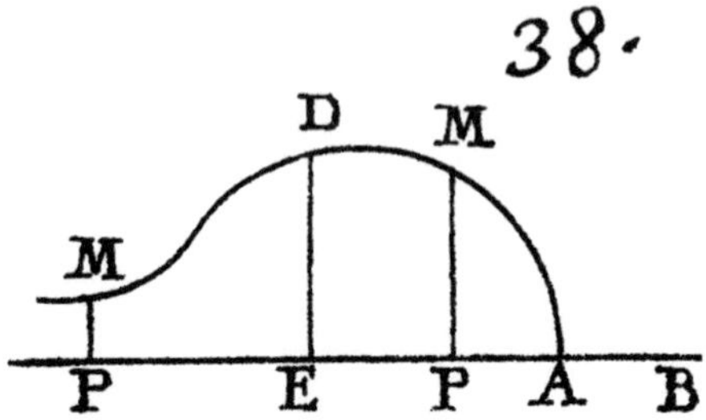

Fig. 3.8 General Problem, $m = 1$ and $n = -2$

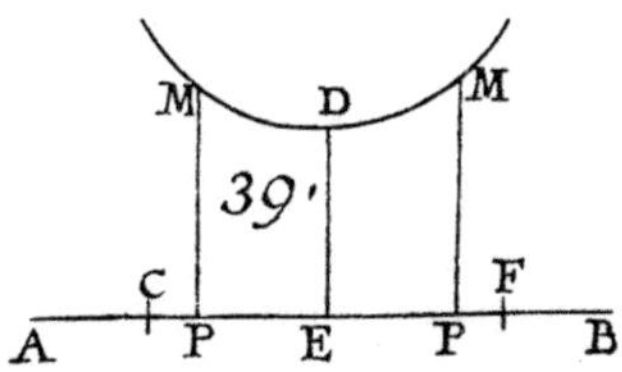

Fig. 3.9 Dividing a Given Line to Minimize a Proportion

MDM is [45] $\frac{xx}{x-a} = y$. In this, if we suppose that $x = a$, the ordinate *PM*, which becomes *BC*, is $\frac{aa}{0}$, that is to say, infinity. Supposing x to be infinite, we have $y = x$, that is to say, that the ordinate is also infinite.

If $m = 1$ and $n = -2$, we have $AE = -a$. From this, it follows that we must therefore express the problem as follows:

Extend the given straight line *AB* on the side of *A* to a point *E* (see Fig. 3.8), so that the quantity $\frac{AE \times \overline{AB}^2}{\overline{BE}^2}$ is greater than any other similar quantity $\frac{AP \times \overline{AB}^2}{\overline{BP}^2}$.

Example V. (§52) If the straight line *AB* is divided into three parts *AC*, *CF*, and *FB* (see Fig. 3.9), we wish to cut the middle part *CF* at the point *E*, so that the ratio of the rectangle $AE \times EB$ to the rectangle $CE \times EF$ is smaller than all other ratios formed in the same way.

Denoting the given quantities *AC* by a, *CF* by b, and *CB* by c, and the unknown *CE* by x, we have $AE = a+x$, $EB = c-x$, $EF = b-x$, and consequently, the ratio of $AE \times EB$ to $CE \times EF$ is $\dfrac{ac + cx - ax - xx}{bx - xx}$, which must be a *minimum*. This is why, if we imagine a curved line *MDM* such that the relationship of the ordinate *PM* (y) to the abscissa *CP* (x) is expressed by the equation $y = \dfrac{aac + acx - aax - axx}{bx - xx}$, then the question reduces to finding a value *CE* for x so that the ordinate *ED* is the least of all similar ordinates *PM*. Therefore, taking differentials and then dividing by $a\,dx$, we form the equation $cxx - axx - bxx + 2acx - abc = 0$, one of the roots of which resolves the question.

If $c = a + b$, then we have $x = \frac{1}{2}b$.

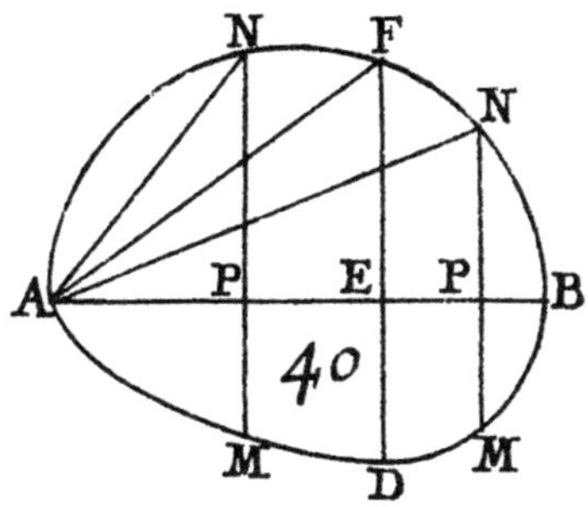

Fig. 3.10 Cones Inscribed in a Sphere

Example VI. (§53) Among all the Cones that may be inscribed [46] in a sphere, we wish to determine the one that has the largest convex surface.[8]

The question reduces to determining the point E on the diameter AB of the semi-circle AFB (see Fig. 3.10) such that, if we draw the perpendicular EF and join AF, the rectangle $AF \times FE$ is the greatest of all similar rectangles $AN \times NP$. For if we imagine that the semi-circle AFB makes a full revolution around the diameter AB, it is clear that it describes a sphere and that the right triangles AEF and APN describe cones inscribed in this sphere, such that the convex surfaces described by the chords AE and AN are to each other as the rectangles $AF \times FE$ and $AN \times NP$.

Therefore let the unknown $AE = x$ and the given quantity $AB = a$, then by the property of the circle we have $AF = \sqrt{ax}$, $EF = \sqrt{ax - xx}$ and consequently $AF \times FE = \sqrt{aaxx - ax^3}$, which must be a *maximum*. This is why we imagine a curved line MDM such that the relationship of the ordinate PM (y) to the abscissa AP (x) is expressed by the equation $\frac{\sqrt{aaxx-ax^3}}{a} = y$. We wish to find the point E, so that the ordinate ED is the greatest among all similar ordinates PM. Taking the differential, we therefore have $\dfrac{2ax\,dx - 3xx\,dx}{2\sqrt{aaxx - ax^3}} = 0$, from which we conclude that $AE\ (x) = \frac{2}{3}a$.

Example VII. (§54) Among all the Parallelepipeds equal to a given cube a^3 and having a given straight line b as one of their sides, we wish to find the one with the least surface area.

Denoting one of the sides that we wish to find by x, the other is $\frac{a^3}{bx}$. Taking the alternate planes of the three sides b, x and $\frac{a^3}{bx}$ of the parallelepiped, their sum, namely $bx + \frac{a^3}{x} + \frac{a^3}{b}$ is half of its surface area, which must be a *minimum*. This is why we imagine, as usual, a curved line that has $\frac{bx}{a} + \frac{aa}{x} + \frac{aa}{b} = y$ as its equation. [47] Taking the differential, we find $\frac{b\,dx}{a} - \frac{aa\,dx}{xx} = 0$, from which we conclude that

[8] I.e., surface area not including the base.

$xx = \frac{a^3}{b}$ and $x = \sqrt{\frac{a^3}{b}}$, so that the three sides of the parallelepiped that satisfy the question are first b, second $\sqrt{\frac{a^3}{b}}$, and third $\sqrt{\frac{a^3}{b}}$. From this, we see that the two sides we wish to find are equal to each other.

Example VIII. (§55) Now we wish to find among all the Parallelepipeds that are equal to a given cube a^3, the one with the least surface area.

Denoting one of the unknown sides by x, it is clear from the preceding example that the other two sides are each $\sqrt{\frac{a^3}{x}}$. Consequently, the sum of the alternate planes, which is half of the surface area, is $\frac{a^3}{x} + 2\sqrt{a^3x}$, which must be a *minimum*. This is why its differential $-\frac{a^3\,dx}{xx} + \frac{a^3\,dx}{\sqrt{a^3x}} = 0$, from which we conclude that $x = a$. Consequently, the two other sides are also $= a$, so that the given cube itself satisfies the question.

Example IX. (§56) Suppose the line AEB is given in position on a plane with two fixed points C and F (see Fig. 3.11) and we have drawn from any of its points P two straight lines CP (u) and PF (z). Let there be given a quantity composed of these indeterminates u and z and as many other given straight lines a, b, etc., as we may wish. We ask what must be the position of the straight lines CE and EF in order that the given quantity that is composed from them is greater or less than this same quantity, when it is composed of the straight lines CP and PF.

Suppose that the lines CE and EF have the required position and, having joined CF, we imagine a curved line DM such that when we draw an arbitrary perpendicular PQM on CF, the ordinate QM expresses the given quantity. It is clear [48] that when the point P falls on the point E, the ordinate QM, which becomes OD, must be the least or the greatest of all its similar ordinates. It is therefore necessary that the differential thus be equal to zero or to infinity. This is why, for

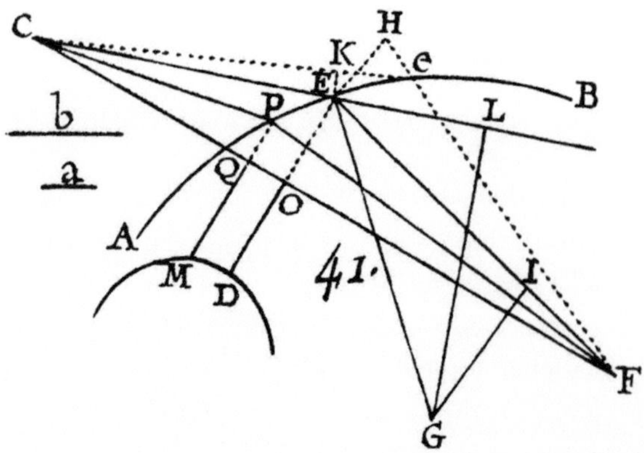

Fig. 3.11 Lines From a Curve to Two Given Points on Opposite Sides

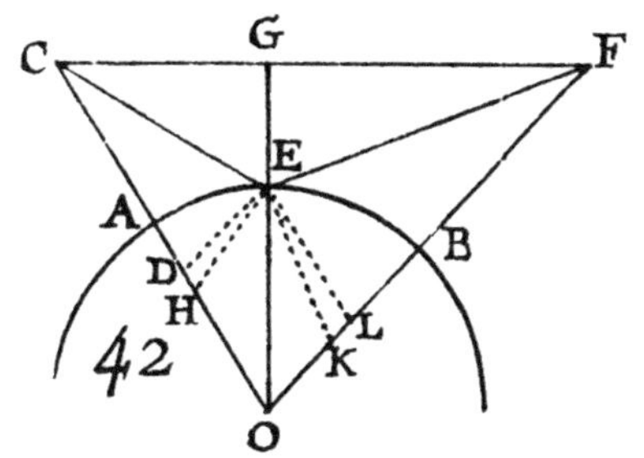

Fig. 3.12 Lines From a Curve to Two Given Points on the Same Side

example, if the given quantity is $au + zz$, we have $a\,du + 2z\,dz = 0$ and consequently $du : -dz :: 2z : a$. From this, we already see that dz must be negative with respect to du, that is to say that the position of the straight lines CE and EF must be such that as u increases, z decreases.

Suppose we now draw EG perpendicular to the line AEB and from any of its points G we draw the perpendiculars GL and GI on CE and EF. Now from the point e, taken infinitely close to E, we draw the straight lines CKe and FeH, and with centers C and F, we draw the little circular arcs EK and EH. We form similar right triangles ELG and EKe and similar right triangles EIG and EHe. For if we remove the same angle LEe from the right angles GEe and LEK, then the remainders LEG and KEe are equal. We prove similarly that the angles IEG and HEe are equal. Therefore, we have that $GL : GI :: Ke\ (du) : He\ (-dz) :: 2z : a$. From this it follows that the position of the straight lines CE and EF must be such that when the perpendicular EG is drawn on the line AEB, the sine GL of the angle GEC is to the sine GI of the angle GEF as the quantities that multiply dz are to those that multiply du. This is what we were required to find.

Corollary. (§57) *If we now wish that the straight line CE is given in position and magnitude, and that the straight line EF is given in magnitude only, and that it is required to find its position, then it is clear that if the angle GEC is given, its sine GL is given as well, and consequently the sine GI of the angle GEF that we wish to find is also given. Thus, if we describe the circle with diameter EG and we apply the value of GI onto its circumference from G to I, the straight line EF that passes through the point I has the required position.*

Let $au + bz$ be the given quantity. We find that $GI = \dfrac{a \times GL}{b}$, from which we see that whatever length we give [49] to EC and EF, the position of this latter is always the same, because they do not enter into the value of GI, which consequently does not change. If $a = b$, it is clear that the position of EF must be on CE, extended on the side of E, because $GL = GI$ when the points C and F fall on opposite sides of the line AEB. However, when they fall on the same side, the angle FEG must be taken equal to the angle CEG (see Fig. 3.12).

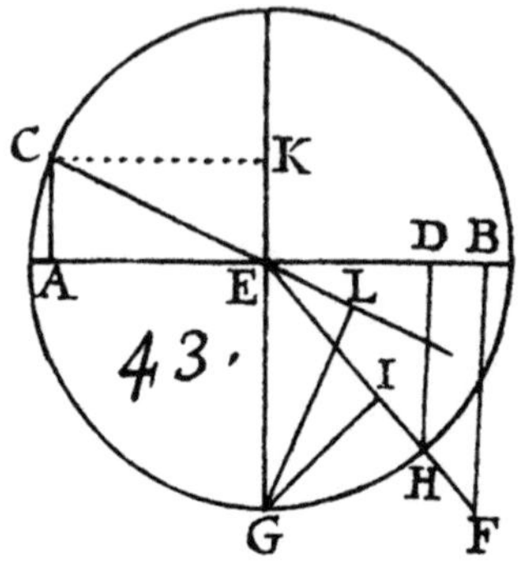

Fig. 3.13 Passing From One Region to Another, With Different Speeds

Example X. (§58) If the circle *AEB* is given in position with the points C and F outside of the circle (see Fig. 3.12), we wish to find the point E on the circumference such that the sum of the straight lines *CE* and *EF* is as small as possible.

Supposing that the point E is the one that we wish to find and that we draw the line *OEG* from the center O, it is clear that it is perpendicular to the circumference *AEB* and therefore (see §57) that the angles *FEG* and *CEG* are equal to each other. If we then draw *EH* so that the angle *EHO* is equal to the angle *CEO*, and similarly, we draw *EK* so that the angle *EKO* is equal to the angle *FEO*, and draw the parallels *ED* and *EL* to *OF* and *OC*, then we form the similar triangles *OCE* and *OEH*, *OFE* and *OEK*, and finally *HDE* and *KLE*. Denoting the given quantities *OE*, *OA* or *OB* by a, *OC* by b and *OF* by c, and the unknowns *OD* or *LE* by x and *DE* or *OL* by y, we have $OH = \frac{aa}{b}$, $OK = \frac{aa}{c}$, and $HD\ (x - \frac{aa}{b}) : DE\ (y) :: KL\ (y - \frac{aa}{c}) : LE\ (x)$. Thus $xx - \frac{aax}{b} = yy - \frac{aay}{c}$, which is the equation of a hyperbola that we can easily construct and which will cut the circle at the point E that we wish to find.

Example XI. (§59)[9] A traveler departing from the location C to go to the location F (see Fig. 3.13) must pass through two regions[10] separated by the straight line *AEB*. We suppose that in the region on the side of C, he covers the distance a in the time c and that in the other, [50] on the side of F, he covers the distance b in the same time c. We wish to find the point E on the straight line *AEB* through which he must pass in order to take the least time possible to make his way from C to F. If we let $a : CE\ (u) :: c : \frac{cu}{a}$ and $b : EF\ (z) :: c : \frac{cz}{b}$, it is clear that $\frac{cu}{a}$ expresses the time that the traveler takes to traverse the straight line *CE*, and similarly that $\frac{cz}{b}$ expresses the time he takes to traverse *EF*, so that $\frac{cu}{a} + \frac{cz}{b}$ must be a *minimum*. From this it follows (see §56) that if we draw *EG* perpendicular to the line *AB*, the sine of the angle *GEC* must be to the sine of the angle *GEF* as a is to b.

[9]Compare to Problem XVI on p. 207.

[10]In L'Hôpital (1696) the word *campagnes* was used, which could mean "countries," or simply "fields."

Given this, suppose that we describe the circle *CGH* with center at the point *E* that we wish to find and with radius *EC*, and that we drop the perpendiculars *CA*, *HD* and *FB* to the straight line *AEB*, and the perpendiculars *GL* and *GI* to *CE* and *EF*, then $a : b :: GL : GI$. Now $GL = AE$ and $GI = ED$, because the right triangles *GEL* and *ECA* are equal and similar, and the right triangles *GEI* and *EHD* are equal and similar, which is easy to prove. This is why, if we denote the unknown *AE* by x, we find that $ED = \frac{bx}{a}$. Denoting the given quantities *AB* by f, *AC* by g and *BF* by h, the similar triangles *EBF* and *EDH* give $EB\ (f - x) : BF\ (h) :: ED\ (\frac{bx}{a}) : DH = \frac{bhx}{af-ax}$. However, because of the right triangles *EDH* and *EAC*, which have equal hypotenuses *EH* and *EC*, we have $\overline{ED}^2 + \overline{DH}^2 = \overline{EA}^2 + \overline{AC}^2$. In analytic terms, that is to say

$$\frac{bbxx}{aa} + \frac{bbhhxx}{aaff - 2aafx + aaxx} = xx + gg.$$

Thus, by removing the fractions and rearranging the equation, we have[11]

$$\begin{array}{llll} aax^4 & -2aafx^3 & + aaffxx & -2aafggx + aaffgg = 0. \\ -bb & +2bbf & + aagg & \\ & & - bbff & \\ & & - bbhh & \end{array}$$

We may also find this equation in the following manner, without having recourse to Example IX.

[51] Denoting, as before, the given quantities *AB* by f, *AC* by g and *BF* by h and the unknown *AE* by x, we make $a : CE\ (\sqrt{gg + xx}) :: c : \frac{c\sqrt{gg+xx}}{a}$, where the last term is equal to the time taken by the traveler in traversing the straight line *CE*. Similarly, $b : EF\ \left(\sqrt{ff - 2fx + xx + hh}\right) :: c : \frac{c\sqrt{ff-2fx+xx+hh}}{b}$, where the last term is equal to the time taken by the traveler in traversing the straight line *EF*. This makes

$$\frac{c\sqrt{gg + xx}}{a} + \frac{c\sqrt{ff - 2fx + xx + hh}}{b}$$

equal to a *minimum*, and consequently its differential

$$\frac{cx\,dx}{a\sqrt{gg + xx}} + \frac{cx\,dx - cf\,dx}{b\sqrt{ff - 2fx + xx + hh}} = 0.$$

[11]This is the notation used in L'Hôpital (1696) for $(a^2 - b^2)x^4 + (-2a^2 f + 2b^2 f)x^3 + (a^2 f {}^+ a^2 g^2 - b^2 f^2 - b^2 h^2)x^2 - 2a^2 fg^2 x + a^2 f^2 g^2 = 0$.

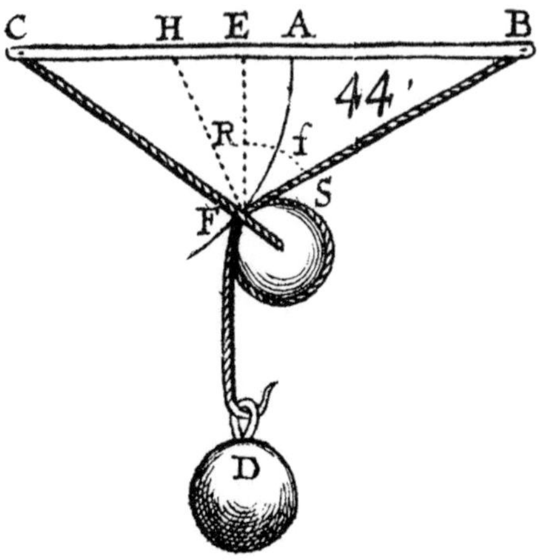

Fig. 3.14 The Pulley Problem

Dividing this latter by $c\,dx$ and removing the incommensurables,[12] we derive the same equation as above, the roots of which provide the desired value for AE.

Example XII. (§60)[13] Let F be a pulley that hangs freely at the end of a rope CF attached at C (see Fig. 3.14), with a weight D suspended by the cord DFB, which passes over the pulley F and which is attached at B, so that the points C and B are situated on the same horizontal line CB. We suppose that the pulley and the cords have no weight and we wish to find the location at which the weight D or the pulley F comes to rest.

It is clear by the principles of Mechanics that the weight D descends as far down as possible below the horizontal CB, from which it follows that the plumb line DFE should be a *maximum*. This is why, if we denote the given quantities CF by a, DFB by b and CB by c, and the unknown CE by x, we have $EF = \sqrt{aa - xx}$, $FB = \sqrt{aa + cc - 2cx}$, and $DFE = b - \sqrt{aa + cc - 2cx} + \sqrt{aa - xx}$, which must be a *maximum*. Consequently, its differential

$$\frac{c\,dx}{\sqrt{aa + cc - 2cx}} - \frac{x\,dx}{\sqrt{aa - xx}} = 0,$$

[12]Two quantities are incommensurable if their ratio is irrational; "removing the incommensurables" means rationalizing the equation.

[13]Compare to Problem XIX on p. 210. For a modern discussion of this problem, see Hahn (1998).

from which we conclude that $2cx^3 - 2ccxx - aaxx + aacc = 0$. [52] Dividing by $x - c$, we have $2cxx - aax - aac = 0$, the roots of which give a value for *CE* such that the perpendicular *ED* passes through the pulley *F* and the weight *D* when they are at rest.

We may also solve this question by a different method than the above.

Denoting *EF* by y and *BF* by z, we have $b - z + y$ equal to a *maximum*, and hence $dy = dz$. Now it is clear that the pulley *F* describes the circle *CFA* about the point *C* as its center. Consequently if we take f infinitely close to *F* and draw *fR* from this parallel to *CB* and *fS* perpendicular to *BF*, then we have $FR = dy$ and $FS = dz$. They are therefore equal to each other and consequently the little right triangles *FRf* and *FSf*, which also have the common hypotenuse *Ff*, are equal and similar. From this we see that the angle *RFf* is equal to the angle *SFf*, that is to say that the point *F* must be so situated on the circumference *FA* that the angles made by the straight lines *EF* and *FB* on the tangents at *F* are equal to each other, or rather (what amounts to the same thing), that the angles *BFC* and *DFC* are equal.

Given this, if we draw *FH* so that the angle *FHC* is equal to the angle *CFB* or *CFD*, the triangles *CBF* and *CFH* are similar. The right triangles *ECF* and *EFH* are also similar, because the angle *CFE* is equal to the angle *FHE*, being supplementary[14] to two equal angles *FHC* and *CFD*. Consequently, we have $CH = \frac{aa}{c}$ and HE $(x - \frac{aa}{c})$: EF (y) :: EF (y) : EC (x). Thus, $xx - \frac{aax}{c} = yy = aa - xx$, by the property of the circle, from which we derive the same equation as before.

Example XIII. (§61)[15] Given the elevation of the pole, we wish to find the day with the smallest crepuscule.[16]

Let *C* be the center of the sphere; *APTOBHQ* the meridian; *HDdO* the horizon; *QEeT* the crepuscular circle parallel [53] to the horizon; *AMNB* the equator; *FEDG* the portion of the parallel to the equator (see Fig. 3.15), which the sun describes on the day of the smallest crepuscule, enclosed between the planes of the horizon and the crepuscular circle; *P* the south pole; *PEM* and *PDN*, the quarters of the circles of declination. The arc *HQ* or *OT* of the meridian included between the horizon and the crepuscular circle, and the arc *OP* of the elevation of the pole are given, and consequently their right sines *CI* or *FL* or *QX*, and *OV*. We wish to find the sine *CK* of the arc *EM* or *DN* of the declination of the sun while it describes the parallel *ED*.

We imagine another portion *fedg* of a parallel to the equator, infinitely close to *FEDG*, with the quarter circles *Pem* and *Pdn*. It is clear that if the time that the sun takes in traversing the arc *ED* is to be a *minimum*, then the differential of the arc

[14] In L'Hôpital (1696), this literally says "the complements within two right angles." We consistently translate this construction as supplementary.

[15] Compare to Problem XX on p. 212.

[16] I.e., the shortest twilight. Evening twilight is defined as the length of time from sunset until the time that the center of the sun is 18° below the horizon. Morning twilight is defined similarly for the period before sunrise.

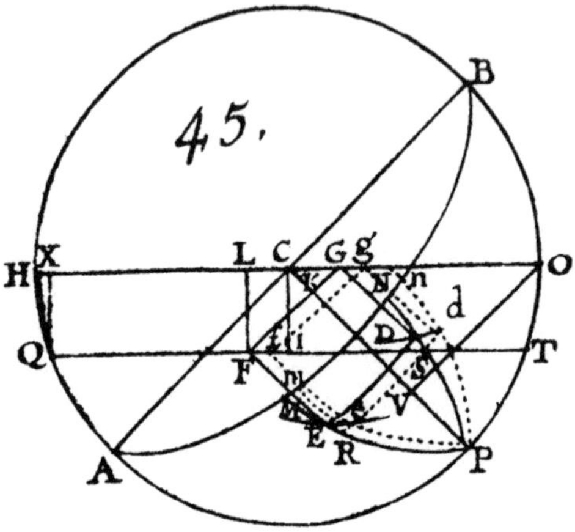

Fig. 3.15 The Celestial Sphere

MN that measures it, and which becomes mn when ED becomes ed, must be null. From this it follows that the little arcs Mm and Nn must be equal, and consequently also the little arcs Re and Sd, are equal to each other. Now the arcs RE and SD, being enclosed between the same parallels ED and ed, are also equal, and the angles at S and R are right angles. Hence the little right triangles ERe and DSd (which we consider to be rectilinear (see §3) because of the infinite smallness of their sides)[17] are equal and similar, and consequently the hypotenuses Ee and Dd are also equal to each other.

Given this, the straight lines DG, EF, dg and ef, common sections of the planes $FEDG$ and $fedg$, parallel to the equator, with the horizon and the crepuscular circle, are perpendicular to the diameters HO and QT, because the planes of all of these circles are each perpendicular to the plane of the meridian. Additionally, the little straight lines Gg ad Ff are equal to each other, because the straight lines FG and fg are parallel. Thus, $\sqrt{\overline{Dd}^2 - \overline{Gg}^2}$ or $DG - dg$ is equal to $\sqrt{\overline{Ee}^2 - \overline{Ff}^2}$ or $fe - FE$. Now it is clear, because we have proved it in §50, that if we draw any two infinitely close ordinates in a semi-circle, the little arc that they enclose is [54] to their differential as the radius is to the abscissa, measured from the center. Here (because of the circles HDO and QET) this gives $CO : CG :: Dd$ or Ee : $DG - dg$ or $fe - FE :: IQ : IF :: CO + IQ$ or $OX : CG + IF$ or GL. However, because of the similar right triangles CVO, CKG and FLG, we have $CO : CG :: OV : GK$ and $GK : GL :: CK : FL$ or QX. Thus $OV : CK :: OX : XQ :: XQ : XH$, by the property of the circle. That is to say, if we take QX as the radius or total sine[18] in the right triangle QXH, whose angle HQX is of 9 degrees, because Astronomers

[17] This right parenthesis was missing in L'Hôpital (1696).

[18] The term "total sine" is a synonym for the radius.

make the arc HQ to be of 18 degrees, then we will have that as the total sine is to the tangent of 9 degrees, so the sine of the elevation of the pole is to the sine of the southern declination of the sun in the time of the smallest crepuscule. From this, it follows that if we subtract 0.8002875 from the logarithm of the sine of the elevation of the pole, the remainder will be the logarithm of the sine that we wish to find. This is what we were required to find.

Chapter 4
Use of the Differential Calculus for Finding Inflection Points and Cusps

[55] Because we will be making use in what follows of second, third, etc., differentials, it is necessary to give an idea of them before going any further.

Definition I. The infinitely small portion by which the differential of a variable quantity continually increases or decreases is called the *differential of the differential* of this quantity, or else its *second differential*. Thus, if we imagine a third ordinate *nq* infinitely close to the second ordinate *mp* (see Fig. 4.1) and we draw *mS* parallel to *AB* and *mH* parallel to *RS*, we call *Hn the differential of the differential Rm*, or else the *second differential* of *PM*.

Similarly, if we imagine a fourth ordinate *of* infinitely close to the third ordinate *nq* and we draw *nT* parallel to *AB* and *nL* parallel to *ST*, we call the difference of the little straight lines *Hn* and *Lo the differential of the second differential*, or else the *third differential* of *PM*. And so forth for the others.

Note. In what follows, we denote each differential by a number of *d*'s that expresses the order or the type. For example, we denote by *dd* the second differential or the differential of the second order, by *ddd* the third differential or the differential of the third order, by *dddd* the fourth differential or the differential of the fourth order, and similarly for the others. Thus, *ddy* denotes *Hn*, *dddy* denotes Lo – Hn, etc.

As for the powers of these differentials, we denote them with numerals placed above and following, as we ordinarily do with powers of whole magnitudes.[1] For example, the square or the cube of *dy* is dy^2 or dy^3, the square or the cube of *ddy* is ddy^2 or *[56]* ddy^3, that of *dddy* is $dddy^2$ or $dddy^3$, that of *ddddy* is $ddddy^2$ or $ddddy^3$, etc.

[1]In (L'Hôpital 1696) the term *grandeurs entières* is used, literally "whole magnitudes," yet even though the comparison seems to be to the use of exponents with finite quantities.

R.E. Bradley et al., *L'Hopital's Analyse des infiniments petits*, Science Networks. Historical Studies 50, DOI 10.1007/978-3-319-17115-9_4

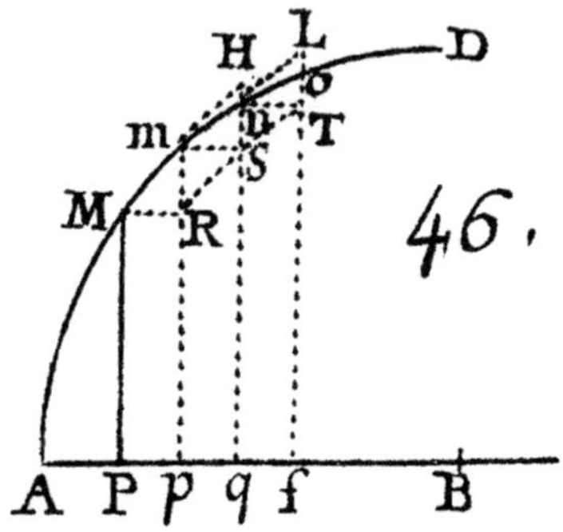

Fig. 4.1 Definition of Higher Order Differentials

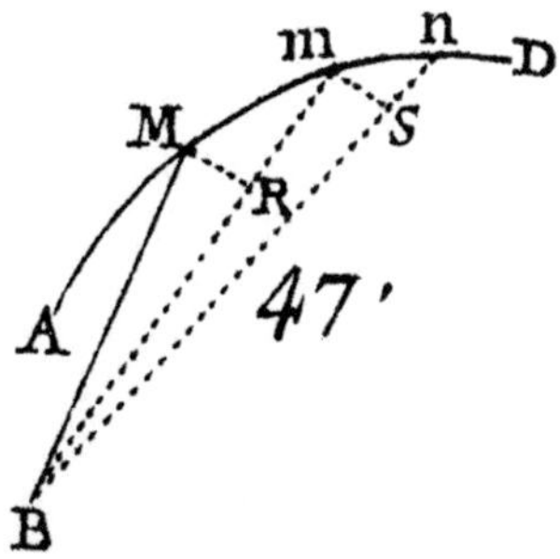

Fig. 4.2 Higher Order Differentials – Ordinates from a Fixed Point

Corollary I. (§62) *If we denote each of the abscissas AP, Ap, Aq, and Af by x, each of the ordinates PM, pm, qn, and fo by y, and each of the curved portions AM, Am, An, and Ao by u, it is clear that dx denotes the differentials Pp, pq and qf of the abscissas, dy denotes the differentials Rm, Sn, and To of the ordinates, and du denotes the differentials Mm, mn, and no of the portions of the curve AMD. Now in order to take, for example, the second differential Hn of the variable PM, we must imagine two little parts Pp and pq on the axis and two others, Mm and mn, on the curve, in order to have the two differentials Rm and Sn. Consequently, if we suppose that the two little parts Pp and pq are equal to each other, it is clear that dx is constant with respect to dy and to du, because Pp, which becomes pq, remains the same while Rm, which becomes Sn, and Mm, which becomes mn, vary. We might suppose that the little parts Mm and mn of the curve are equal to each other and therefore that du is constant with respect to dx and dy. Finally, if we suppose that Rm and Sn are equal, then dy is constant with respect to dx and du and its differential HN (ddy) is null.*

Similarly, to take the third differential of *PM*, or the differential of the second differential *Hn*, we must imagine three little parts *Pp*, *pq*, and *qf* on the axis, three others, *Mm*, *mn*, and *no* on the curve, and also three others, *Rm*, *Sn*, and *To*, on their ordinates. We then have *dx* or *du* or *dy* as constant, according to whether we suppose that either Pp, pq and qf, or Mm, mn and no, or Rm, Sn and To are equal to one another. It is similar for the fourth differentials, the fifth, and so on.

All of this should also include curves AMD, whose ordinates BM, Bm and Bn all emanate from a fixed point B (see Fig. 4.2), because to have, for example, the second differential of BM, we must imagine two other ordinates Bm and Bn, which make infinitely small angles MBm and mBn. Then, describing the small circular arcs MR and mS, with center B, the difference [57] of the little straight lines Rm and Sn is the second differential of BM. We may take the little arcs MR and mS to be constant, or the little portions Mm and mn of the curve, or finally the little straight lines Rm and Sn. It is similar for third differentials of the ordinate BM, the fourth, and so on.

Remark. (§63) We must carefully remark that:

1. There are different orders of the infinitely small: Rm, for example (see Fig. 4.1), is infinitely small with respect to PM and infinitely large with respect to Hn. Similarly the region $MPpm$ is infinitely small with respect to the region APM and infinitely large with respect to the triangle MRm.
2. The entire differential Pf is also infinitely small with respect to AP, because all quantities that are the sum of a finite number of quantities, such as Pp, pq, and qf, that are infinitely small with respect to some other quantity AP, always remain infinitely small with respect to that same quantity. In order for it to become of the same order, it is necessary that the number of quantities of the lower order that composed it be infinite.

Corollary II. (§64) *We may indicate all possible cases of second order differentials in the following way.*

1. In curves in which the ordinates mR and nS are parallel to each other (see Figs. 4.3, 4.4), we prolong the little straight line Mm to H, where it meets the ordinate Sn. We describe the arc nk with center m and radius mn and we draw the little straight lines nl, li and kcg, parallel to mS and to Sn. Given this, if we wish dx to be constant, that is to say that MR be equal to mS, then it is clear that the triangle mSH is similar and equal to the triangle MRm and, furthermore, that Hn is ddy, that is to say, the difference of Rm and Sn, and that $Hk = ddu$.

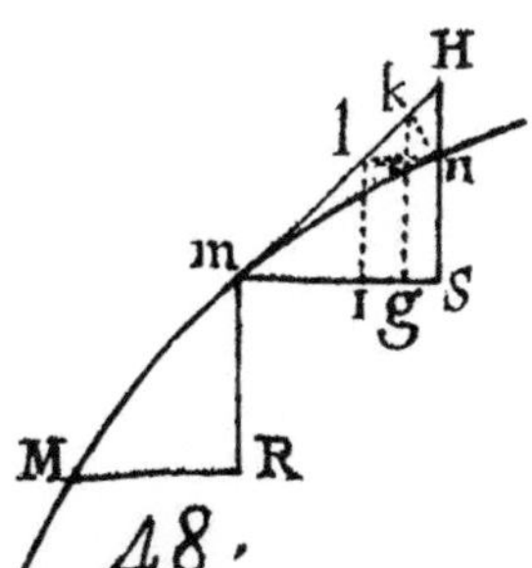

Fig. 4.3 Parallel Ordinates – First Case

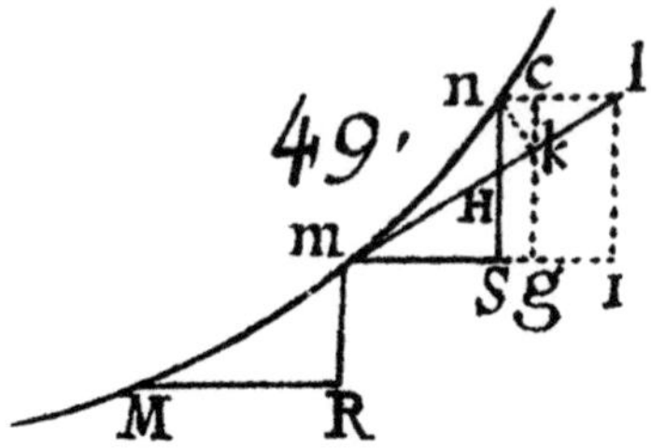

Fig. 4.4 Parallel Ordinates – Second Case

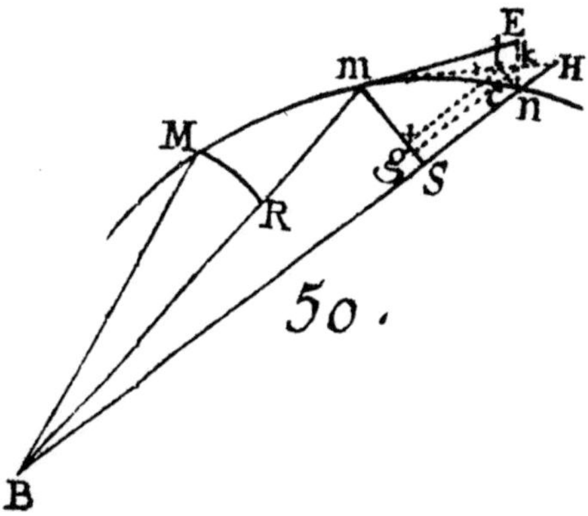

Fig. 4.5 Ordinates from a Fixed Point – First Case

However, if we suppose that du is constant, that is to say that $Mm = mn$ or $= mk$, it is then clear that the triangle mgk is similar and equal to the triangle MRm, and thus that $kc = ddy$ and [58] Sg or $cn = ddx$. Finally, if we take dy as constant, that is to say that $mR = nS$, it follows that the triangle mil is equal and similar to the triangle MRm, and therefore that iS or $nl = ddx$ and that $lk = ddu$.

2. In curves in which the ordinates BM, Bm and Bn emanate from the same point B (see Figs. 4.5, 4.6), we describe the arcs MR and mS from the center B, which we regard (see §3) as little straight lines, perpendicular to Bm and Bn. If we now prolong Mm to E and describe the little arc nkE with center m and radius mn, we make the angle $EmH = mBn$. We draw the little straight lines nl, li and kcg parallel to mS and Sn. Given this, because the angle at S in the triangle BSm is right, the angle $BmS + mBn$ or $+EmH$ makes a right angle. Consequently, the angle BmE is equal to a right angle $+SmH$, which is also equal to the right angle $MRm + RMm$, because it is external to the triangle RMm. Thus, the angle $SmH = RMm$.

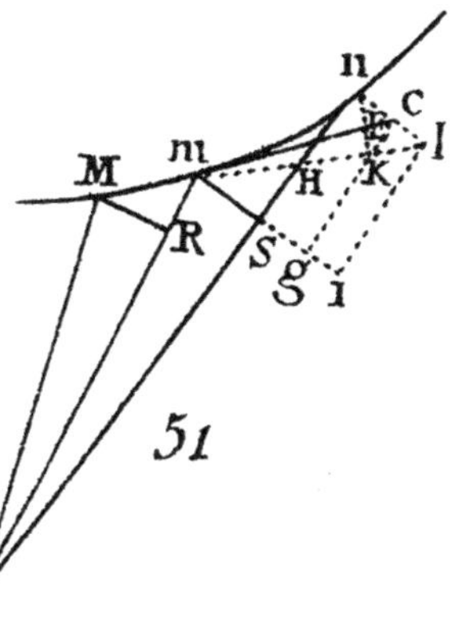

Fig. 4.6 Ordinates from a Fixed Point – Second Case

It follows from this that:

1. If we wish dx to be constant, that is that the little arcs MR and mS are equal to each other, the triangle SmH is similar and equal to the triangle RMm and hence $HN = ddy$ and $Hk = ddu$.
2. If we take du as constant, the triangle gmk is similar and equal to the triangle RMm and hence kc expresses ddy and Sg or cn expresses ddx.
3. Finally, if we take dy as constant, the triangles iml and RMm are equal and similar and thus iS or $ln = ddx$ and $lk = ddu$.

Proposition I.

Problem. (§65) *To take the differential of a quantity composed of any differentials.*

We take whatever differential we may wish as constant and treat the others as variable quantities, making use of the rules prescribed in the first chapter.

The differential of $\frac{y\,dy}{dx}$ is $\frac{dy^2+y\,ddy}{dx}$, taking dx as constant, and $\frac{dx\,dy^2-y\,dy\,ddx}{dx^2}$, taking dy as constant.

[59] The differential of $\frac{z\sqrt{dx^2+dy^2}}{dx}$, taking dx as constant, is $dz\sqrt{dx^2+dy^2} + \frac{z\,dy\,ddy}{\sqrt{dx^2+dy^2}}$, all divided by dx, that is to say,

$$\frac{dz\,dx^2 + dz\,dy^2 + z\,dy\,ddy}{dx\sqrt{dx^2+dy^2}}.$$

Taking dy as constant, it is $dz\,dx\sqrt{dx^2+dy^2} + \frac{z\,dx^2\,ddx}{\sqrt{dx^2+dy^2}} - z\,ddx\sqrt{dx^2+dy^2}$, all divided by dx^2, that is to say,

$$\frac{dz\,dx^3 + dz\,dx\,dy^2 - z\,dy^2\,ddx}{dx^2\sqrt{dx^2+dy^2}}.$$

The differential of $\frac{y\,dy}{\sqrt{dx^2+dy^2}}$, taking dx as constant, is $\overline{dy^2 + y\,ddy}\sqrt{dx^2 + dy^2} - \frac{y\,dy^2\,ddy}{\sqrt{dx^2+dy^2}}$, all divided by $dx^2 + dy^2$, that is to say,

$$\frac{dx^2\,dy^2 + dy^4 + y\,dx^2\,ddy}{\overline{dx^2 + dy^2}\sqrt{dx^2 + dy^2}}.$$

Taking dy as constant, it is

$$\frac{dx^2\,dy^2 + dy^4 - y\,dy\,dx\,ddx}{\overline{dx^2 + dy^2}\sqrt{dx^2 + dy^2}}.$$

The differential of $\frac{\overline{dx^2+dy^2}\sqrt{dx^2+dy^2}}{-dx\,ddy}$, or $\frac{\overline{dx^2+dy^2}^{\frac{3}{2}}}{-dx\,ddy}$, taking dx as constant, is

$$\frac{-3\,dx\,dy\,ddy^2\,\overline{dx^2 + dy^2}^{\frac{1}{2}} + dx\,dddy\,\overline{dx^2 + dy^2}^{\frac{3}{2}}}{dx^2\,ddy^2}.$$

However, we must observe that in this last case, we are not free to take dy as constant, because under this assumption, the differential ddy is null and, consequently, it must not be present in the given quantity.

Definition II. When a curved line AFK is in part concave and in part convex towards a straight line AB (see Figs. 4.7, 4.8) or towards a fixed point B (see Figs. 4.9, 4.10), the point F that separates the concave part from the convex part and, consequently, is the end of one and the beginning of the other, is called an

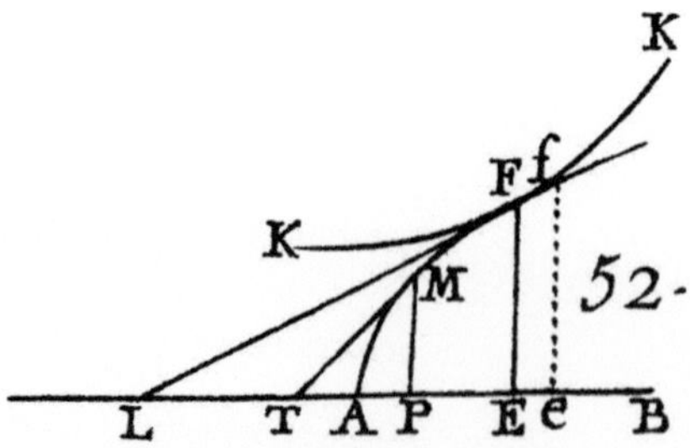

Fig. 4.7 Inflection Point

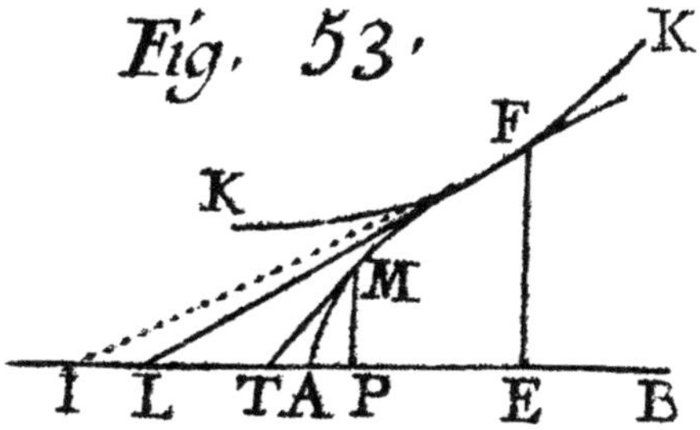

Fig. 4.8 Cusp

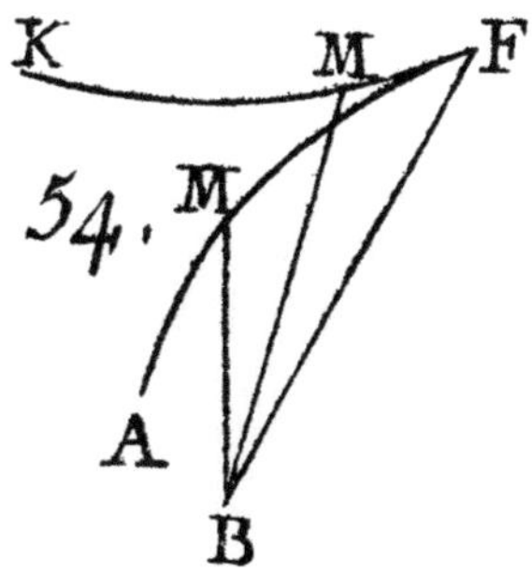

Fig. 4.9 Cusp

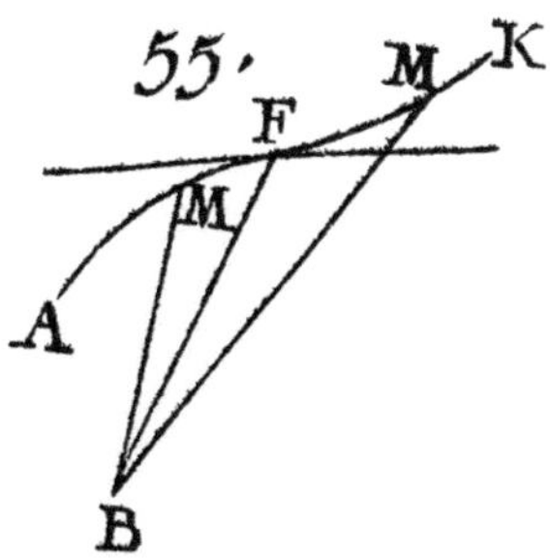

Fig. 4.10 Inflection Point

inflection point[2] when the curve, having arrived at F, continues along its path in the same direction, and a *cusp*[3] when it turns back on its path on the side of its origin.

---

[2]Compare this to the definition presented in Problem XXI on p. 217.

[3]Following Bernoulli [letter 22], L'Hôpital used the term "*point de rebroussement*," literally a "point of turning back." The mathematical term "turning point" has a different meaning from this, so we use the term "cusp," which is the standard English term for this type of point.

Proposition II.

General Problem. [60] (§66) *When the nature of the curved line AFK is given, we wish to determine the inflection point or the cusp F.*

In the first case, suppose that the curved line AFK has a straight line AB as its diameter and that its ordinates PM, EF, etc., are all parallel to one another (see Figs. 4.7, 4.8). If we draw an ordinate FE with the tangent FL from a point F, as well as an ordinate MP with a tangent MT from any point M on the portion AF, then it is clear that:

1. In curves that have an inflection point, that as the abscissa AP increases continually, the portion AT of the diameter, intercepted between the origin of the x's and the point where it meets the tangent, also increases up to when the point P coincides with E, after which it decreases. From this we see that AT, when applied to the point P, must become a *maximum* AL when the point P coincides with the point E that we wish to find.
2. In curves that have a cusp, as the portion AT increases continually, the abscissa AP also increases up to when the point where T coincides with L, after which it decreases. From this we see that AP, when applied to the point T, must become a *maximum* AE when the point T coincides with L.

Now if we denote AE by x and EF by y, then we have $AL = \frac{y\,dx}{dy} - x$, the differential of which is $\frac{dy^2\,dx - y\,dx\,ddy}{dy^2} - dx$ (assuming that dx is constant) which, being divided by dx, the differential of AE, must be null or infinity (see §47). This gives $-\frac{y\,ddy}{dy^2} = 0$ or is equal to infinity. Multiplying by dy^2 and dividing by $-y$ we have $ddy = 0$ or is equal to infinity, which we make use of in what follows as the general formula for finding the inflection point or the cusp F. Given the nature of the curve AFK, we have the value of dy in terms of dx and taking the differential of this value, supposing that dx [61] is constant, we will find the value of ddy in terms of dx^2. When this is first set equal to zero and then to infinity, we may use one or the other of these assumptions to find a value of AE such that the ordinate EF cuts the curve AFK in either an inflection point or a cusp F.

The origin A of the x's may be situated so that $AL = x - \frac{y\,dx}{dy}$, instead of $\frac{y\,dx}{dy} - x$, and such that AL or AE is a *minimum* instead of a *maximum*. However, as the consequence of this is still the same and because this causes no difficulty, I will not dwell on this case.

The same thing may also be found in a different way. It is clear that if we take dx as constant and we assume that the ordinate y increases, then Sn is smaller than SH or than Rm in the concave part (see Figs. 4.3, 4.4), and greater in the convex part. From this we see that the values of Hn (ddy) must go from positive to negative at the inflection point or cusp F. Consequently, it must be either null or infinity (see §47).

In the second case, suppose that the curve AFK has as its ordinates the straight lines BM, BF, and BM,[4] which all emanate from the same point B (see Figs. 4.9, 4.10). We draw any ordinate BM we may wish, with a tangent MT that meets BT, which is perpendicular to BM, at the point T (see Figs. 4.11, 4.12). Now take the point m to be infinitely close to M and draw the ordinate Bm, the tangent mt and the perpendicular Bt on Bm, which meets MT at O. Assuming that the ordinate BM increases as it becomes Bm, it is clear that in the concave part, Bt is greater than BO and, on the contrary, it is smaller in the convex part, so that at the inflection point or cusp F, the value of Ot must go from positive to negative.

Given this, if we describe the little circular arcs MR and TH with center B (see Fig. 4.11), we make similar triangles mRM, MBT, and THO and similar little sectors BMR and BTH. Denoting BM by y and MR by dx, we have $mR\ (dy) : RM\ (dx) :: BM\ (y) : BT = \frac{y\,dx}{dy} :: MR\ (dx) : TH = \frac{dx^2}{dy} :: TH \left(\frac{dx^2}{dy}\right) : HO = \frac{dx^3}{dy^2}$. Now if we take the differential of BT [62] $\left(\frac{y\,dx}{dy}\right)$, assuming dx as constant, we have $Bt - BT$ or $Ht = \dfrac{dx\,dy^2 - y\,dx\,ddy}{dy^2}$. Consequently, $OH + Ht$ or $Ot = \dfrac{dx^3 + dx\,dy^2 - y\,dx\,ddy}{dy^2}$, from which it follows, multiplying by dy^2 and dividing by dx, that the value of $dx^2 + dy^2 - y\,ddy$ is null or infinity at the inflection point or cusp F. Now, given the nature of the curve AFK (see Figs. 4.9, 4.10), we have the values of dy in terms of dx and of ddy in terms of dx^2. When substituted into $dx^2 + dy^2 - y\,ddy$, these give

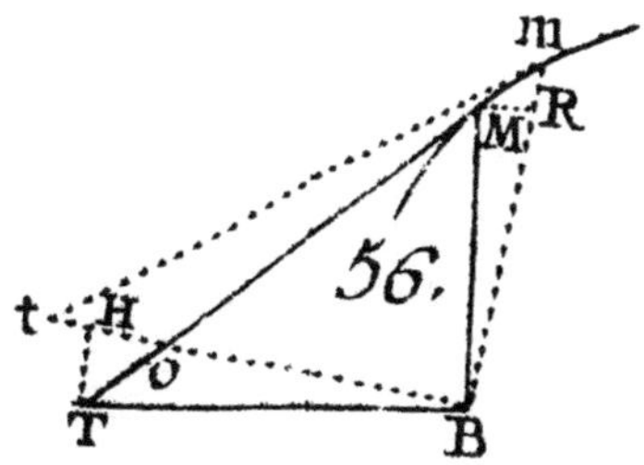

Fig. 4.11 Ordinates from a Fixed Point – Concave

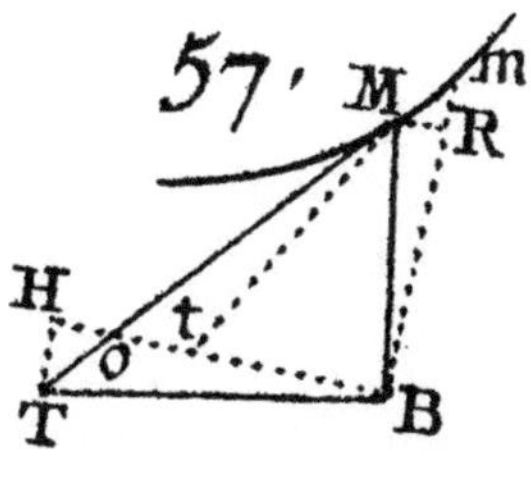

Fig. 4.12 Ordinates from a Fixed Point – Convex

[4]In figures 4.9 and 4.10, the author makes use of the same letter M for two different points.

a quantity which, when set first to zero and then to infinity, is used to find a value of BF such that, when we describe a circle with center B and this radius, we cut the curve AFK at the inflection point or cusp F. This is what was proposed.

To find the same thing by another method, we must consider that in the concave part, the angle BmE is greater than the angle Bmn (see Figs. 4.5, 4.6), while on the other hand, in the convex part it is smaller. Consequently, the angle $BmE - Bmn$ or Emn, that is to say the arc En that measures it, goes from positive to negative at the point F that we wish to find. Now, taking dx as constant, the similar right triangles HmS and Hnk give $Hm\ (du) : mS\ (dx) :: Hn\ (-ddy) : nk = -\frac{dx\,ddy}{du}$, where we should observe that the value of Hn is negative, because as $BM\ (y)$ increases, $Rm\ (dy)$ decreases. However, because the sectors BmS and mEk are similar, we have $Bm\ (y) : mS\ (dx) :: mE\ (du) : Ek = \frac{dx\,du}{y}$. Consequently, $Ek + kn$ or $En = \dfrac{dx\,du^2 - y\,dx\,ddy}{y\,du}$. From this it follows, in multiplying by $y\,du$ and dividing by dx that $du^2 - y\,ddy$ or $dx^2 + dy^2 - y\,ddy$ must go from positive to negative at the point F (see Figs. 4.9, 4.10) that we wish to find.

If we suppose that y becomes infinite, then the terms dx^2 and dy^2 are null with respect to the term $y\,ddy$. Consequently, the formula $dx^2 + dy^2 - y\,ddy = 0$ or infinity becomes the formula $-y\,ddy = 0$ or infinity. That is to say, dividing by $-y$, that $ddy = 0$ or infinity, which is the formula in the first case. This must also happen because [63] the ordinates BM, BF, and BM thereby become parallel.

Corollary. (§67) *When $ddy = 0$, it is clear that the differential of AL must be null with respect to that of AE (see Fig. 4.7) and consequently that the two infinitely close tangents FL and fL must coincide with each other, making only a single straight line fFL. However, when ddy = infinity (see Fig. 4.8), the differential of AL must be infinitely large with respect to that of AE, or (what is the same thing) the differential of AE is infinitely small with respect to that of AL. Consequently, we may draw two tangents FL and Fl from the same point F, which make and infinitely small angle LFl between themselves.*

Likewise, when $dx^2 + dy^2 - y\,ddy = 0$, it is clear that Ot must become infinitely small with respect to MR (see Figs. 4.11, 4.12). Thus, the two infinitely close tangents MT and mt must coincide with each other, when the point M becomes an inflection point or a cusp. However, on the contrary, when $dx^2 + dy^2 - y\,ddy =$ infinity, Ot must be infinite with respect to MR, or (what is the same thing) MR is infinitely small with respect to Ot. Consequently, the point m must fall on the point M, that is to say that we may draw two tangents from the same point M that make an infinitely small angle between themselves, when this point becomes an inflection point or a cusp.

It is clear that the tangent to an inflection point or a cusp F, when prolonged, touches and cuts the curve AFK at the same point.

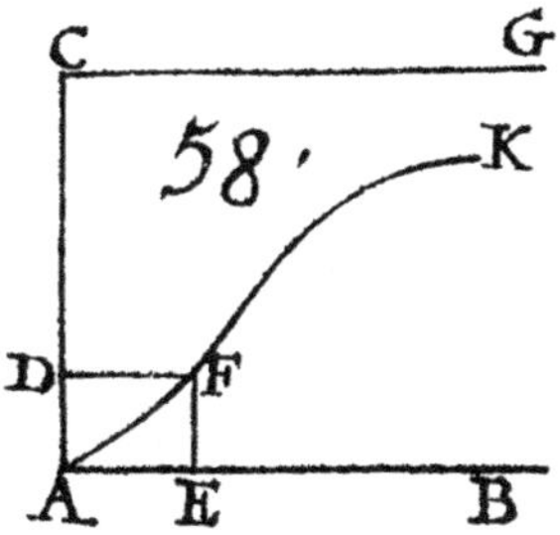

Fig. 4.13 A Curve Related to the "Witch of Agnesi"

Example I. (§68)[5] Let AFK be a curved line that has the straight line AB as diameter (see Fig. 4.13), and is such that the relationship of the abscissa AE (x) to the ordinate EF (y) is expressed by the equation[6] $axx = xxy + aay$. We wish to find a value of AE such that the ordinate EF meets the curve AFK at an inflection point F.

[64] The equation of the curve is $y = \frac{axx}{xx+aa}$ and consequently $dy = \frac{2a^3x\,dx}{\overline{xx+aa}^2}$. Taking the differential of this quantity, supposing dx to be constant, then setting it equal to zero, we find

$$\frac{2a^3\,dx^2 \times \overline{xx+aa}^2 - 8a^3xx\,dx^2 \times \overline{xx+aa}}{\overline{xx+aa}^4} = 0.$$

Multiplying this by $\overline{xx+aa}^4$ and dividing by $2a^3\,dx^2 \times \overline{xx+aa}$ gives $xx+aa-4xx = 0$, from which we conclude that $AE\ (x) = a\sqrt{\frac{1}{3}}$.

If we substitute the value $\frac{1}{3}aa$ for xx in the equation of the curve $y = \frac{axx}{xx+aa}$, we find that $EF\ (y) = \frac{1}{4}a$, so that we may determine the inflection point F without supposing that the curve AFK is described.

If we draw AC parallel to the ordinate EF equal to the given straight line a and we draw CG parallel to AB, it is asymptotic to the curve AFK. This is because if we suppose x to be infinite, we may take xx for $xx + aa$ and consequently the equation of the curve $y = \frac{axx}{xx+aa}$ becomes $y = a$.

Example II. (§69) Let $y - a = \overline{x-a}^{\frac{3}{5}}$. Thus $dy = \frac{3}{5}\overline{x-a}^{-\frac{2}{5}}$ and

$$ddy = -\frac{6}{25}\overline{x-a}^{-\frac{7}{5}}dx^2 = \frac{-6\,dx^2}{25\sqrt[5]{\overline{x-a}^7}},$$

---

[5]Compare this to the example on p. 218.

[6]This is essentially the curve that later became known as the "Witch of Agnesi." This curve considered here is that curve reflected in the line $y = \frac{a}{2}$; see p. xxxvi for more about this curve.

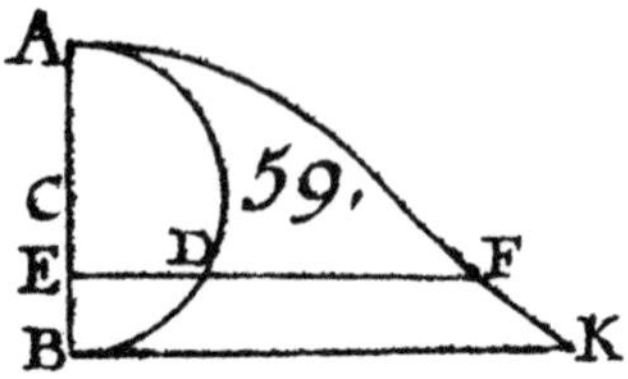

Fig. 4.14 Prolate Cycloid

taking dx as constant. Now, if we assume this fraction is equal to zero, we find that $-6\,dx^2 = 0$, which can never happen. We must assume that it is infinitely large, and consequently its denominator $25\sqrt[5]{x-a}^{\,7}$ is infinitely small or zero. From this, the unknown AE $(x) = a$.

Example III. (§70) Consider a prolate half-roulette AFK (see Fig. 4.14), whose base BK is greater than the semi-circumference ADB of the generating circle with center C. We wish to determine [65] the point E on the diameter AB, such that the ordinate EF meets the cycloid at the inflection point F.

Denoting the given quantities ADB by a, BK by b and AB by $2c$ and the unknowns AE by x, ED by z, the arc AD by u and EF by y, then by the property of the roulette, we have $y = z + \frac{bu}{a}$ and consequently $dy = dz + \frac{b\,du}{a}$. Now by the property of the circle, we have $z = \sqrt{2cx - xx}$, $dz = \frac{c\,dx - x\,dx}{\sqrt{2cx-xx}}$ and $du(\sqrt{dx^2 + dy^2}) = \frac{c\,dx}{\sqrt{2cx-xx}}$. Thus, substituting the values of dz and du, we find

$$dy = \frac{ac\,dx - ax\,dx + bc\,dx}{a\sqrt{2cx - xx}},$$

whose differential (taking dx as constant) gives

$$\frac{\overline{bcx - acc - bcc} \times dx^2}{\overline{2cx - xx} \times \sqrt{2cx - xx}} = 0,$$

from which we conclude that AE $(x) = c + \frac{ac}{b}$ and $CE = \frac{ac}{b}$.

It is clear that in order for there to be an inflection point F, b must be greater than a, because if it were less than a, then CE would be greater than CB.

Example IV. (§71)[7] We wish to find the inflection point F of the Conchoid AFK of *Nicomedes* (see Fig. 4.15), which has the point P as pole and the straight line BC as asymptote. Its property is such that if we draw the straight line PF, from the pole P to any of its points F, which meets the asymptote BC at D, then the part DF is always equal to the same given straight line a.

[7]Compare this to the example given on p. 72.

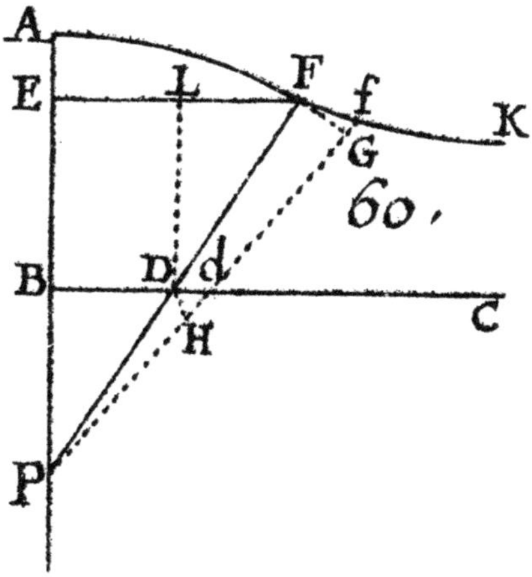

Fig. 4.15 Conchoid of Nicomedes

If we draw *PA* perpendicular to *BC* and *FE* parallel to *BC*, then we denote the given quantities *AB* or *FD* by a and *BP* by b, and the unknowns *BE* by x and *EF* by y. If we draw *DL* parallel to *BA*, the similar triangles *DLF* and *PEF* give $DL\ (x) : LF\ \sqrt{aa-xx} :: PE\ (b+x) : EF\ (y) = \frac{\overline{b+x}\sqrt{aa-xx}}{x}$, whose differential is[8] $dy = \frac{x^3\,dx+aab\,dx}{xx\sqrt{aa-xx}}$. If we then take the differential of this quantity and set it equal to zero, we form the equation

$$\frac{\overline{2a^4b - aax^3 - 3aabxx} \times dx^2}{\overline{aax^3 - x^5} \times \sqrt{aa-xx}} = 0,$$

[66] which reduces to $x^3 + 3bxx - 2aab = 0$, one of the roots of which gives the value of *BE* that we wish to find.

If $a = b$, the previous equation changes to $x^3 + 3axx - 2a^3 = 0$, which, when divided by $x + a$, gives $xx + 2ax - 2aa = 0$. Consequently $BE\ (x) = -a + \sqrt{3aa}$.

In Another Way. We take the lines *PF* that emanate from the pole P as the ordinates and use the formula (see §66) $yddy = dx^2 + dy^2$, in which dx is assumed to be constant. We imagine another ordinate *Pf* that makes an infinitely small angle *FPf* with *PF* and we describe the little arcs *FG* and *DH* with center P. If we denote the given quantities *AB* by a and *BP* by b, and the unknowns *PF* by y and *PD* by z, then we have $y = z + a$ from the property of the conchoid, which gives $dy = dz$. Now,

[8]If one applies the rules of the differential calculus, the value of dy would be the negative of the given value. However, the coordinates in this problem are set up so that as y increases, x decreases, so the author adjusts the sign of dy as described earlier (see §8). In any case, the value of ddy that follows has the correct sign.

because DBP is a right triangle, we have $DB = \sqrt{zz - bb}$. Because of the similar triangles DBP and dHD, we have DB $(\sqrt{zz - bb})$: BP (b) :: dH (dz) : $HD = \frac{b\,dx}{\sqrt{zz-bb}}$. Also, because of the similar triangles PDH and PFG, we have PD (z) : PF $(z + a)$ :: HD $\left(\frac{b\,dz}{\sqrt{zz-bb}}\right)$: FG $(dx) = \frac{bx\,dz + ab\,dz}{z\sqrt{zz-bb}}$. From this, we conclude dz or $dy = \frac{z\,dx\sqrt{zz-bb}}{bz+ab}$, the differential of which (supposing dx to be constant) is

$$ddy = \frac{\overline{bz^3 + 2abzz - ab^3} \times dz\,dx}{\overline{bz + ab}^2 \sqrt{zz - bb}}$$
$$= \frac{\overline{bz^4 + 2abz^3 - ab^3z} \times dx^2}{\overline{bz + ab}^3},$$

substituting the value of dz. Thus, if we substitute $z + a$ in place of y in the general formula (see §66) $y\,ddy = dx^2 + dy^2$, as well as the values that we have just found in terms of dx and dx^2 for dy and ddy, we form the following equation

$$\frac{\overline{z^4 + 2az^3 - abbz} \times dx^2}{\overline{bz + ab}^2} = \frac{\overline{z^4 + 2abbz + aabb} \times dx^2}{\overline{bz + ab}^2},$$

which reduces to $2z^3 - 3bbz - abb = 0$, one of the roots of which, when added to a, gives the value of the unknown PF.

If $a = b$, then we have $2z^3 - 3aaz - a^3 = 0$, which when divided by $z + a$, gives $zz - az - \frac{aa}{2} = 0$, whose solution gives PF $(z + a) = \frac{3}{2}a + \frac{1}{2}a\sqrt{3} = \frac{3a+a\sqrt{3}}{2}$.

Example V. [67] (§72)[9] Let AFK (see Fig. 4.15) be another kind of conchoid, such that if we draw the straight line PF from any of its points F to the pole P, which meets the asymptote BC at D, then the rectangle $PD \times DF$ is always equal to the same rectangle $PB \times BA$. We wish to find the inflection point F.

If we denote the unknowns BE by x and EF by y and the given quantities AB by a and BP by b, then we have $PD \times DF = ab$. The parallels BD and EF give $PD \times DF$ (ab) : $PB \times BE$ (bx) :: $\overline{PF}^2$ $(bb + 2bx + xx + yy)$: $\overline{PE}^2$ $(bb + 2bx + xx)$. Thus $bbx + 2bxx + x^3 + yyx = abb + 2abx + axx$, or

$$yy = \frac{abb + 2abx + axx - bbx - 2bxx - x^3}{x}$$

[9]Compare this to the example given on p. 223.

and

$$y = \overline{b+x}\sqrt{\frac{a-x}{x}} = \sqrt{ax-xx} + b\sqrt{\frac{a-x}{x}},$$

the differential of which gives[10]

$$dy = \frac{-ax\,dx + 2xx\,dx + ab\,dx}{2x\sqrt{ax-xx}}.$$

Taking the differential again, we form the equation[11]

$$\frac{\overline{3aab - aax - 4abx} \times dx^2}{\overline{4axx - 4x^3}\sqrt{ax-x^2}} = 0,$$

which reduces to $x = \frac{3ab}{a+4b}$, the value of the unknown *BE*.

If we set the value of dy, $\frac{-ax\,dx+2xx\,dx+ab\,dx}{2x\sqrt{ax-xx}}$ equal to zero, we have $xx - \frac{1}{2}ax + \frac{1}{2}ab = 0$, the two roots of which are $\frac{a+\sqrt{aa-8ab}}{4}$ and $\frac{a-\sqrt{aa-8ab}}{4}$. When a is greater than $8b$, these give two values of *BH* and *BL* (see Fig. 4.16), such that the ordinate *HM* is less than its neighbors and the ordinate *LN* is greater. That is to say, the tangents at *M* and *N* are parallel to the axis *AB* and thus the point *E* falls between the points *H* and *L*.

However, if $a = 8b$, the lines *BH*, *BE*, and *BL* are all equal to $\frac{1}{4}a$ and thus the tangent at the inflection point *F* is parallel to the axis *AB* (see Fig. 4.17). Finally, if

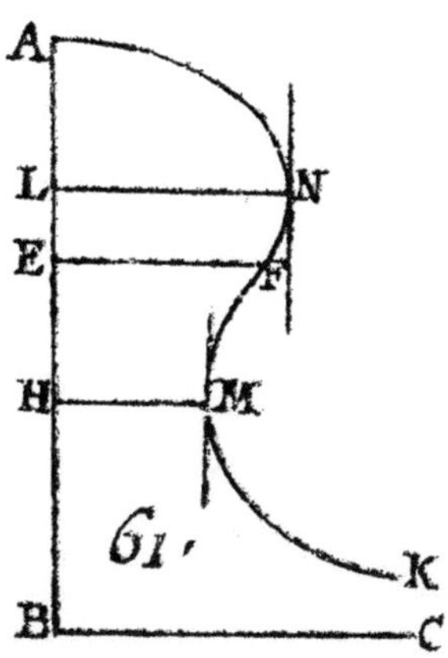

Fig. 4.16 Conchoid with Two Extrema

[10] As in Example IV, in this chapter, this value of dy has its sign adjusted.

[11] In (L'Hôpital 1696), the denominator contained $\overline{4ax - 4x^3}$, with a note in the Errata to replace $-4x^3$ with $-4xx$. In fact, that term $-4x^3$ is correct, whereas for $4ax$ should have been replaced with $4axx$.

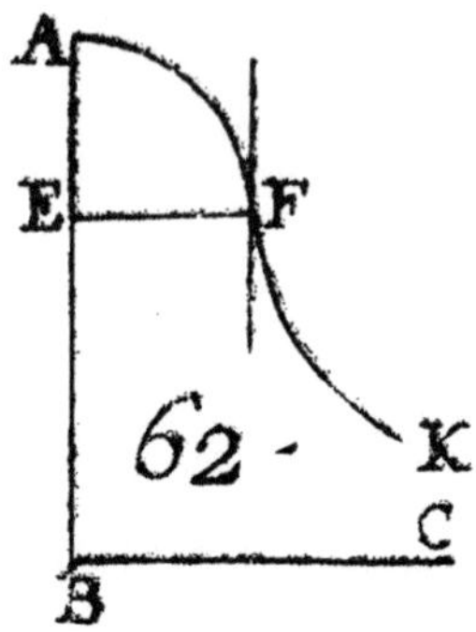

Fig. 4.17 Conchoid with Tangent Parallel to the Axis at the Inflection Point

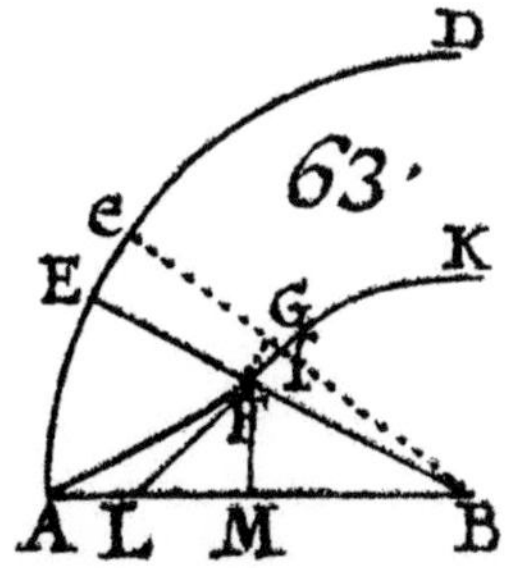

Fig. 4.18 Parabolic Spiral

a is less than $8b$, the two roots are imaginary and consequently there are no tangents that can be parallel to the axis.

[68] We may also resolve this question by taking the lines PF and Pf that emanate from the pole P as ordinates (see Fig. 4.15) and making use of the formula $y\,ddy = dx^2 + dy^2$, as we have done in the previous example.

Example VI. (§73)[12] Let AED be a circle with the point B as its center (see Fig. 4.18), with a curved line AFK, such that if we draw the radius BFE at will, the square on FE is equal to the rectangle of the arc AE with a given straight line b.[13] We wish to determine the inflection point F of this curve.

Denoting the arc AE by z, the radius BA or BE by a and the ordinate BF by y, we have $bz = aa - 2ay + yy$ and (taking differentials) $\frac{2y\,dy - 2a\,dy}{b} = dz = Ee$. Now, because of the similar sectors BEe and BFG, we have $BE\ (a) : BF\ (y) :: Ee\ \left(\frac{2y\,dy-2a\,dy}{b}\right) : FG\ (dx) = \frac{2yy\,dy-2ay\,dy}{ab}$, whose differential, taking dx as constant, gives $4y\,dy^2 - 2a\,dy^2 + 2yy\,ddy - 2ay\,ddy = 0$, and consequently $y\,ddy = \frac{a\,dy^2 - 2y\,dy^2}{y-a}$.

[12]Compare this to the example given on p. 230.

[13]I.e., the product of the lengths of the arc AE and the line b.

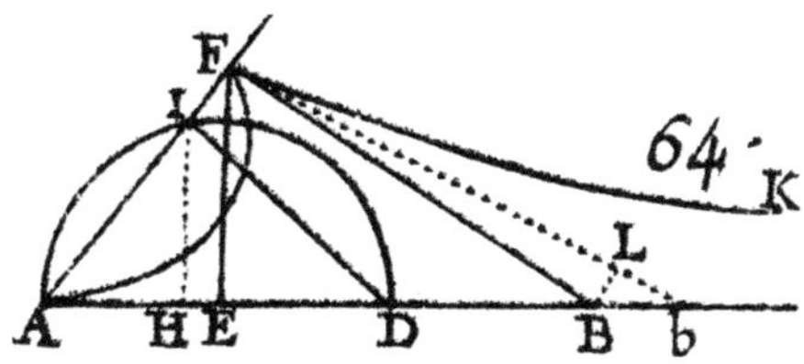

Fig. 4.19 Example of a Cusp

Thus, if we substitute the values of dx^2 and $y\,ddy$ in terms of dy^2 in the general formula $y\,ddy = dx^2 + dy^2$ (see §66), we form the equation

$$\frac{a\,dy^2 - 2y\,dy^2}{y - a} = \frac{4y^2\,dy^2 - 8ay^3\,dy^2 + 4aayy\,dy^2 + aabb\,dy^2}{aabb}.$$

This reduces to $4y^5 - 12ay^4 + 12aay^3 - 4a^3yy + 3aabby - 2a^3bb = 0$, the solution of which gives the value of BF that we wish to find.

It is clear that the curve AFK, which we may call a *parabolic Spiral*,[14] must have an inflection point F. Because the circumference AED does not initially differ noticeably from the tangent at A, it follows from the nature of the parabola that it must initially be concave towards this tangent and that subsequently, when the curvature of the circumference about its center becomes noticeable, it must become concave towards this center.

Example VII. [69] (§74) Let AFK be a curved line that has the straight line AB (see Fig. 4.19) as its axis with the following property: if we draw any tangent FB that meets AB in the point B, the intercepted part AB is always to the tangent BF in a given ratio of m to n. We wish to determine the cusp F.

If we denote the unknown and variable quantities AE by x and EF by y, we have $EB = -\frac{y\,dx}{dy}$ (because as x increases, y decreases) and $FB = \frac{y\sqrt{dx^2+dy^2}}{dy}$. Now, by the property of the curve, $AE + EB$ or AB $\left(\frac{x\,dy - y\,dx}{dy}\right) : BF$ $\left(\frac{y\sqrt{dx^2+dy^2}}{dy}\right) :: m : n$. Therefore, $m\sqrt{dx^2 + dy^2} = \frac{nx\,dy}{y} - n\,dx$ and its differential gives

$$\frac{m\,dy\,ddy}{\sqrt{dx^2 + dy^2}} = \frac{-ny\,dx\,dy + nxy\,ddy - nx\,dy^2}{yy},$$

taking dx as constant and negative, from which we conclude

$$ddy = \frac{\overline{-ny\,dx\,dy - nx\,dy^2}\sqrt{dx^2 + dy^2}}{myy\,dy - nxy\sqrt{dx^2 + dy^2}}.$$

[14] This curve is closely related to Spiral of Fermat. See p. xxxvi for further discussion of this curve.

If we now set this fraction equal to zero, we find that $-y\,dx - x\,dy = 0$, which tells us nothing. This is why we must suppose that this fraction is equal to infinity, that is to say its denominator is equal to zero. This gives $\sqrt{dx^2 + dy^2} = \frac{my\,dy}{nx} = \frac{nx\,dy - ny\,dx}{my}$, because of the equation of the curve. From this we conclude that $dx = \frac{nnxx\,dy - mmyy\,dy}{nnxy}$. Now squaring both sides of the equation $my\,dy = nx\sqrt{dx^2 + dy^2}$ we also find that $dx = \frac{dy\sqrt{mmyy - nnxx}}{nx} = \frac{nnxx\,dy - mmyy\,dy}{nnxy}$. Finally, we conclude that $y\sqrt{mm - nn} = nx$, which gives the following construction.

Let a semi-circle *AID* be described with diameter $AD = m$. If we take the chord $DI = n$ and extend *AI* indefinitely, then I say that it meets the curve *AFK* in the cusp F.

[70] This is because, if we draw *IH* perpendicular to *AB*, the similar right triangles *DIA*, *IHA*, and *FEA* give $DI\ (n) : IA(\sqrt{mm - nn}) :: IH : HA :: FE\ (y) : EA\ (x)$. Consequently, $y\sqrt{mm - nn} = nx$, which is the location to be constructed.

It is clear that *BF* is parallel to *DI* because $AB : BF :: AD\ (m) : DI\ (n)$, from which it follows that the angle *AFB* is a right angle. Consequently, the lines *AB*, *BF*, and *BE* are in continued proportion.[15]

We may find this same property without any calculation if we imagine (see §67) two tangents *FB* and *Fb* drawn from the same cusp F, which make an infinitely small angle *BFb* between themselves. This is because if we describe the little arc *BL* with center F, we have $m : n :: Ab : bF :: AB : BF :: Ab - AB$ or $Bb : bF - BF$ or $bL :: BF : BE$, because of the similar right triangles *BbL* and *FBE*. Therefore, etc.

If $m = n$, it is clear that the straight line *AF* is evidently perpendicular to the axis *AB* and thus the tangent *FB* is parallel to this axis. This is what we already knew must happen, because in this case the curve *AF* must be a semi-circle which has its diameter perpendicular to the axis *AB*. However, if m is less than n, it is clear that there is no cusp, because then the equation $y\sqrt{mm - nn} = nx$ contains a contradiction.

[15] I.e., *BE* is the third proportional to *AB* and *BF*, or $AB : BF :: BF : BE$.

Chapter 5
Use of the Differential Calculus for Finding Evolutes

Definition. [71] We imagine any curved line *BDF* (see Fig. 5.1), which is concave towards the same side, and is covered or wrapped by a thread *ABDF*, one of the extremities of which is fixed at *F* and the other of which is pulled along the tangent *BA*. If we make the extremity *A* move by always holding the thread taut and by continually evolving[1] the curve *BDF*, it is clear that the extremity *A* of this thread describes a curved line *AHK* by this motion.

Given this, the curve *BDF* is called the *Evolute* of the curve[2] *AHK*.

The straight parts *AB*, *HD*, and *KF* of the thread *ABDF* are called the *radii of the evolute.*[3]

Corollary I. (§75) *As long as the length of the thread ABDF always remains the same, it follows that the portion BD of the curve is equal to the difference of the radii DH and BA, which emanate from its extremities. Furthermore, the portion DF is equal to the difference of the radii FK and DH, and the entire curve BDF is equal to the difference of the radii FK and BA. From this, we see that if the radius BA of the curve were null, that is to say that if the extremity A of the thread fell on the origin B of the curve BDF, then the radii DH and FK of the evolute would be equal to the portions BD and BDF of the curve BDF.*

Corollary II. (§76) *If we consider the curve BDF to be a polygon BCDEF with an infinity of sides, then it is clear that the extremity A of the thread ABCDEF*

---

[1]In L'Hôpital (1696) the verb *déveloper* is used to describe the peeling away of the thread from the curve *BDF*. We translate this as "evolving." Conversely, the term *enveloper* is used, e.g. §110, Footnote 4, to describe the inverse process of laying the thread back onto the curve *BDF*.

[2]In L'Hôpital (1696) the term *Dévelopée* is used for the curve *AHK*. We translate this with the standard modern term "evolute" (Huygens 1673, p. 74).

[3]The modern term for this is "radii of curvature," but L'Hôpital (1696) makes no mention of curvature here, referring instead to *rayons de la dévelopée*.

R.E. Bradley et al., *L'Hopital's Analyse des infiniments petits*, Science Networks. Historical Studies 50, DOI 10.1007/978-3-319-17115-9_5

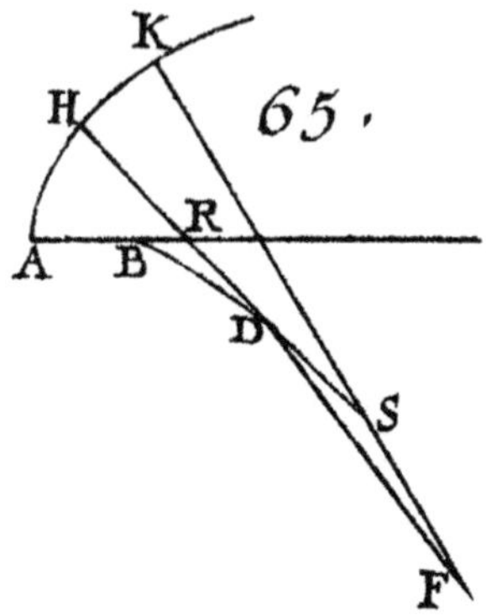

Fig. 5.1 Definition of Involute and Evolute

describes the little arc AG, which has [72] the point C for its center, right up until the radius CG makes a straight line with the little side CD neighboring CB. Furthermore, it describes the little arc GH, which has the point D as its center, right up until the radius DH makes a straight line with the little side DE, and so on, until the curve BCDEF is entirely evolved. The curve AHK may therefore be considered as an assemblage of an infinity of little circular arcs AG, GH, HI, IK, etc., which have the points C, D, E, F, etc., as centers. From this it follows that:

1. The radii of the evolute touch the curve continually as *DH* at *D*, *KF* at *F*, etc., and they are all perpendicular to the curve *AHK* that they describe, as *DH* at *H*, *FK* at *K*, etc. This is because *DH*, for example, is perpendicular to the little arc *GH* and to the little arc *HI*, because it passes through their centers *D* and *E*. From this we see that:[4]

   (a) The evolute *BDF* (see Fig. 5.1) bounds the space where all the perpendiculars to the curve *AHK* fall.

   (b) If we prolong any radius *HD* that cuts the radius *AB* at *R* until it meets any other radius *KF* at *S*, we may always draw two perpendiculars to the curve *AHK* from all of the points of the part *RS*, except from the point of tangency *D*, from which we may only draw one, namely *DH*. For it is clear that the intersection *R* of the radii *AB* and *DH* traverses all the points of the part *RS* while the radius *AB* describes, by its extremity *A*, the line *AHK* to which it is continually perpendicular and that the radii *AB* and *HD* do not coincide, except when the intersection *R* falls on the point of tangency *D*.

2. If we prolong (see Fig. 5.2) the little arcs *HG* to *l*, *IH* to *m*, *KI* to *n*, etc., towards the origin *A* of the evolution, each little arc such as *IH* will touch outside its

[4]Rather than using letters (a) and (b), the author numbered the following two points 1° and 2°, embedded within a paragraph already numbered 1. We have chosen to use letters instead for clarity.

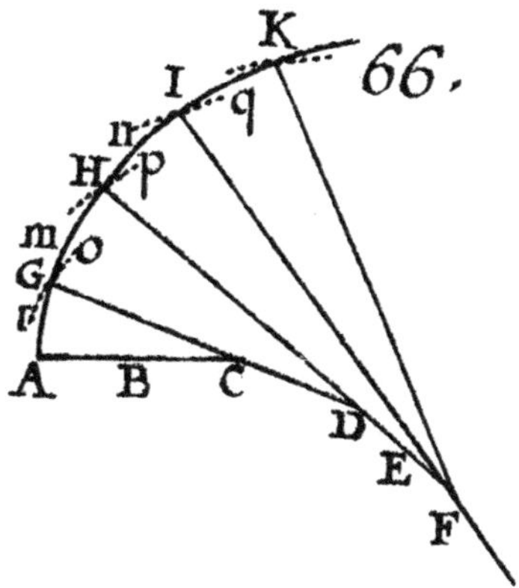

Fig. 5.2 Evolute as a Polygonal Curve

neighbor *HG*, because the radii *CA*, *DG*, *EH*, and *FI* always increase, as long as the little arcs that make up the curve *AHK* move away from the point *A*. For the same reasons, if we prolong the little arcs *AG* to *o*, *GH* to *p*, [73] and *HI* to *q* in the direction opposite to the point *A*, each little arc such as *HI* will touch inside its neighbor *IK*. Now, because the points *H* and *I*, as well as *D* and *E*, may be considered as falling on each other, owing to the infinite smallness of both the arc *HI* and the side *DE*, it follows that if we describe a circle *mHp* with any point *D* in the middle of the evolute *BDF* as its center and with *DH* as its radius, it touches outside the part *HA*, which falls entirely within the circle, and touches inside the other part *HK*, which falls entirely outside the same circle. That is to say, it both touches and cuts the curve *AHK* at the same point *H*, in the same way as the tangent at an inflection point cuts the curve at the same point.

3. The radius *HD* of the little arc *HG* differs from the radius *CG* and *EH* of the neighboring arcs *GA* and *HI* by only the infinitely small quantity *CD* or *DE*. From this it follows that no matter how little we decrease the radius *DH*, it will be less than *CG* and therefore that its circle touches inside the part *HA* and, on the other hand, no matter how little we increase it, it will exceed *HE* and therefore its circle touches outside the part *HK*. Consequently, the circle *mHp* is the smallest of all those that touch outside of the part *HA* and, on the other hand, is the largest of all those that touch inside the part *HK*. That is to say, we cannot put another circle between this one and the curve.[5]
4. Because the curvature of circles increases proportionately as their radii decrease, it follows that the curvature of the little arc *HI* is to the curvature of the little arc *AG* reciprocally as the radius *BA* or *CA* of the latter is to the radius *DH* or *EH*. That is to say, the curvature of the curve *AHK* at *H* is to the curvature at *A* as the radius *BA* is to the radius *DH*. Similarly, the curvature at *K* is to the curvature

[5]This is an implicit reference to the osculating circle, which will be mentioned in Chapter 10.

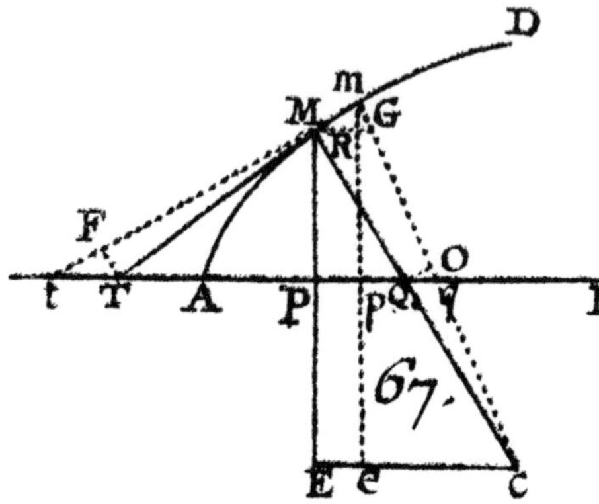

Fig. 5.3 Construction of the Radius of the Evolute – Perpendicular Ordinates

at H as the radius *DH* is to the radius *FK*. From this we see that the curvature of the line *AHK* continually decreases as the line *BDF* is evolved, so that at the point A, where the evolution begins, it is as large as possible [74] and at the point K where I suppose that it ceases, it is the smallest.

5. The points of the evolute are nothing more than the meeting points of the perpendiculars drawn from the extremities of the little arcs that make up the curve *AHK*. For example, the point D or E is the intersection point of the perpendiculars *HD* and *IE* of the little arc *HI*. Therefore, if the curve *AHK* is given, along with the position of one of its perpendiculars *HD*, then to find the point D or E, where it touches the evolute, we need to only find the point of intersection of the infinitely close perpendiculars *HD* and *IE*. We will show how to do this in the Problem that follows.

Proposition I.

General Problem. (§77) *Given the nature of the curved line AMD* (see Fig. 5.3) *with any one of its perpendiculars MC, we wish to find the length of the radius MC of its evolute. That is to say, we wish to find the intersection of the infinitely close perpendiculars MC and mC.*

Suppose, in the first place, that the curved line *AMD* has the straight line *AB* as axis, on which the ordinates *PM* are perpendicular. We imagine another ordinate *mp*, which is infinitely close to *MP*, because we suppose the point m to be infinitely close to M. From the point of intersection C, we draw *CE* parallel to the axis *AB*, which meets the ordinates *MP* and *mp* in the points E and e. Finally, drawing *MR* parallel to *AB*, we form right triangles *MRm* and *MEC*, which are similar because the angles *EMR* and *CMm* are right angles and the angle *CMR* is common to them, so the angle *EMC* is equal to the angle *RMm*.

Therefore, if we denote the given quantities *AP* by x and *PM* by y, and the unknown *ME* by z, we have *Ee* or *Pp* or $MR = dx$, $Rm = dy = dz$, $Mm =$

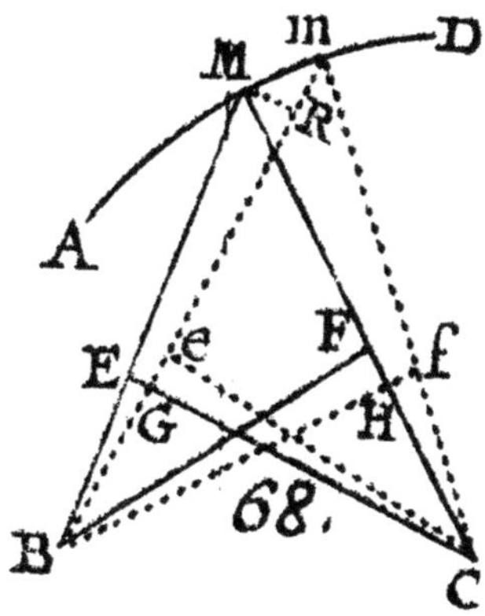

Fig. 5.4 Construction of the Radius of the Evolute – Polar Ordinates

$\sqrt{dx^2 + dy^2}$ and MR (dx) : Mm ($\sqrt{dx^2 + dy^2}$) :: ME (z) : $MC = \frac{z\sqrt{dx^2+dy^2}}{dx}$. Now because the point C is the center of the little arc Mm, its radius CM, which becomes Cm when [75] EM increases by its differential Rm, remains the same. Its differential is therefore null, which gives

$$\frac{dz\,dx^2 + dz\,dy^2 + z\,dy\,ddy}{dx\,\sqrt{dx^2 + dy^2}} = 0$$

(supposing that dx is constant). From this we conclude that

$$ME\ (z) = \frac{dz\,dx^2 + dz\,dy^2}{-dy\,ddy} = \frac{dx^2 + dy^2}{-ddy},$$

substituting the value dy for dz.

Suppose, in the second place, that the ordinates BM and Bm (see Fig. 5.4) all emanate from the same point B. If, from the point C that we wish to find, we drop the perpendiculars CE and Ce to these ordinates, which I suppose to be infinitely close, and describe the little arc MR with center B, then we form the similar right triangles RMm and EMC, and the similar triangles BMR, BEG, and CeG. If we denote BM by y, ME by z, and MR by dx, then we have $Rm = dy$, $Mm = \sqrt{dx^2 + dy^2}$, CE or $Ce = \frac{z\,dy}{dx}$ and $MC = \frac{z\sqrt{dx^2+dy^2}}{dx}$. Next we find, as in the first case, that

$$z = \frac{dz\,dx^2 + dz\,dy^2}{-dy\,ddy}.$$

Now

$$BM\ (y) : CE\ \left(\frac{z\,dy}{dx}\right) :: MR\ (dx) : Ge = \frac{z\,dy}{y},$$

and

$$me - ME \text{ or } Rm - Ge = dz = \frac{y\,dy - z\,dy}{y}.$$

Thus, substituting this value in place of dz, we have

$$ME\ (z) = \frac{y\,dx^2 + y\,dy^2}{dx^2 + dy^2 - y\,ddy}.$$

If we suppose that y is infinite, the terms dx^2 and dy^2 are null with respect to $y\,ddy$ and consequently this last formula changes to the one in the preceding case. This must also happen, because then the ordinates become parallel to each other, and the arc MR becomes a straight line perpendicular to these ordinates.

If the nature of the curve AMD is now given, we find the values of dy^2 and ddy in terms of dx^2, or of dx^2 and ddy in terms of dy^2, which, when substituted into the preceding formulas, give a value for ME freed of all differentials and entirely known. If we draw EC perpendicular to ME, it cuts the perpendicular to the curve MC at the point C that we wish to find. This is what was proposed.

Corollary I. [76] (§78) *Because of the similar right triangles MRm and MEC (see Figs. 5.3, 5.4), we have*

$$MC = \frac{\overline{dx^2 + dy^2}\,\sqrt{dx^2 + dy^2}}{-dx\,ddy}$$

in the first case and

$$MC = \frac{\overline{y\,dx^2 + y\,dy^2}\,\sqrt{dx^2 + dy^2}}{dx^3 + dx\,dy^2 - ydx\,ddy}$$

in the second case.

Remark. (§79) There are several other ways to find the radii of the evolute. I will include some of them here, in order to give different approaches to those who have not yet mastered this calculus.

The first case, for curves whose ordinates are perpendicular to the axis.

First Method. Let MR (see Fig. 5.3) be prolonged to G, where it meets the perpendicular mC. The right angles MRm and MmG give $RG = \frac{dy^2}{dx}$ and consequently

$MG = \frac{dx^2+dy^2}{dx}$. Now because the triangles MRm and MPQ are similar (where the points Q and q are the intersections of the infinitely close perpendiculars MC and mC with the axis AB), it follows that $MQ = \frac{y\sqrt{dx^2+dy^2}}{dx}$ and $PQ = \frac{y\,dy}{dx}$. Consequently $AQ = x + \frac{y\,dy}{dx}$, the differential of which is $Qq = dx + \frac{dy^2+y\,ddy}{dx}$ (taking dx to be constant). Because the triangles CMG and CQq are similar, we have

$$MG - Qq\left(\frac{-y\,ddy}{dx}\right) : MG\left(\frac{dx^2+dy^2}{dx}\right) :: MQ\left(\frac{y\sqrt{dx^2+dy^2}}{dx}\right) : MC = \frac{\overline{dx^2+dy^2}\sqrt{dx^2+dy^2}}{-dx\,ddy}.$$

Second Method. If we describe the little arc QO with center C, then the little right triangles QOq and MRm are similar, because Mm and QO are parallel and MR and Qq are parallel. Consequently

$$Mm\ \left(\sqrt{dx^2+dy^2}\right) : MR\ (dx) :: Qq\ \left(\frac{dx^2+dy^2+y\,ddy}{dx}\right) : QO = \frac{dx^2+dy^2+y\,ddy}{\sqrt{dx^2+dy^2}}.$$

Now the similar sectors CMm [77] and CQO give

$$Mm - QO\ \left(\frac{-y\,ddy}{\sqrt{dx^2+dy^2}}\right) : Mm\ \left(\sqrt{dx^2+dy^2}\right) :: MQ\ \left(\frac{y\sqrt{dx^2+dy^2}}{dx}\right) : MC = \frac{\overline{dx^2+dy^2}\sqrt{dx^2+dy^2}}{-dx\,ddy}.$$

Third Method. If we draw the infinitely close tangents MT and mt, we have $PT - AP$ or $AT = \frac{y\,dx}{dy} - x$, the differential of which gives $Tt = -\frac{y\,dx\,ddy}{dy^2}$. If we describe the little arc TH with center m, we form the right triangle HTt, which is similar to RmM, because the angles HtT and RMm, or PTM, are equal, since they differ from each other only by the angle Tmt, which is infinitely small. This gives

$$Mm\ \left(\sqrt{dx^2+dy^2}\right) : mR\ (dy) :: Tt\left(-\frac{y\,dx\,ddy}{dy^2}\right) : TH = \frac{-y\,dx\,ddy}{dy\sqrt{dx^2+dy^2}}.$$

Now the sectors TmH and MCm are similar, because the angle $Tmt + MmC$ is right and the angle $MmC + MCm$ is also right, because of the triangle CMm, considered to be a right triangle at M. Thus,

$$TH\ \left(-\frac{y\,dx\,ddy}{dy\sqrt{dx^2+dy^2}}\right) : Mm\ \left(\sqrt{dx^2+dy^2}\right) :: Tm \text{ or } TH\ \left(\frac{y\sqrt{dx^2+dy^2}}{dy}\right) : MC = \frac{\overline{dx^2+dy^2}\sqrt{dx^2+dy^2}}{-dx\,ddy}.$$

Fourth Method. We consider (see §64) the second differentials, taking dx to be constant. The similar triangles (see Fig. 5.5) HmS and Hnk give Hm or $Mm\ \left(\sqrt{dx^2+dy^2}\right) : mS$ or $MR\ (dx) :: Hn\ (-ddy) : nk = -\frac{dx\,ddy}{\sqrt{dx^2+dy^2}}$. Now the angle kmn is equal to the angle that the tangents at the points M and m make between themselves and consequently, as we have just proven, is equal to the angle MCm, from which it follows that sectors nmk and MCm are similar. Therefore

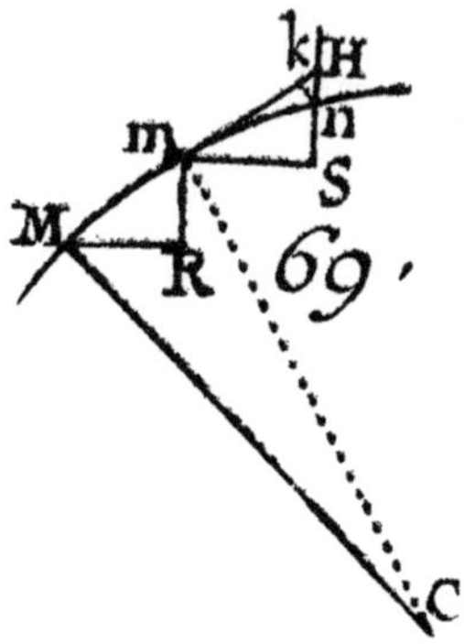

Fig. 5.5 Second Differential of the Radius of the Evolute – Perpendicular Ordinates

$nk\ \left(-\frac{dx\,ddy}{\sqrt{dx^2+dy^2}}\right)$: mk or Mm (see §2) $\left(\sqrt{dx^2+dy^2}\right)$ :: Mm $\left(\sqrt{dx^2+dy^2}\right)$: $MC = \frac{\overline{dx^2+dy^2}\sqrt{dx^2+dy^2}}{-dx\,ddy}$. We take mH or Mm for mk, because they differ from each other by only the little straight line Hk, which is infinitely smaller than them. Similarly, Hn is infinitely smaller than Rm or Sn.

The second case, for curves whose ordinates emanate from the same fixed point.

[78] First Method. From the fixed point B (see Fig. 5.4), drop perpendiculars BF and Bf to the infinitely close radii CM and Cm. The right triangles mMR and BMF, which are similar (because when we add the same angle FMR to the angles mMR and BMF they each make a right angle), give MF or $MH = \frac{y\,dx}{\sqrt{dx^2+dy^2}}$ and $BF = \frac{y\,dy}{\sqrt{dx^2+dy^2}}$, the differential of which (taking dx to be constant) is[6]

$$Bf - BF \text{ or } Hf = \frac{dx^2\,dy^2 + dy^4 + y\,dx^2\,ddy}{\overline{dx^2+dy^2 \times \sqrt{dx^2+dy^2}}}.$$

Now, because the sectors CMm and CHf are similar, we form the proportion $Mm - Hf : Mm :: MH : MC$ and consequently

$$MC = \frac{\overline{y\,dx^2 + y\,dy^2}\sqrt{dx^2+dy^2}}{dx^3 + dx\,dy^2 - y\,dx\,ddy}.$$

[6] In L'Hôpital (1696) the overline in the denominator of the following fraction extended to cover the multiplication symbol.

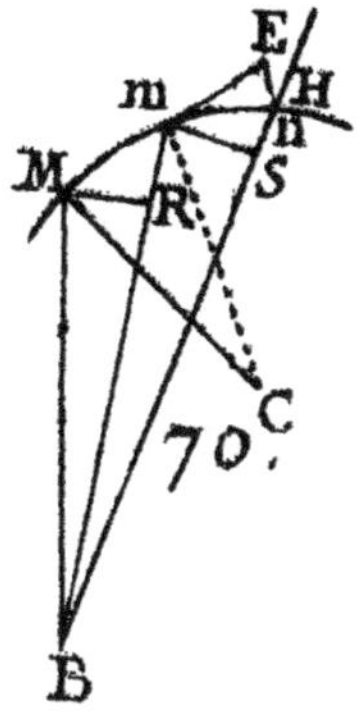

Fig. 5.6 Second Differential of the Radius of the Evolute – Polar Ordinates

Second Method. We consider (see §64) the second differentials, supposing dx to be constant. The similar sectors (see Fig. 5.6) BmS and mEk give

$$Bm\ (y) : mS\ (dx) :: mE\ \left(\sqrt{dx^2+dy^2}\right) : Ek = \frac{dx\sqrt{dx^2+dy^2}}{y}.$$

Now because the right triangles HmS and Hnk are similar, we have

$$Hm \text{ or } Mm\ \left(\sqrt{dx^2+dy^2}\right) : mS \text{ or } MR\ (dx) :: Hn\ (-ddy) : nk = -\frac{dx\,ddy}{\sqrt{dx^2+dy^2}}.$$

Consequently

$$En = \frac{dx^3 + dx\,dy^2 - y\,dx\,ddy}{y\sqrt{dx^2+dy^2}}$$

and, taking a third proportional to En and Em or Mm, the similar sectors Emn and MCm give the same value as before for MC.

If we denote Mm $\left(\sqrt{dx^2+dy^2}\right)$ by du and if we take dy as constant in place of dx, we find in the first case that $MC = \frac{du^3}{dy\,ddx}$ and in the second case that $MC = \frac{y\,du^3}{dx\,du^2+y\,dy\,ddx}$. Finally, if we take du as constant, it follows in the first case $MC = \frac{dx\,du}{-ddy}$ or $\frac{dy\,du}{ddx}$, because the differential of $dx^2+dy^2 = du^2$ is [79] $dx\,ddx+dy\,ddy = 0$ and hence that $\frac{dx}{-ddy} = \frac{dy}{ddx}$. It follows in the second case that $MC = \frac{y\,dx\,du}{dx^2-y\,ddy}$ or $\frac{y\,dy\,du}{dx\,dy+y\,ddx}$.

Corollary II. (§80) *Because we find but a single value for ME or MC (see Fig. 5.8), it follows that a curved line AMD may have only one evolute BCG.*

Corollary III. (§81) *If the value of ME (see Figs. 5.3, 5.4)* $\left(\frac{dx^2+dy^2}{-ddy}\right)$ *or* $\left(\frac{y\,dx^2+y\,dy^2}{dx^2+dy^2-y\,ddy}\right)$ *is positive, we must take the point E on the same side of the axis or of the point B, as we have supposed in making our calculations. From this we see that the curve is therefore concave towards this axis or this point. However, if the value of ME is negative, we must take the point E on the opposite side, from which we see that the curve is therefore convex. Hence, at an inflection point or cusp that separates the convex part from the concave part, the value of ME must go from positive to negative and consequently the infinitely close or contiguous perpendiculars must go from convergent to divergent. Now this can only happen in two ways: either they must increase as long as they approach the inflection point or cusp, and it must be the case that they become parallel, that is to say that the radius of the evolute is infinite, or else they decrease, and it must necessarily be the case that they coincide with each other, that is to say that the radius of the evolute is zero. All of this is in perfect accord with what we have demonstrated in the previous chapter.*

Remark. (§82) Because up until now we have believed that the radius of the evolute is always infinitely large at an inflection point, [80] this is the proper place to show that there is, so to speak, an infinity of kinds of curves that all have the radius of the evolute at their inflection point equal to zero, whereas there is only one kind in which this radius is infinite.

Let *BAC* (see Fig. 5.7) be one of the curves that has an infinite radius of the evolute at its inflection point *A*. If we evolve the parts *BA* and *AC* beginning from the point *A*, then it is clear that we form a curved line *DAE* that also has an inflection point at the same point *A*, whose radius of the evolute at that point is equal to zero. Moreover, if we were to form, in the same way, a third curve by the evolution of the second curve *DAE* and a fourth curve by the evolution of the third one, and so on to infinity, it is clear that the radius of the evolute at the inflection point *A* in all these curves is always equal to zero. Thus, etc.

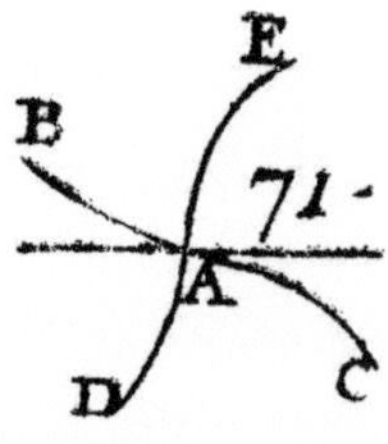

Fig. 5.7 Evolute at an Inflection Point

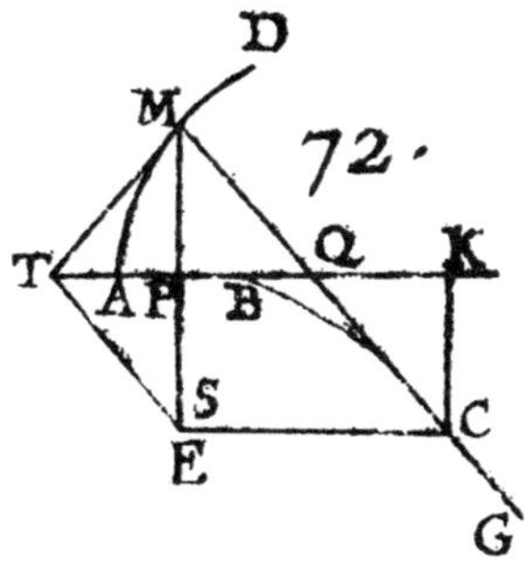

Fig. 5.8 Finding the Origin of the Evolute

Proposition II.

Problem. (§83) *In curves AMD* (see Fig. 5.8)*, for which the axis AB makes a right angle with the tangent at A, we wish to find the point B where this axis touches the evolute BCG.*

If we suppose that the point M becomes infinitely close to the vertex A, it is clear that the perpendicular MQ meets the axis at the point B that we wish to find. From this it follows that, if we find the general value of[7] PQ $\left(\frac{y\,dy}{dx}\right)$ in terms of x or of y and if we then make x or $y = 0$, we will make the point P fall on the point A and the point Q fall on the point B that we wish to find. That is to say, PQ will therefore become equal to the AB that we wish to find. This will be clarified by the following examples.

Example I. (§84) Let the curve AMD (see Fig. 5.8) be a Parabola which has the given straight line a as its [81] parameter. The equation of the parabola is $ax = yy$, whose differential gives $dy = \frac{a\,dx}{2y} = \frac{a\,dx}{2\sqrt{ax}}$. Taking the differential of this latter equation, supposing that dx is constant, we find that $ddy = \frac{-a\,dx^2}{4x\sqrt{ax}}$. Finally, substituting these values in place of dy and ddy in the formula $\frac{dx^2+dy^2}{-ddy}$, we have (see §77)

$$ME = \frac{\overline{a+4x}\sqrt{ax}}{a} = \sqrt{ax} + \frac{4x\sqrt{ax}}{a}.$$

This gives the following construction.

[7]The expression that follows is the first time the ratio $\frac{dy}{dx}$ appears in L'Hôpital (1696).

Let T be the point where the tangent MT meets the axis and let the line TE be drawn parallel to MC. I say that it meets the prolonged MP at the point E that we wish to find. Because the angles MPT and MTE are right, we have

$$MP\ (\sqrt{ax}) : PT\ (2x) :: PT\ (2x) : PE = \frac{4xx}{\sqrt{ax}} = \frac{4x\sqrt{ax}}{a}$$

and consequently $MP + PE = \sqrt{ax} + \frac{4x\sqrt{ax}}{a}$.

Furthermore, because of the [similar] right triangles[8] MPQ and MEC, we have

$$PM\ (\sqrt{ax}) : PQ\ \left(\frac{1}{2}a\right) :: ME\ \left(\sqrt{ax} + \frac{4x\sqrt{ax}}{a}\right) : EC \text{ or } PK = \frac{1}{2}a + 2x$$

and consequently $QK = 2x$. This gives us the following new construction.

Let QK be taken to be the double of AP or (what amounts to the same thing) let PK be taken equal to TQ and let KC be drawn parallel to PM. It will meet the perpendicular MC at the point C, which is on the evolute BCG.

Alternate Method. $yy = ax$ and $2y\,dy = a\,dx$, whose differential (supposing dx to be constant) gives $2dy^2 + 2y\,ddy = 0$, from which we conclude $-ddy = \frac{dy^2}{y}$. Substituting this value into the formula $\frac{dx^2+dy^2}{-ddy}$ (see §77), we find $ME = \frac{y\,dy^2+y\,dx^2}{dy^2}$ and consequently EC or $PK = \frac{y\,dy^2+y\,dx^2}{dy\,dx} = \frac{y\,dy}{dx} + \frac{y\,dx}{dy} = PQ + PT$ or TQ. This gives the same constructions as above, because $MP : PT :: dy : dx :: PT\ \left(\frac{y\,dx}{dy}\right) : PE = \frac{y\,dx^2}{dy^2} = \frac{4x\sqrt{ax}}{a}$.

[82] Now, we desire to find the point B where the axis AB touches the evolute BCG. We have $PQ\ \left(\frac{y\,dy}{dx}\right) = \frac{1}{2}a$. However, because this quantity is constant, it always remains the same, no matter what the location of the point M. Hence, when it falls on the vertex A, we still have the same PQ, which in this case becomes $AB = \frac{1}{2}a$.

To find the nature of the evolute BCG in the manner of *Descartes*.[9] We denote the abscissa BK by u and the ordinate KC or PE by t. From this we have $CK\ (t) = \frac{4x\sqrt{ax}}{a}$ and $AP + PK - AB\ (u) = 3x$. Thus, substituting its value $\frac{1}{3}u$ for x in the equation $t = \frac{4x\sqrt{ax}}{a}$, we have the new equation $27att = 16u^3$, which expresses the relationship of BK to KC. From this we see that the evolute BCG of

[8] In L'Hôpital (1696) the similarity of these triangles is never mentioned, although it is used the proportional relation that follows.

[9] The phrase "in the manner of Descartes" simply means to give an equation for the curve; evolutes are not discussed by Descartes.

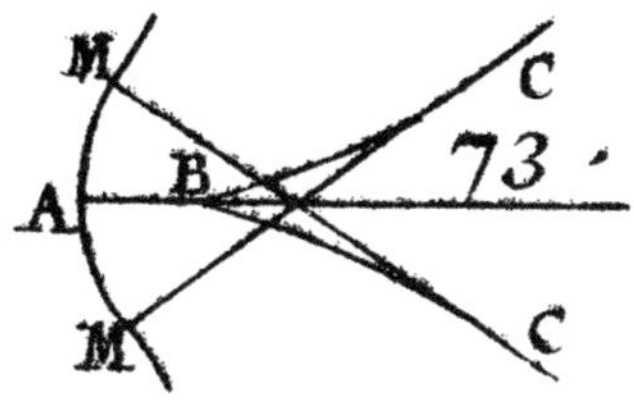

Fig. 5.9 Evolute of the Parabola

the ordinary parabola is a semi-cubical parabola,[10] whose parameter is equal to $\frac{27}{16}$ of the parameter of the given parabola.

It is clear that the evolute *CBC* (see Fig. 5.9) of the common entire parabola *MAM* has two parts, *CB* and *BC*, which have their convexities opposite to each other, so that they form a cusp at the point *B*.

Note. By *geometric curves*, we understand curves such as *AMD* and *BCG (see Fig. 5.8)*, for which the relationship between the abscissas *AP* and *BK* and the ordinates *PM* and *KC* may be expressed by an equation in which there are no differentials.[11] By *geometric*, we understand all that may be done by means of such lines. Here, we suppose that the abscissas and ordinates are straight lines.

Corollary. (§85) *When the given curve AMD is geometric, it is clear that we may always find (as in this example) an equation to express the nature of [83] its evolute BCG and that therefore this evolute is also geometric. Furthermore, I say that it is rectifiable,*[12] *that is that we may geometrically find straight lines equal to any one of its portions BC. For it is evident (see §75) that, with the assistance of the line AMD, which is geometric, we may determine a point M on the tangent CM to the portion BC, such that CM differs from the portion BC only by the given straight line AB.*

Example II. (§86) Let the given curve *MDM* (see Fig. 5.10) be a hyperbola between its asymptotes, that has the equation $aa = xy$.

We have $\frac{aa}{y} = x$ and $\frac{-aa\,dy}{yy} = dx$. Supposing dx to be constant, we have (see §1)

$$\frac{-aayy\,ddy + 2aay\,dy^2}{y^4} = 0,$$

[10] In L'Hôpital (1696) this is called the *une seconde Parabole cubique*, literally "a second cubical parabola," but we use the traditional English term for this curve.

[11] As noted in the Preface, "geometric" was used by Descartest in the same way we use "algebraic" in modern terminology.

[12] In modern usage, a curve is said to be rectifiable if it has a finite arc length. The meaning here is somewhat different: that we may determine the arc length of any portion of a curve geometrically in the sense of Descartes. This will be further clarified in §108.

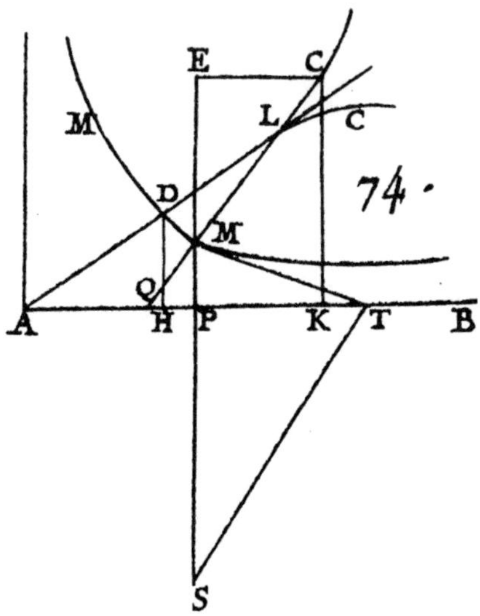

Fig. 5.10 Evolute of the Hyperbola

from which we conclude $ddy = \frac{2\,dy^2}{y}$. Substituting this value in $\frac{dx^2+dy^2}{-ddy}$, it follows (see §77) that[13] $ME = \frac{y\,dx^2+y\,dy^2}{-2dy^2}$, so that EC or $PK = -\frac{y\,dy}{2\,dx} - \frac{y\,dx}{2\,dy}$. This gives the following constructions.

From the point T, where the tangent MT meets the asymptote AB, draw the line TS parallel to MC, which meets the prolonged MP at S. Let ME be taken equal to half of MS on the other side of the asymptote (which we consider here to be the axis), because the value of ME is negative. Alternately, let PK be taken equal to half of TQ on the same side as the point T. I say that if we draw EC parallel or KC perpendicular to the axis, they will cut the straight line MC at the required point C. For it is clear that $MS = \frac{y\,dx^2+y\,dy^2}{dy^2}$ and that $TQ = \frac{y\,dy}{dx} + \frac{y\,dx}{dy}$.

If we pay some attention to the figure of the hyperbola MD, we will see that the evolute CLC must have a cusp L, as did the evolute of the parabola. To find it, I note that the radius DL of the evolute is smaller than any other radius [84] MC. From this it follows that the differential of its expression (see §78)

$$\frac{\overline{dx^2+dy^2}\sqrt{dx^2+dy^2}}{-dx\,ddy} \quad \text{or} \quad \frac{\overline{dx^2+dy^2}^{\frac{3}{2}}}{-dx\,ddy}$$

[13] In L'Hôpital (1696), the equal sign was missing from this expression.

is zero or infinite (see Ch. 3). Taking dx to be constant, as usual, this gives

$$\frac{-3dx\,dy\,ddy^2\,\overline{dx^2+dy^2}^{\frac{1}{2}} + dx\,d\,d\,dy\,\overline{dx^2+dy^2}^{\frac{3}{2}}}{dx^2\,ddy^2} = 0 \quad \text{or} \quad \infty.$$

Dividing this by $\overline{dx^2+dy^2}^{\frac{1}{2}}$ and then multiplying by $dx^2\,ddy^2$, we derive the equation $dx^2\,d\,d\,dy + dy^2\,d\,d\,dy - 3dy\,ddy^2 = 0$ or ∞, which allows us to find a value AH for x such that, when we draw the ordinate HD and the radius of the evolute DL, the point L will be the desired cusp.

In this example, we have $y = \frac{aa}{x}$, $dy = \frac{-aa\,dx}{xx}$, $ddy = \frac{2aa\,dx^2}{x^3}$ and $d\,d\,dy = \frac{-6aa\,dx^3}{x^4}$. Therefore, when we substitute these values in the preceding equation, we find that AH $(x) = a$. From this it follows that the point D is the vertex of the hyperbola and that the lines AD and DL form the same straight line AL, which is the axis of the hyperbola.

Example III. (§87) Let the general equation be $y^m = x$ (see Figs. 5.8, 5.10), which expresses the nature of all of the Parabolas to infinity when the exponent m denotes either a whole or a fractional positive number and all of the Hyperbolas when it denotes a negative number.

We have $my^{m-1}\,dy = dx$, whose differential, taking dx to be constant, gives $\overline{mm-my}\,y^{m-1}\,dy^2 + my^{m-1}\,ddy = 0$. Dividing this by my^{m-1}, it follows that $-ddy = \frac{\overline{m-1}\,dy^2}{y}$. Substituting this value in $\frac{dx^2+dy^2}{-ddy}$ we conclude (see §77) that $ME = \frac{y\,dx^2+y\,dy^2}{\overline{m-1}\,dy^2}$ and consequently that EC or $PK = \frac{y\,dy}{\overline{m-1}\,dx} + \frac{y\,dx}{\overline{m-1}\,dy}$. This gives the following general constructions.

From the point T where the tangent MT meets the axis AP, draw the line TS parallel to MC, which meets [85] the prolonged MP at the point S. Let $ME = \frac{1}{m-1}MS$ or else $PK = \frac{1}{m-1}TQ$ be taken. It is clear that if we draw a parallel to the axis through the point E, or a perpendicular to the axis through the point K, they meet MC at the point C that we wish to find.

If m is negative, as occurs in hyperbolas (see Fig. 5.10), the value of ME is negative and consequently they are convex towards their axis, which is therefore an asymptote. However, for the parabolas, where m is positive, two cases may occur. Either m is less than 1 (see Fig. 5.11) and thus they are convex on the side of their axis, which is a tangent to the vertex, or m is greater than 1 (see Fig. 5.8) and thus they are concave towards their axis, which is perpendicular to the vertex.

In this last case, to find the point B where the axis AB touches the evolute. We have PQ $\left(\frac{y\,dy}{dx}\right) = \frac{y^{2-m}}{m}$, which gives three different cases. For either $m = 2$, which only occurs in the ordinary parabola, and hence because the exponent of y is zero, this unknown vanishes, and consequently $AB = \frac{1}{2}$, that is to say half of the parameter. Alternately, if m is less than 2, then because the exponent of y is positive, it is found in the numerator, which makes it (setting it equal (see §83) to zero) the null fraction, that is to say the point B falls on the point A (see Fig. 5.12)

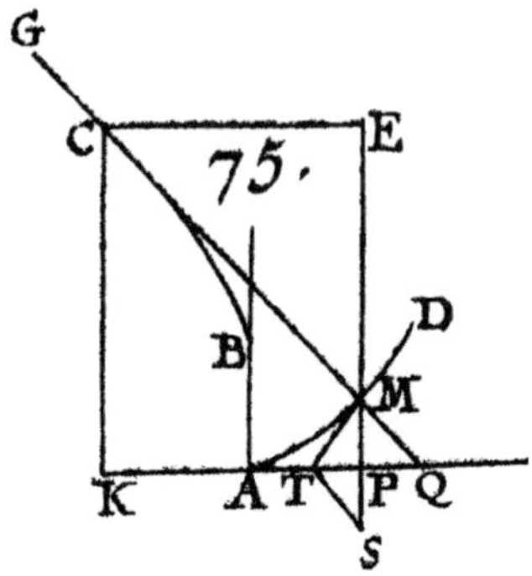

Fig. 5.11 Evolute of the Parabola, $m < 1$

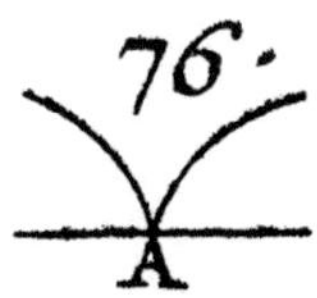

Fig. 5.12 The Semi-cubical Parabola

in this case, as in the semi-cubical parabola $axx = y^3$. Finally, if m is greater than 2, then because the exponent of y is negative, it is in the denominator, which makes it (when it becomes zero) the infinite fraction, that is to say the point B is infinitely far from the point A, or (what is the same thing) that the axis AB is an asymptote to the evolute, as in the cubical parabola[14] $aax = y^3$ (see Fig. 5.13). We may remark in this last case that the evolute CLO of the half-parabola ADM has a cusp L, so that by the evolution of the part LO, continued to infinity, the point D describes only the determinate portion DA, [86] whereas by the evolution of the other part LC, also continued to infinity, it describes the infinite portion DM.

We determine the point L in the same way as for the hyperbola. For example, if we let $aax = y^3$ or $y = x^{\frac{1}{3}}$, we have $dy = \frac{1}{3}x^{-\frac{2}{3}}\,dx$, $ddy = -\frac{2}{9}x^{-\frac{5}{3}}\,dx^2$ and $dddy = \frac{10}{27}x^{-\frac{8}{3}}\,dx^3$. When we substitute these values in the equation $dx^2\,dddy + dy^2\,dddy - 3dy\,ddy^2 = 0$, we find (see §86) $AH\ (x) = \sqrt[4]{\frac{1}{91125}}$. It is similar for the others.

Remark. (§88) If we suppose that m is greater than 1, so that the parabolas are always concave on the side of their axes, there are different cases that may arise. For if the numerator of the fraction denoted by m is even and the denominator odd, all such parabolas fall on one side and the other of their axes (see Fig. 5.9) in a position

[14]In L'Hôpital (1696) this is called *la premiere parabole cubique*, literally "the first cubical parabola."

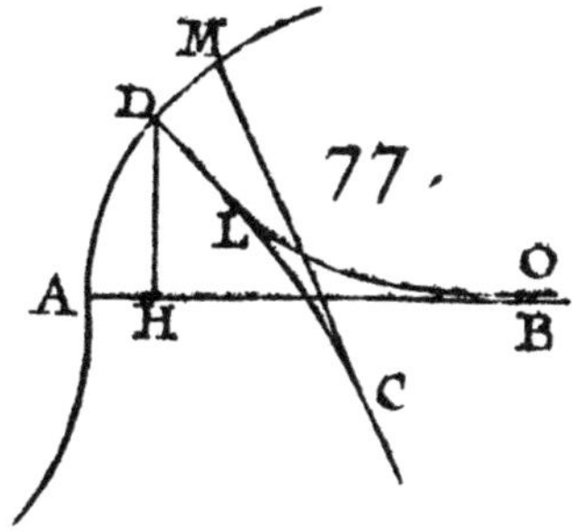

Fig. 5.13 Evolute of the Parabola, $m > 2$

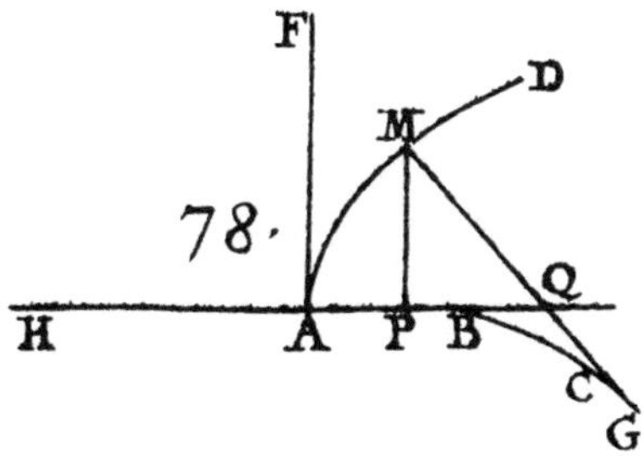

Fig. 5.14 The Hyperbola

similar to that of the ordinary parabola. However, if the numerator and denominator are each odd, they have a reversed position on one side and the other of their axes, so that their vertex A is an inflection point, as in the cubic parabola (see Fig. 5.13) $x = y^{\frac{3}{1}}$ or $aax = y^3$. Finally, if the numerator is odd and the denominator is even, they have a reversed position on the same side of their axes, so that the vertex A is a cusp, as in the semi-cubical parabola (see Fig. 5.12) $x = y^{\frac{3}{2}}$ or $axx = y^3$. All of this follows from the fact that an even power may never have a negative value. Given this, it is evident that

1. For an inflection point A (see Fig. 5.13), the radius of the evolute may be infinitely large, as in $aax = y^3$ or infinitely small, as in $aax^3 = y^5$.
2. For a cusp A (see Fig. 5.12), the radius of the evolute may be infinite, as in $a^3xx = y^5$ or zero, as in $axx = y^3$.
3. [87] It does not follow that if the radius of the evolute is either infinite or zero, the curves therefore have an inflection point or a cusp. For in $a^3x = y^4$ it is infinite, and in $ax^3 = y^4$ it is null, but nevertheless these parabolas fall on one side and the other of their axes, in a position similar to that of the ordinary parabola.

Example IV. (§89) Let the curve *AMD* be a Hyperbola or an Ellipse (see Figs. 5.14, 5.15) that has *AH* (a) as its axis and *AF* (b) as its parameter.

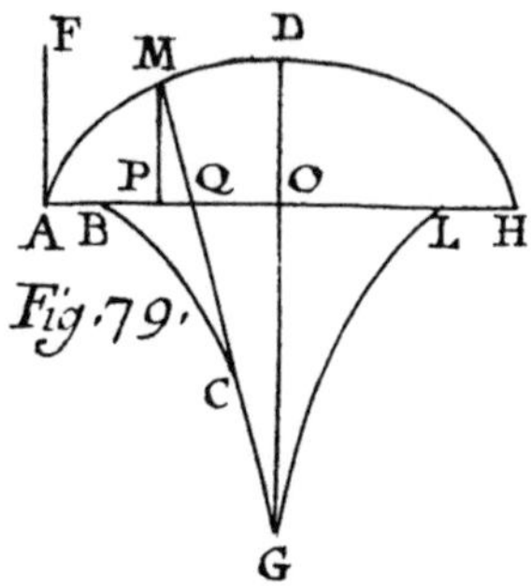

Fig. 5.15 The Ellipse

By the property of these lines, we have

$$y = \frac{\sqrt{abx \mp bxx}}{\sqrt{a}},$$

$$dy = \frac{ab\,dx \mp 2bx\,dx}{2\sqrt{aabx \mp abxx}} \quad \text{and}$$

$$ddy = \frac{-a^3bb\,dx^2}{\overline{4aabx \mp 4abxx}\sqrt{aabx \mp abxx}}.$$

If we then put these values in $\frac{\overline{dx^2+dy^2}\sqrt{dx^2+dy^2}}{-dx\,ddy}$, the general expression for MC (see §78), we find that in these two curves,

$$MC = \frac{\overline{aabb \mp 4abbx + 4bbxx + 4aabx \mp 4abxx}\sqrt{aabb \mp 4abbx + 4bbxx + 4aabx \mp 4abxx}}{2a^3bb}$$

$$= \frac{4MQ^3}{bb},$$

because, in both cases

$$MQ\left(\frac{y\sqrt{dx^2+dy^2}}{dx}\right) = \frac{\sqrt{aabb \mp 4abbx + 4bbxx + 4aabx \mp 4abxx}}{2a}.$$

This gives the following construction, which also works for the Parabola.

Let MC be taken as four times the fourth continued proportional[15] of the parameter AF to the perpendicular MQ, terminated by the axis. Then the point C is on the evolute.

[15]The fourth continued proportion to a and b is the value of y such that $a : b :: b : x :: x : y$.

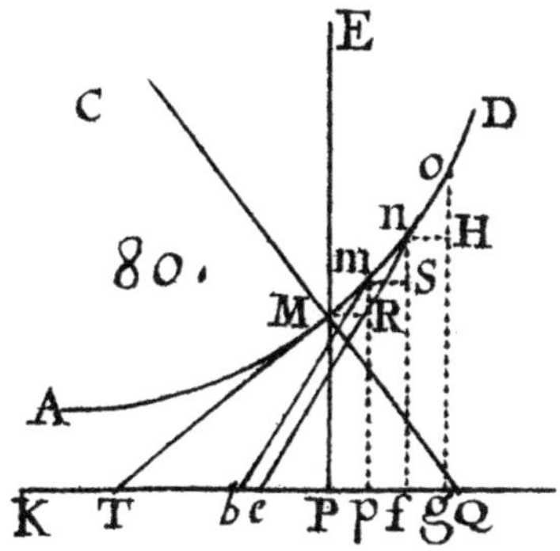

Fig. 5.16 The "Logarithmic" or Exponential Curve

If we let $x = 0$, we have (see §83) $AB = \frac{1}{2}b$. Also, if for the Ellipse we let $x = \frac{1}{2}a$, we find that $DG = \frac{a\sqrt{ab}}{2b}$, that is to say equal to half the parameter of the minor axis. From this we see that in the ellipse, the evolute *BCG* terminates at a point *G* on the minor axis *DO*, where it makes a cusp, whereas in parabola and the hyperbola it extends to infinity.

[88] If $a = b$ in the Ellipse, we have that[16] $MC = \frac{1}{2}a$, from which it follows that all the radii of the evolute are equal to one another and consequently that it is a single point. That is to say, in this case the ellipse becomes a circle, which has its center as its evolute. We already knew this to be true from elsewhere.

Example V. (§90) Let the curve *AMD* be an ordinary logarithmic (see Fig. 5.16), whose nature is such that if we drop the perpendicular *MP* from any of its points *M* to the asymptote[17] *KP* and draw the tangent *MT*, then the subtangent *PT* is always equal to the same given straight line a.

We therefore have $PT \left(\frac{y\,dx}{dy}\right) = a$, from which we conclude that $dy = \frac{y\,dx}{a}$, whose differential, taking dx to be constant, is $ddy = \frac{dy\,dx}{a} = \frac{y\,dx^2}{aa}$. Substituting these values in $\frac{dx^2+dy^2}{-ddy}$ we find (see §77) that $ME = \frac{-aa-yy}{y}$ and consequently that EC or $PK = \frac{-aa-yy}{a}$. This gives the following construction.

Let *PK* be taken equal to *TQ* on the same side as *T*, because its value is negative, and let *KC* be drawn parallel to *PM*. I say that it meets the perpendicular *MC* in the point *C* that we wish to find, because $TQ = \frac{aa+yy}{a}$.

If we wish that point *M* is the one with the greatest curvature, we make use of the formula $dx^2\,dddy + dy^2\,dddy - 3dy\,ddy^2 = 0$, which we found (see §86) in the

[16] In L'Hôpital (1696) the equal sign was missing in the next expression.

[17] In L'Hôpital (1696) the asymptote was given as *BF*, but the correction *KP* was noted in the *Errata*.

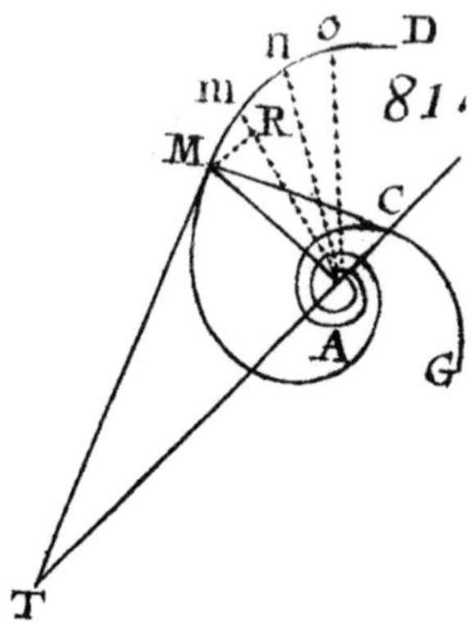

Fig. 5.17 The Logarithmic Spiral

second example. Substituting for dy, ddy, and $dddy$ their values $\frac{y\,dx}{a}$, $\frac{y\,dx^2}{aa}$, and $\frac{y\,dx^3}{a^3}$, we find that $PM\ (y) = a\sqrt{\frac{1}{2}}$.

If we take dx to be constant, it is clear that the ordinates y are to one another as their differentials dy or $\frac{y\,dx}{a}$, from which it follows that they also make a geometric progression. For if we imagine that the asymptote or the axis PK is divided into an infinite number of little equal parts Pp or MR, pf or mS, fg or nH, etc., contained between the [89] ordinates PM, pm, fn, go, etc., we have $PM : pm :: Rm : Sn :: PM + Rm$ or $pm : pm + Sn$ or Fn. Similarly, we prove that $pm : fn :: fn : go$, and so on. The ordinates PM, pm, fn, go, etc., are therefore in a geometric progression.

Example VI. (§91) Let the curve AMD be a logarithmic spiral (see Fig. 5.17), whose nature is such that when we draw the straight line MA from any of its points M to a fixed point A, which is its center, and the tangent MT, the angle AMT is always the same.

Because the angle AMT or AmM is constant, the ratio of mR (dy) to RM (dx) is also constant. Therefore we must have that the differential of $\frac{dy}{dx}$ is null, which gives $ddy = 0$, supposing that dx is constant. This is why in erasing the term $y\,ddy$ in the general expression $\frac{y\,dx^2 + y\,dy^2}{dx^2 + dy^2 - y\,ddy}$ for ME in the case where the ordinates all emanate from the same point, we find that $ME = y$, that is to say $ME = AM$. This gives the following construction.

Let AC be drawn perpendicular to AM, which meets the straight line MC perpendicular to the curve at the point C. The point C is on the evolute[18] ACG.

The angles AMT and ACM are equal, because when we add the same angle AMC to both of them, they make a right angle. The evolute ACG is therefore the same logarithmic spiral as the given AMD, and it differs only in its position.

[18] In L'Hôpital (1696), this was given as ACB, but there is no B in figure 5.17.

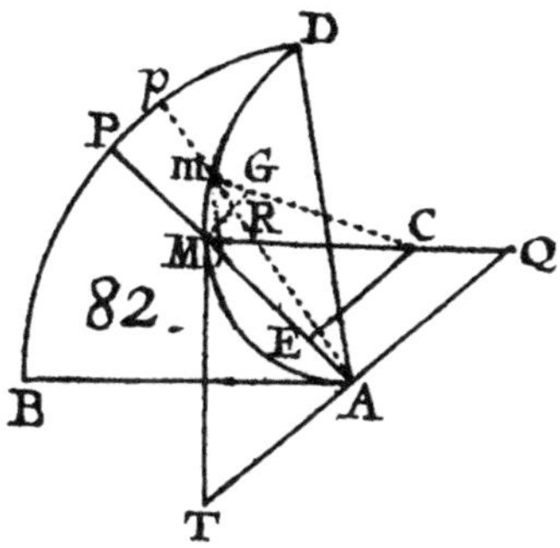

Fig. 5.18 The Spiral of Archimedes

If we suppose that, given the point C on the evolute ACG, it is required to determine the length CM of the radius of the evolute at this point, which (see §75) is equal to the portion AC that makes an infinity of revolutions before reaching A, it is clear that we need to only draw AM perpendicular to AC. Therefore, if we draw AT perpendicular to [90] AM, the tangent MT is also equal to the portion AM of the given logarithmic spiral AMD. If we imagine an infinite number of ordinates AM, Am, An, Ao, etc., that make infinitely small and equal angles amongst themselves, it is clear that the triangles MAm, mAn, nAo, etc., are similar, because the angles at A are equal. By the property of the logarithmic, the angles at m, n, o, etc., are also equal. Consequently, $AM : Am :: Am : An$. Also, $Am : An :: An : Ao$, and so on. From this we see that the ordinates AM, Am, An, Ao, etc., make a geometric progression when the angles they make amongst themselves are equal.

Example VII. (§92) Let the curve AMD be one of the spirals (see Fig. 5.18) up to infinity, formed in the sector BAD with the property that if we draw any radius AMP and denote the entire arc BPD by b, its part BP by z, the radius AB or AP by a and its part AM by y, we have this proportion[19] $b : z :: a^m : y^m$.

The equation of the spiral is $y^m = \frac{a^m z}{b}$, whose differential gives $my^{m-1}\,dy = \frac{a^m\,dz}{b}$. Because the sectors AMR and APp are similar, we have $AM\ (y) : AP\ (a) :: MR\ (dx) : Pp\ (dz) = \frac{a\,dx}{y}$. Hence, substituting this value in place of dz in the equation we have just found, we have $my^m\,dy = \frac{a^{m+1}\,dx}{b}$, whose differential (taking dx as constant) is $mmy^{m-1}\,dy^2 + my^m\,ddy = 0$. Dividing by my^{m-1}, we conclude that $-y\,ddy = m\,dy^2$ and consequently (see §77)

$$ME\ \left(\frac{y\,dx^2 + y\,dy^2}{dx^2 + dy^2 - y\,ddy}\right) = \frac{y\,dx^2 + y\,dy^2}{dx^2 + \overline{m+1}\,dy^2}.$$

This gives the following construction.

[19]When $m = 1$, this is the Spiral of Archimedes. When $m = 2$, this is the Spiral of Fermat.

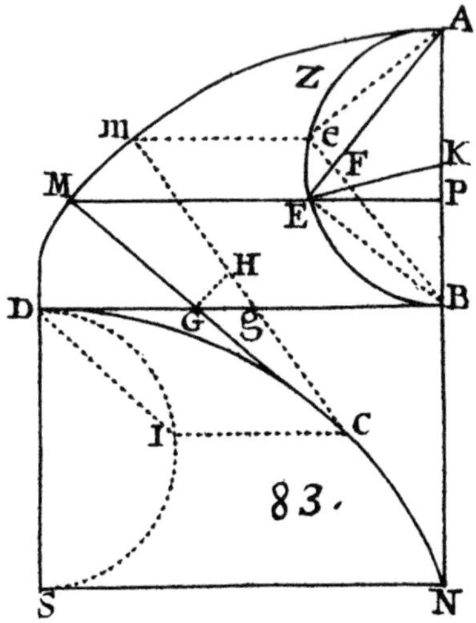

Fig. 5.19 Evolute of the simple Half-Roulette (Half-Cycloid)

Let the straight line TAQ be drawn through the center A perpendicular to AM, which meets the tangent MT at T and the perpendicular MQ at Q. Let [91] $TA + \overline{m+1}AQ : TQ :: MA : ME$. I say that if we draw EC parallel to TQ, it will meet MQ at a point C, which is on the evolute.

This is so, because MRG and TAQ are parallel, so that we have $MR\ (dx) + \overline{m+1}RG\ \left(\frac{dy^2}{dx}\right) : MG\ \left(dx + \frac{dy^2}{dx}\right) :: TA + \overline{m+1}AQ : TQ :: AM\ (y) : ME = \frac{y\,dx^2 + y\,dy^2}{dx^2 + \overline{m+1}\,dy^2}$.

Example VIII. (§93) Let AMD be a simple *half-roulette* (see Fig. 5.19) whose base BD is equal to the semi-circumference BEA of the generating circle.

Denoting AP by x, PM by y, the arc AE by u and the diameter AB by $2a$, we have, by the property of the circle $PE = \sqrt{2ax - xx}$ and by that of the roulette $y = u + \sqrt{2ax - xx}$, the differential of which gives

$$dy = du + \frac{a\,dx - x\,dx}{\sqrt{2ax - xx}}$$

$$= \frac{2a\,dx - x\,dx}{\sqrt{2ax - xx}} \quad \text{or} \quad dx\sqrt{\frac{2a - x}{x}},$$

substituting for du its value $\frac{a\,dx}{\sqrt{2ax-xx}}$. Supposing dx to be constant, we have

$$ddy = \frac{-a\,dx^2}{x\sqrt{2ax - xx}}$$

and substituting these values in $\frac{\overline{dx^2+dy^2}\sqrt{dx^2+dy^2}}{-dx\,ddy}$, it follows (see §78) that $MC = 2\sqrt{4aa - 2ax}$, that is to say $2BE$ or $2MG$.

If we let $x = 0$, we have $AN = 4a$ for the radius of the evolute at the vertex A. However, if we let $x = 2a$, we find that the radius of the evolute at the point D becomes null or zero, from which we see that the evolute has its origin at D and that it ends at N in such a way that $BN = BA$.

To understand the nature of this evolute, we need only complete the rectangle BS, describe the semi-circle DIS that has diameter DS and draw DI parallel to MC or to BE. With this done, it is clear that the angle BDI is equal to the angle EBD and consequently that the arcs DI and BE are equal to each other, from which it follows that their chords DI, and BE or GC, are [92] also equal. Thus, if we join IC, it will be equal and parallel to DG, which by the generation of the roulette is equal to the arc BE or DI. Consequently the evolute DCN is a half-roulette that has as its base the straight line NS, which is equal to the semi-circumference DIS of its generating circle. That is to say it is the same half-roulette $AMDB$ placed in a reversed position.

Corollary. (§94) *It is clear (see §75) that the portion DC of the roulette is the double of its tangent CG, or of the corresponding chord DI. Also, the half-roulette DCN is double the diameter BN or DS of its generating circle.*

Alternate Solution. (§95) We may also find the length of the radius MC without any calculus, in the following way. If we imagine another perpendicular mC infinitely close to the first one, another parallel me, another chord Be and we describe the little arcs GH and EF with centers C and B, then we form the right triangles GHg and EFe, which are equal and similar. For $Cg = Ee$, because BG or ME is equal to the arc AE and similarly, Bg or me is equal to the arc Ae. Furthermore, Hg or $mg - MG = Fe$ or $Be - BE$ and hence GH is equal to EF. Now because the perpendiculars MC and mC are parallel to the chords EB and eB, the angle MCm is equal to the angle EBe. Hence, because the arcs GH and EF, which measure these angles, are equal, it follows that their radii CG and BE are also equal. Consequently, MC should be taken to be the double of MG or of BE.[20]

Lemma. (§96) *If there is any number of quantities a, b, c, d, e, etc., either finite or infinite, then whether these quantities are lines, surfaces, or solids, then the sum $a - b + b - c + c - d + d - e$, etc., of all of their differences is equal to the largest a, minus the smallest* [93] *e, or simply the largest, when the smallest is zero. This is evident.*

Corollary I. (§97) *Because the sectors CMm and CGH are similar, it is clear that Mm is twice GH or EF, which is equal to it, and because this always happens in whatever location we suppose the point M to be, it follows that the sum of all the little arcs Mm, that is the portion Am of the half-roulette AMD, is double the sum of all the little arcs EF. Now the little arc EF is part of the chord AE, perpendicular to BE, and is the difference of the chords AE and Ae, because the little straight line*

[20]§94 and §95 give solutions to Pascal's problems on the cycloid. It is conceivable that §95 is the solution that Fontenelle attributes to L'Hôpital and age 15, see p. 295.

eF, which is perpendicular to AE, may be considered as a little arc described with center A. Consequently, the sum of all the little arcs EF in the arc AZE is the sum of the differences of all the chords AE, Ae, etc., in this arc. That is to say by the Lemma, it is equal to the arc AE. Therefore it is clear that the portion AM of the half-roulette AMD is the double of the corresponding chord AE.

Corollary II. (§98) *The space MGgm or (see §2) the trapezoid* $MGHm = \frac{1}{2}Mm + \frac{1}{2}GH \times MG = \frac{3}{2}EF \times BE$*, that is to say, it is the triple of the triangle EBF or EBe. From this it follows that the space MGBA – the sum of all the trapezoids – is the triple of the circular space BEZA – the sum of all these triangles.*

Corollary III. (§99) *If we denote BP by z, the arc AZE or EM or BG by u and the radius KA by a, we have the parallelogram* $MGBE = uz$*. Now the region of the roulette* $MGBA = 3BEZA = 3EKB + \frac{3}{2}au$ *and consequently the space AMEB enclosed by the portion of the roulette AM, the parallel ME, the chord BE and the diameter AB is* $= 3EKB + \frac{3}{2}au - uz$*. From this it follows that if we take [94] BP* $(z) = \frac{3}{2}a$*, the space AMEB is the triple of the corresponding triangle EKB and consequently we have its quadrature, independent of that of the circle. Mr.* Huygens *was the first to note this. Here is another kind of space that has the same property.*

If we cut the segment *BEZA* out of the space *AMEB*, there remains the space $AZEM = 2EKB + au - uz$. From this we see that if the point *P* falls on the center *K*, the space *AZEM* is equal to the square of the radius. It is clear that among all the spaces *AMEB* and *AZEM*, only the two that we have just determined have their absolute quadrature independent of that of the circle.

Example IX. (§100) Let the half-roulette[21] *AMD* (see Fig. 5.20) be described by the revolution of the semi-circle *AEB* about another immobile circle *BGD*. We wish

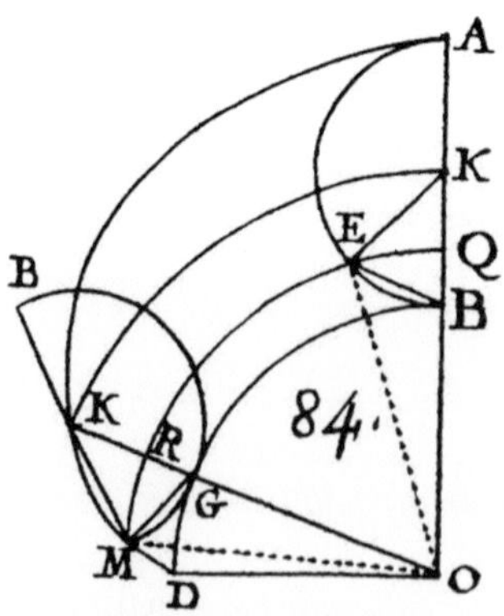

Fig. 5.20 The Epicycloid

[21]In L'Hôpital (1696) the term *demi-roulette* is used here, as in §15 and in the previous example. Unlike those cases the curve being described here is not a cycloid but rather an epicycloid.

to determine the point on a perpendicular *MG* given in position that touches the evolute.

In order to use the general formulas, we must use straight lines perpendicular to the axis *OA* as the ordinates of the curve *AMD* and then seek an equation that expresses the relationship between the abscissas and their ordinates, or of their differentials. However, because these calculations would be quite laborious, it would be much better in these sorts of situations to attempt a solution by making use of the generation itself.

When the semi-circle *AEB* is brought to the position *MGB* in which it touches the base[22] *BD* at *G* and the describing point *A* falls on the point *M* of the half-roulette *AMD*, it is clear that

1. The arc *GM* is equal to the arc *GD*, because the arc *GB* on the mobile circle is equal to the arc *GB* of the immobile circle.
2. *MG* is perpendicular (see §43) to the curve, for if we consider the semi-circumference *MGB* or *AEB* and the base *BGD* as an assemblage of an infinity of equal little straight lines [95] each one in correspondence, it is manifest that the half-roulette *AMD* is an assemblage of an infinity of little arcs that have as their centers successively all the points touching *G* and that are each described from the same point *M* or *A*.
3. If we describe the concentric arc *ME* from the center *O* of the immobile circle, then the arcs *MG* and *EB* of the mobile circle are equal to each other, as well as their chords *MG* and *EB* and the angles *OGM* and *OBE*. This is because the straight lines *OK* and *OK* that join the centers of these two circles are equal, because they pass through the points of contact *B* and *G*. For this reason, if we draw the radii *OM*, *OE*, and *KE*, we form the triangles *OKM* and *OKE*, which are equal and similar. Because the angle *OKM* is therefore equal to the angle *OKE*, the arcs *MG* and *BE* of the equal semi-circles *MGB* and *BEA*, which measure these angles, are equal, as also are their chords *MG* and *BE*. From this it follows that the angles *OGM* and *OBE* are also equal.

Given this, let it be understood that there is another perpendicular *mC* (see Fig. 5.21) infinitely close to the first one, another concentric arc *me* and another chord *Be*. Let the little arcs *GH* and *EF* be described from centers *C* and *B*. The right triangles *GHg* and *EFe* are equal and similar, because Gg or $Dg - DG = Ee$ or to the arc *Be* – the arc *BE* and also Hg or $mg - MG = Fe$ or to $Be - BE$. The little arc *GH* is therefore equal to the little arc *EF*, from which it follows that the angle *GCH* is to the angle *EBF* as *BE* is to *CG*. Therefore all of the difficulty is reduced to finding the ratio of these angles. This is done in the following way.

If we draw the radii *OG*, *Og*, *KE*, and *Ke* and denote *OG* or *OB* by b and *KE* or *KB* or *KA* by a, it is clear that the angle $EBe = OBe - OBE = Ogm - OGM =$ (drawing *GL* and *GV* parallel to *Cm* and *Og*) $LGM - OGV = GCH - Gog$. Thus

[22] In this and subsequent articles, the immobile circle is sometimes referred to as "the base."

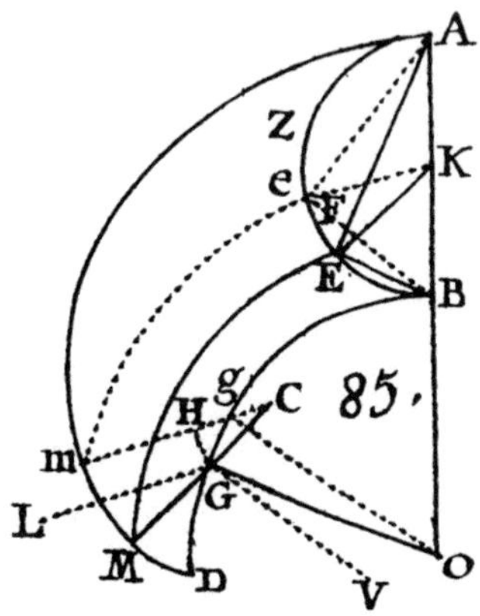

Fig. 5.21 The Epicycloid and its Differentials

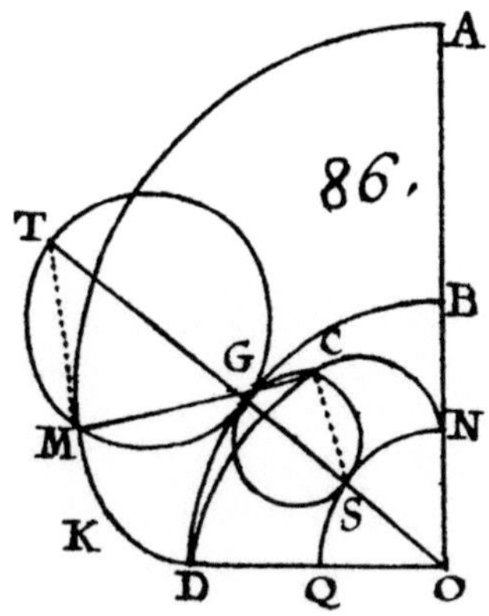

Fig. 5.22 The Epicycloid – Construction of its Evolute

we have the angle $GCH = GOg + EBF$. Now because the arcs Gg and Ee are equal we also have $GOg : EKe$ or $2EBF :: KE\ (a) : OG\ (b)$ and consequently the angle $GOg = \frac{2a}{b}EBF$ and $GCH = \frac{2a+b}{b}EBF$. Thus $GCH : EBF$ or $BE : CG :: \frac{2a+b}{b} : 1$ and consequently the [96] unknown $CG = \frac{b}{2a+b}BE$ or MG. This gives the following construction.

If we make $OA\ (2a + b) : OB\ (b) :: MG : GC$ (see Fig. 5.22), then point C is on the evolute. It is clear that

1. This evolute begins at the point D and that it touches the base BGD, because the arc GM becomes infinitely small at this point.
2. It terminates at the point N, so that $OA : OB :: AB : BN :: OA - AB$ or $OB : OB - BN$ or ON. That is to say OA, OB, and ON are in continued proportion.
3. If we now describe the circle NSQ with center O, I say that the evolute DCN is formed by the revolution of the mobile circle GCS, which has GS or BN as

its diameter, around the immobile circle *NSQ*. That is to say, it is a half-roulette similar to the given one, or of the same kind (because the diameters *AB* and *BN* of the mobile circles are in the same ratio as the radii *OB* and *ON* of the immobile circles) and positioned in an opposite situation so that its vertex is at D. To prove this, suppose that the diameters of the mobile circles are found on the straight line *OT*, drawn at will from the center O. It passes through the points of contact S and G and if we make AB or $TG : BN$ or $GS :: MG : GC$, the point C is on the evolute and furthermore on the circumference of the circle *GCS*. This is because the angle *GMT* is right, so that the angle *GCS* is also right. Now because the angles *MGT* and *CGS* are equal, the arc *TM* or *GB* is to the arc *CS* as the diameter *GT* is to the diameter $GS :: OG : OS :: GB : NS$ and consequently the arcs *CS* and *SN* are equal. Hence, etc.

Corollary I. (§101) *It is clear (see §75) that the portion DC of the roulette is equal to the straight line CM and consequently that DC is to its tangent* $CG :: AB + BN : BN :: OB+ON : ON$*, that is to say as the sum of the diameters of the two generating circles, or of the mobile and immobile circles, is to the radius of the immobile circle. This truth is also discovered by the [97] method that follows. Because the triangles CMm and CGH are similar, we have* $Mm : GH$ *or* $EF :: MC : GC :: OA + OB$ $(2a+b) : OB\ (b)$*. From this it follows (as in §97) that the portion AM of the roulette is to the corresponding chord AE as the sum of the diameters of the generating circle and of the base is to the radius of the base.*

Corollary II. (§102) *The trapezoid (see Fig. 5.21)* $MGHm = \overline{\tfrac{1}{2}GH + \tfrac{1}{2}Mm} \times MG$*. Now*

$$CG\left(\frac{b}{2a+b}MG\right) : CM\left(\frac{2a+2b}{2a+b}MG\right) :: GH : Mm = \frac{2a+2b}{b}GH.$$

Therefore, because $GH = EF$ *and* $MG = EB$*, we have* $MGHm = \frac{2a+3b}{2b}EF \times EB$*, that is to say the trapezoid* $MGHm$ *is always to the corresponding triangle* $EBF ::$ $2a + 3b : b$*.*

From this it follows that the space MGBA enclosed by the perpendiculars MG and AB of the roulette, by the arc BG and by the portion MA of the roulette is to the corresponding segment of the circle BEZA $:: 2a + 3b : b$*.*

Corollary III. (§103) *It is clear that the indefinite quadrature of the roulette (see Fig. 5.23) depends on the quadrature of the circle. However, if we take OQ to the mean proportional between OK and OA and if we describe the arc QEM from this radius, I say that the space* $ABEM$ *enclosed by the diameter AB, the chord BE, the*

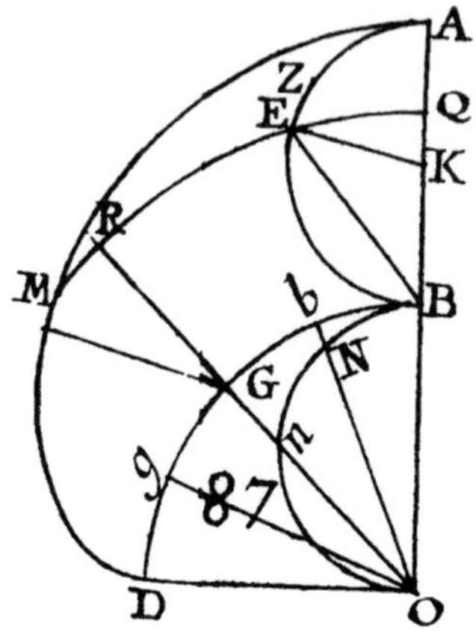

Fig. 5.23 The Epicycloid – Quadrature

arc EM and by the portion AM of the roulette, is to the triangle EKB :: $2a + 3b : b$. *This is because, if we denote the arc AE or GB by u and the radius OQ by z, we have* $OB\ (b) : OQ\ (z) :: GB\ (u) : RQ$ *or* $ME = \frac{uz}{b}$ *and consequently the space* $RGBQ$ *or* $MGBE$*, that is to say*

$$\overline{\frac{1}{2}GB + \frac{1}{2}RQ} \times BQ = \frac{zzu - bbu}{2b}.$$

Now (see §102) the space of the roulette

$$MGBA = \frac{2a+3b}{b} \times BEZA = \frac{2a+3b}{b} \times EKB + \frac{2a+3b}{b} \times KEZA \left(\frac{au}{2}\right).$$

Therefore, if we remove this space from the previous one, what remains is

$$ABEM = \frac{2aau + 3abu + bbu - zzu}{2b} + \frac{2a+3b}{b} \times EKB = \frac{2a+3b}{b}EKB,$$

[98] because $zz = 2aa + 3ab + bb$ *by the construction. From this we see that the quadrature of this space is independent of the quadrature of the circle and it is the only one among all similar spaces.*

Here is another space that has the same property. If we remove the segment $BEZA$ $\left(\frac{1}{2}au + EKB\right)$ from the space $ABEM$, what remains is the space

$$AZEM = \frac{2aau + 2abu + bbu - zzu}{2b} + \frac{2a+2b}{b}EKB = \frac{2a+2b}{b}EKB,$$

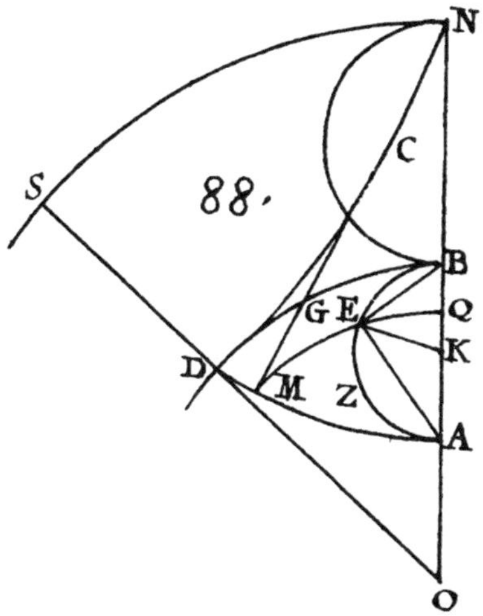

Fig. 5.24 The Hypocycloid

making $zz = 2aa + 2ab + bb$. That is to say, if we divide the semi-circumference equally into two parts at the point E, the space *AZEM* is to the double of the triangle *EKB*, that is to say to the square of the radius :: *OK* $(a + b)$: *OB* (b).

Corollary IV. (§104) *If the mobile circle AEB (see Fig. 5.24) rolls inside the immobile circle BGD, its diameter AB becomes negative, from the positive it had previously been, and thus we must change the sign of the terms where it has an odd dimension. From this it follows that:*

1. If we draw the perpendicular *MG* at will on the roulette and we make *OA* $(b - 2a)$: *OB* (b) :: *MG* : *GC*, the point *C* is (see §100) on the evolute *DCN* described by the revolution of the circle with diameter *BN*, inside the circumference *NS*, concentric with *BD*.
2. If we describe the arc *ME* with center O, the portion *AM* of the roulette is (see §101) to the chord *AE* :: $2b - 2a : b$.
3. The space *MGBA* is (see §102) to the segment *BEZA* $= 3b - 2a : b$.
4. If we take $OQ = \sqrt{2aa - 3ab + bb}$, that is to say the mean proportional between *OK* and *OA*, then the space *ABEM* enclosed by the portion *AM* of the roulette, the arc *ME*, the chord *EB* and the diameter *AB* is (see §103) to the triangle *EKB* :: $3b - 2a : b$. However, if we make *OQ* or $OE = \sqrt{2aa - 2ab + bb}$, that is to say the arc *AE* is a quarter of the circumference, then the space *AZEM* enclosed by the portion *AM* of the roulette and the arcs *ME* and *AE* is (see §103) to the triangle *EKB*, which in this case is half the square of the radius, :: $2b - 2a : b$.

Corollary V. [99] (§105) *If we imagine that the radius OB of the immobile circle (see Figs. 5.22, 5.24) becomes infinite, the arc BGD becomes a straight line and the curve AMD becomes the ordinary roulette. Now because in this case the diameter*

AB of the mobile circle is null with respect to that of the immobile circle, it follows that:

1. $MG : GC :: b : b$, because $b \pm 2a = b$, that is to say that $MG = GC$. Therefore, if we take $BN = AB$ and we draw the straight line *NS* parallel to *BD*, the evolute *DCN* is formed by the revolution of the circle, which has diameter *BN*, on the base *NS*.
2. The portion of the roulette *AM* (see Figs. 5.21, 5.24) is to the corresponding chord $AE :: 2b : b$.
3. The space *MGBA* is to the segment $BEZA :: 3b : b$.
4. Because *BQ* or $\pm OQ \mp OB$, (see Figs. 5.23, 5.24) which I denote by x, is $= \mp b \pm \sqrt{2aa \pm 3ab + bb}$, from which we conclude (by removing the incommensurables) that $xx \pm 2bx = 2aa \pm 3ab$. We have $x = \frac{3}{2}a$ by erasing the terms in which b does not appear, because they are null with respect to the others. That is to say that if we take $BP = \frac{3}{4}AB$ (see Fig. 5.19) in the ordinary roulette and we draw the straight line *PEM* parallel to the base *BD*, the space *AMEB* is the triple of the triangle *EKB*. Operating in the same way, we find that if the point P falls on the center K, the space *AZEM* enclosed by the portion *AM* of the roulette, the straight line *ME*, and the arc *AE*, is equal to the square of the radius. This is what we have already demonstrated above in §99.

Remark. (§106) Because the arcs *DG* and *GM* (see Fig. 5.20) are always equal to each other, it follows that the angle *DOG* is always to the angle $GKM :: GK : OG$. This is why if the origin D of the roulette *DMA*, the radii *OG* and *GK* of the generating circles, and the point of contact G are given, and if, in this position, we wish to determine the point M that describes the roulette, we need to only [100] draw the radius *KM* so that the angle *GKM* is to the given angle $DOG :: OG : GK$. I say now that this can always be done geometrically, when the ratio of these radii may be expressed by numbers, and consequently the roulette *DMA* is therefore geometric.

Suppose, for example, that $OG : GK :: 13 : 5$. It is clear that the angle *MKG* must contain twice the given angle *DOG* and another $\frac{3}{5}$ of this angle. All difficulties therefore reduce to dividing the angle *DOG* into five equal parts. Now this is something that is known among the Geometers, that we can always divide a given angle or arc into as many equal parts as we may wish, because we always end up with some equation that involves only straight lines. Thus, etc.

I say furthermore that the roulette *DMA* is mechanical or, what is the same thing, that we cannot determine its points M geometrically when the ratio of *OG* to *KG* cannot be expressed by numbers, that is to say when it is a surd.[23]

For every line, be it mechanical or geometrical, either returns to itself or extends to infinity, (see Fig. 5.25) because we may always continue its generation. Thus,

[23] Although in modern usage the term "surd" is the irrational root of an integer, in the seventeenth century it was synonymous with an irrational number.

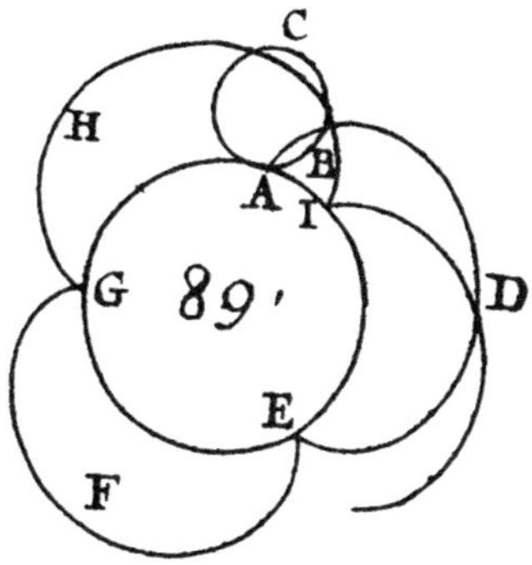

Fig. 5.25 Epicycloid – Irrational Case

if the mobile circle *ABC* describes the roulette *ADE* by its point *A* in the first revolution, this roulette is not yet finished, and in continuing to roll, it will describe a second revolution *EFG*, then a third one *GHI*, and so on, until the describing point *A* once again falls, after many revolutions, on the same point from which it departed. And from there, if we begin to roll the mobile circle *ABC* again, it will describe the same curved line once again, so that all of these roulettes taken together make up but a single curve *ADEFGHI*, etc. Now if the radii of these generating circles are incommensurable, their circumferences are also and consequently the describing point *A* of the mobile circle *ABC* can never fall again on the point *A* of the immobile circle, from which it had departed, no matter how great [101] the number of revolutions may be. Thus, there is an infinity of roulettes that nevertheless form but a single curved line *ADEFGHI*, etc. If we now draw an indefinite straight line across the immobile circle, it is clear that it will cut the curve, continued to infinity, in an infinite number of points. Now, because the equation that expresses the nature of a geometric line must have at least as many dimensions as the number of different points in which a straight line may cut this line, it follows that the equation that expresses the nature of this curve has an infinity of dimensions. Because this cannot be, we evidently see that the curve must be mechanical or transcendental.

Proposition III.

Problem. (§107) *Given the curved line BFC* (see Fig. 5.26), *to find an infinity of lines AM, BN, EFO, of which it is the common evolute.*

If we evolve the curve *BFC* beginning at the point *A*, it is clear that all the points *A*, *B*, *F*, of the thread *ABFC* describe in this motion the curved lines *AM*, *BN*, *FO*, which all have the given curve *BFC* as their common evolute. However, we must observe that because the line *FO* only has the part *FC* as its evolute, its origin is not *F* and, in order to find it, it is necessary to evolve the remaining part *BF* beginning at the point *F* so as to describe the portion *EF* of the curve *EFO*, whose origin is at *E*, and whose evolute is the entire curve *BFC*.

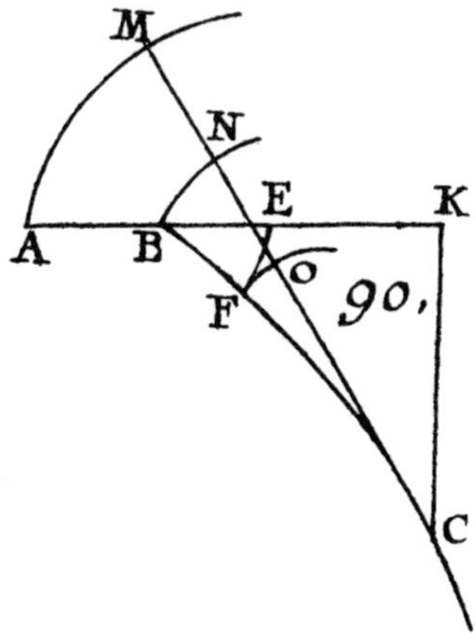

Fig. 5.26 Finding Multiple Involutes

If we wish to find the points M, N, and O without making use of the thread $ABFC$, we need to only take, on any tangent *CM* other than *BA*, the parts *CM*, *CN*, and *CO* equal to *ABFC*, *BFC*, and *FC*.

Corollary. [102] (§108) *It is clear that*

1. *The curves AM, BN, and EFO have quite different natures from one another, because the curve AM has a radius of the evolute equal to AB at its vertex A, unlike that of the curve BN, which is null. It is also evident from the figure itself of the curve EFO that it is quite different from the curves AM and BN.*
2. *The curves AM, BN and EFO are only geometric when the given BFC is geometric and also rectifiable. For if it were not geometric, then by taking BK as the abscissa, we would not find the ordinate KC to be geometric. Furthermore, if it were not rectifiable, then if we draw the tangent CM, we would not be able to determine geometrically the points M, N, and O on the curves AM, BN, and EFO, because we would not be able to find straight lines equal to the curved line BFC and to its parts BF and FC geometrically.*

Remark. (§109) If we evolve a curved line *BAC* (Fig. 5.27) that has an inflection point at *A*, beginning at a point *D* other than the inflection point, we form the part *DEF* by the evolution of the part *BAD* and the remaining part *DG* by the evolution of the part *DC*, so that *FEDG* is the entire curve formed by the evolution of *BAC*. Now it is clear that this curve turns back on itself at the points D and E, with the difference that at the cusp D, the parts *DE* and *DG* have their convexity opposed to each other, whereas at the point E, the parts *DE* and *EF* are concave towards the same side. In the previous chapter, we showed how to find cusps such as D. It is

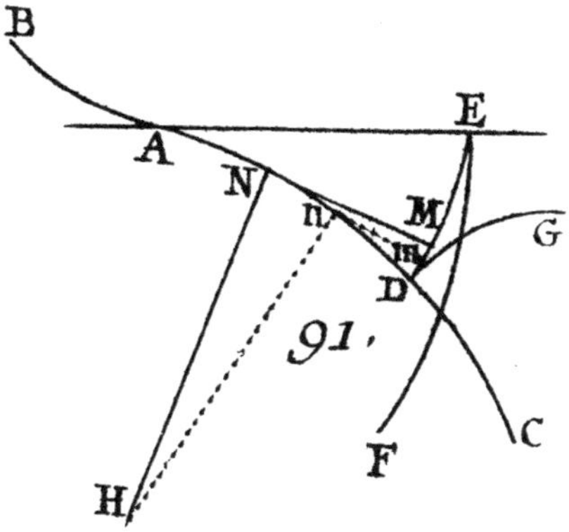

Fig. 5.27 Cusps of the First and Second Kind

now a matter of determining the points E, which we may call cusps of the second kind,[24] and which nobody, as far as I know, has yet considered.

To arrive at this goal, we erect [103] two perpendiculars *MN* and *mn* at will on the part *DE*, terminated by the involute at the points N and n, on which we erect two other perpendiculars *NH* and *nH* on the former lines *NM* and *nm*. These form two little sectors *MNm* and *NHn*, which are similar because the angles *MNm* and *NHn* are equal. Therefore, we have $Nn : Mm :: NH : NM$. Now at the inflection point A, the radius *NH* becomes (see §81) infinite or zero and the radius *MN*, which becomes *AE*, remains of finite magnitude. Thus, at a cusp E of the second kind, it must be the case that the ratio of the differential *Nn* of the radius *MN* of the evolute to the differential *Mm* of the curve becomes either infinitely large or infinitely small. Consequently, because (see §86)

$$Nn = \frac{-3\,dx\,ddy^2\,dy^2\,\overline{dx^2+dy^2}^{\frac{1}{2}} + dx\,d\,d\,dy\,\overline{dx^2+dy^2}^{\frac{3}{2}}}{dx^2\,ddy^2}$$

and $Mm = \sqrt{dx^2+dy^2}$, we have

$$\frac{dx^2\,dddy + dy^2\,dddy - 3\,dy\,ddy^2}{dx\,ddy^2} = 0 \text{ or } \infty.$$

[24]Cusps of this kind are sometimes called *ramphoid*, meaning beak-like, as opposed to cusps of the first kind, such as D, which are called *keratoid*, meaning horn-like.

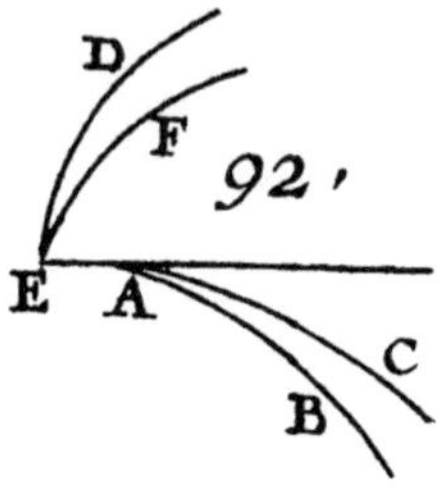

Fig. 5.28 Cusp of the Second Kind

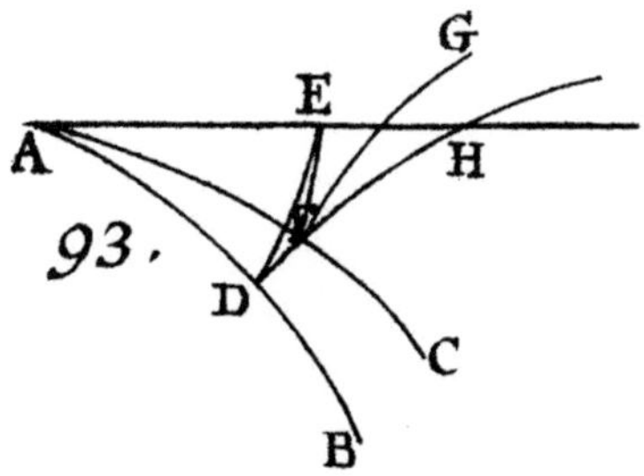

Fig. 5.29 Cusp of the Second Kind

Multiplying this by $dx\,ddy^2$, we find the formula

$$dx^2\,dddy + dy^2\,dddy - 3\,dy\,ddy^2 = 0 \text{ or } \infty,$$

which we may use to determine cusps of the second kind.

We may also conceive of curves *DEF* or *HDEFG* (see Figs. 5.28, 5.29) that turn back in the shape of a cusp of the second kind, which have as evolute another curve *BAC* that turns back in the shape of a cusp of the second kind, such that its cusp A corresponds to the cusp E, that is to say it is situated on the radius of the evolute that corresponds to the point E. Now it is clear that under this assumption, the radius *EA* of the evolute is always a *maximum* or a *minimum* and consequently that the differential of the general expression $\frac{\overline{dx^2+dy^2}^{\frac{3}{2}}}{-dx\,ddy}$ of the radius of the evolute (see §78) must be either null or infinite at the point E that we wish to find. This gives the same formula as before, so that it is the general formula for finding cusps of the second kind.

Chapter 6
Use of the Differential Calculus for Finding Caustics by Reflection

Definition I. [104] If we imagine that an infinity of rays *BA*, *BM*, and *BD* *(see Figs. 6.1, 6.2)*, which emanate from a radiant point[1] *B*, are reflected when they encounter a curved line *AMD*, so that the angles of reflection are equal to the angles of incidence, then the line *HFN*, which touches the reflected rays or their prolongations *AH*, *MF*, and *DN*, is called *the Caustic by reflection.*[2]

Corollary I. (§110) *If we prolong HA to I* (see Fig. 6.1), *so that* $AI = AB$*, and if we evolve*[3] *the caustic HFN beginning at the point I, we describe the curve ILK so that the tangent FL is continually (see §75) equal to the portion FH of the caustic plus the straight line HI. Moreover, if we imagine two incident and reflected rays Bm and mF, infinitely close to BM and MF, and if we prolong Fm to l and describe the little arcs MO and MR with centers F and B, then we form the little right triangles MOm and MRm, which are similar and equal. This is because the angles* $OmM = FmD = RmM$*, and furthermore the hypotenuse Mm is common, and so the little sides Om and Rm are equal to each other. Now, because Om is the differential of LM, and Rm is the differential of BM, and because this always happens no matter where we take the point M, it follows that* $ML - IA$ *or* $AH + HF - MF$*, the sum (see §96) of all the differentials Om in the portion AM of the curve, is* $= BM - BA$*, the sum (see §96) of the of all the differentials Rm in the same portion AM. Therefore, the portion HF of the caustic HFN is equal to* $BM - BA + MF - AH$.

[1] In L'Hôpital (1696) the term *point lumineux* was used, literally meaning "luminous point." We adopt the modern term "radiant point," e.g. Lockwood (1971, pp. 183–185).

[2] A caustic by reflection is sometimes called a "Catacaustic."

[3] I.e., describe the involute of *HFN*; see Chapter 5.

R.E. Bradley et al., *L'Hopital's Analyse des infiniments petits*, Science Networks. Historical Studies 50, DOI 10.1007/978-3-319-17115-9_6

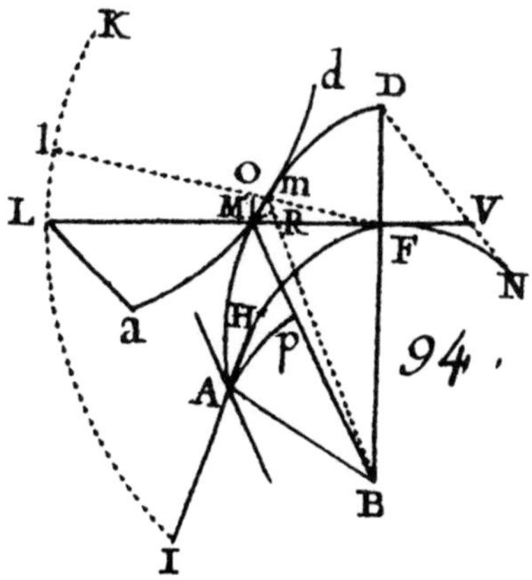

Fig. 6.1 Caustic by Reflection, Concave Case

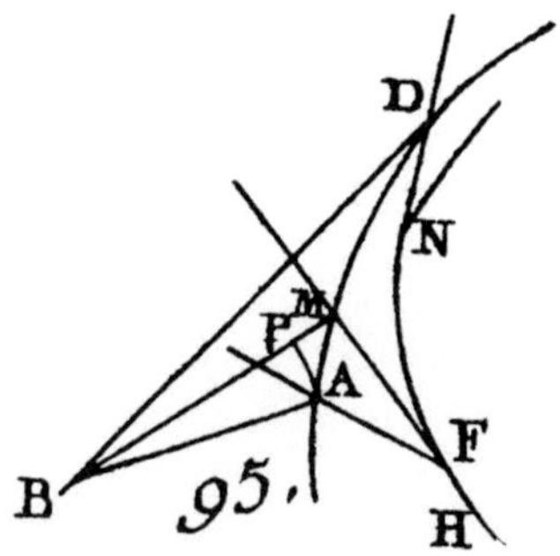

Fig. 6.2 Caustic by Reflection, Convex Case

There could be different cases, according to whether the incident ray *BA* is greater or less than *BM*, and whether the [105] reflected ray *AH* evolves or envelops[4] the portion *HF* to arrive at *MF*. However, we will still prove, as we have just done, that the difference of the incident rays is equal to the difference of the reflective rays, by joining to one of them the portion of the caustic that it evolves before falling on the other. For example *(see Fig. 6.2)*, $BM - BA = MF + FH - AH$; from which we conclude that $FH = BM - BA + AH - MF$.

If we describe the circular arc of *AP* with center *B* *(see Figs. 6.1, 6.2)*, then it is clear that *PM* is the difference of the incident rays *BM* and *BA*. Moreover, if we suppose that the radiant point *B* becomes infinitely distant from the curve *AMD*, the incident rays *BA* and *BM* *(see Fig. 6.3)* become parallel and the arc *AP* becomes a straight line perpendicular to these rays.

[4]In L'Hôpital (1696) the verb *enveloper* is used to mean the reverse of the process of describing the involute. For example, in Fig. 6.2, the tangential thread *HA* is laid on the curve *HFN* as one moves from the tangent *HA* to the tangent *FN*, instead of being peeled away in the usual evolution.

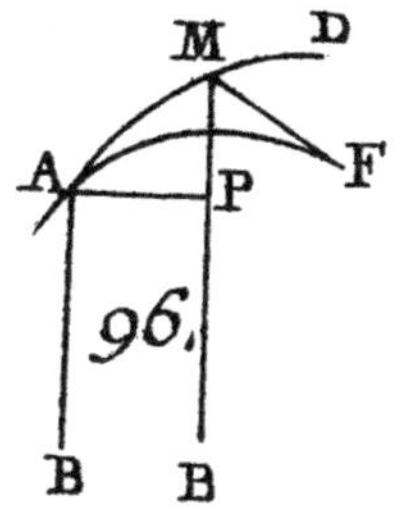

Fig. 6.3 Caustic by Reflection, Infinitely Distant Radiant Point

Corollary II. (§111) *If we imagine that the figure BAMD* (see Fig. 6.1) *is reflected in the same plane, so that the point B falls on the point* I*, and that therefore the tangent at A to the curve AMD in its first position, still touches it in this new position, and if we make the curve aMd roll on the curve AMD, that is to say on itself, so that the portions aM and AM are always equal, then I say that the point B describes by this motion a kind of roulette ILK whose evolute is the caustic HFN.*

This is because it follows from the generation that:

1. The line *LM* drawn from the describing point L to the point of contact M is (see §43) perpendicular to the curve *ILK*.
2. *La* or $IA = BA$, and $LM = BM$.
3. The angles made by the straight lines *ML* and *BM* on the common tangent at M are equal, and consequently if we prolong *LM* to F, the ray *MF* is the reflected ray of the incident ray *BM*.

From this we see that the perpendicular *LF* touches the caustic *HFN*, and because this always happens whenever we take the point L, it follows that the curve *ILK* is formed by the evolution of the caustic *HFN*, plus the straight line *HI*.

It follows from this that the portion *FH* or $FL-HI = BM+MF-BA-AH$. [106] This is what we have just demonstrated in another way in the preceding Corollary.

Corollary III. (§112) *If the tangent DN becomes infinitely close to the tangent FM, it is clear that the point of contact* N*, and the point of intersection* V *will coincide with the other point of contact* F*. Thus, to find the point* F *where the reflected ray MF touches the caustic HFN, we need only find the point of intersection of the infinitely close reflected rays MF and mF. Indeed, if we imagine an infinity of incidence rays infinitely close to one another, we will see a polygon with an infinity of sides, the assemblage of which makes up the caustic HFN born of the intersections of the reflected rays.*

Proposition I.

General Problem. (§113) *Given the nature of the curve AMD* (see Fig. 6.4), *the radiant point* B*, and the incident ray BM, we wish to find the point* F *on the reflected ray MF, given in position, where it touches the caustic.*

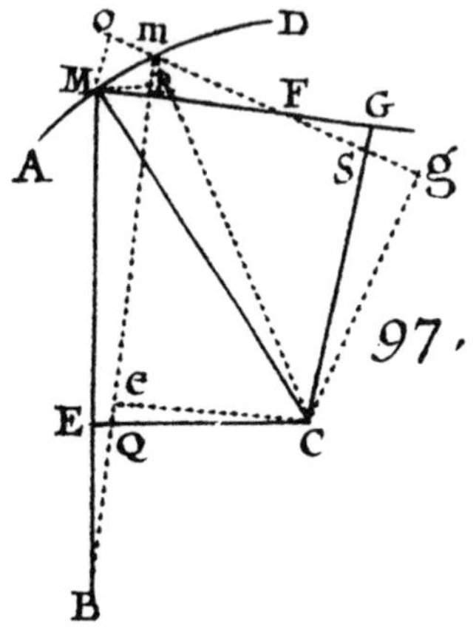

Fig. 6.4 Intersection of Reflected Ray with the Caustic

By the previous chapter, we find the length *MC* of the radius of the evolute at the point M. We take the infinitely small arc *Mm*, and draw the straight lines *Bm*, *Cm*, and *Fm*. We describe the little arcs *MR* and *MO* with centers B and F, and we drop the perpendiculars *CE*, *Ce*, *CG*, and *Cg* on the incident and reflected rays. We then denote the given quantities *BM* by y, and *ME* or *MG* by a.

Given this, we prove, as in the first Corollary (see §110), that the triangles *MRm* and *MOm* are similar and equal, and thus $MR = MO$. Now, because of the equality of the angles of incidence and reflection, we also have $CE = CG$ and $Ce = Cg$, and consequently $CE - Ce$ or $EQ = CG - Cg$ or SG. Therefore, because of the similar triangles *BMR* and *BEQ*, and *FMO* and *FGS*, we have [107] $BM + BE(2y - a)$: $BM(y) :: MR + EQ$ or $MO + GS : MR$ or $MO :: MG(a) : MF = \frac{ay}{2y-a}$.

If the radiant point B falls on the other side of the point E with respect to the point M, or (what is the same thing) if the curve *AMD* is convex towards the radiant point B, then y changes from positive to negative, and consequently we have $MF = \frac{-ay}{-2y-a}$ or $\frac{ay}{2y+a}$.

If we suppose that y becomes infinite *(see Fig. 6.3)*, that is to say that the point B is infinitely distant from the curve *AMD*, the incident rays are parallel to one another, and we have $MF = \frac{1}{2}a$, because a is null with respect to $2y$.

Corollary I. (§114) *Because we have found only one value for MF* (see Figs. 6.1, 6.2) *along the radius of the evolute, it follows that a curved line AMD may have but a single caustic HFN by reflection, because (see §80) it has but a single evolute.*

Corollary II. (§115) *When AMD is geometric* (see Fig. 6.4), *it is clear (see §85) that its evolute is also geometric, that is to say that we find all of its points C geometrically. From this it follows that all points F of its caustic are also determined geometrically* (see Figs. 6.1, 6.2), *that is to say that the caustic HFN is geometric. However, I say furthermore, that this caustic is always rectifiable, because it is* *clear (see §110) that we may find straight lines equal to any of its portions with the assistance of the curve AMD, which we assume to be geometric.*

Corollary III. (§116) *If the curve AMD* (see Fig. 6.4) *is convex towards the radiant point B, then the value of MF* $\left(\frac{ay}{2y+a}\right)$ *is always positive, and consequently we must take the point F on [108] the same side as the point C with respect to the point M, as we have assumed in making the calculations. From this we see that infinitely close reflected rays are divergent.*

However, if the curve *AMD* is concave towards the radiant point B, the value of *MF* $\left(\frac{ay}{2y-a}\right)$ is positive when y is greater than $\frac{1}{2}a$, negative when it is less than, and infinite when it is equal. From this it follows that if we describe a circle that has half of the radius of the evolute *MC* as its diameter, then the infinitely close reflecting rays are convergent when the radiant point B falls outside of its circumference, they are divergent when falls inside, and finally they are parallel when it falls on the circumference.

Corollary IV. (§117) *If the incident ray BM touches the curve AMD at the point M, we have $ME(a) = 0$, and consequently $MF = 0$. Because the reflected ray is therefore in the direction of the incident ray, and because the nature of the caustic consists of touching all reflected rays, it follows that it will also touch the incident ray BM at the point M. That is to say that the caustic and the given curve have the same tangent at the point M, which is common to them.*

If the radius *MC* of the evolute is null, we also have $ME(a) = 0$, and consequently $MF = 0$. From this we see that the given curve and the caustic make an angle equal to the angle of incidence between them at their common point M.

If the radius *CM* of the evolute is infinite, then the little arc *Mm* becomes a straight line, and we have $MF = \mp y$, because when $ME(a)$ is infinite, y is null with respect to a. Now, because this value is negative when the point B falls on the same side as the point C with respect to the line *AMD*, and positive when it falls on the opposite side, it follows that the infinitely close reflected rays are always divergent when the line *AMD* is straight.

Corollary V. [109] (§118) *It is clear that given any two of the three points B, C, and F, we easily find the third.*

1. Let the curve *AMD (see Fig. 6.5)* be a Parabola, which has the radiant point B as its focus. It is clear by the elements of the conic sections, that all reflected rays are parallel to the axis, and consequently *MF* is always infinite wherever we suppose the point M to be. Therefore, we have $a = 2y$. From this it follows that if we take *ME* to be twice *MB*, and if we draw the perpendicular *EC*, it cuts *MC* perpendicular to the curve *AMD* at a point C, which is on the evolute of this curve.
2. Let the curve *AMD (see Fig. 6.6)* be an Ellipse, which has the radiant point B as one of its foci. Again, it is clear that all reflected rays *MF* meet in the same point F, which is the other focus. If we denote *MF* by z, we have (see §113)

$z = \frac{ay}{2y-a}$, from which we find the quantity that we wish to find, $ME(a) = \frac{2yz}{y+z}$. However, if the curve *AMD (see Fig. 6.7)* is a Hyperbola, the focus F falls on the other side, and consequently $MF(z)$ becomes negative. From this it follows that we therefore have $ME(a) = \frac{-2yz}{y-z}$ or $\frac{2yz}{z-y}$. This gives the following construction, which also serves for the Ellipse.

Let *ME (see Figs. 6.6, 6.7)* be taken as the fourth proportional to the transverse semi-axis,[5] the incident ray, and the reflected ray. Let *EC* be drawn perpendicular; it cuts the line *MC*, which is perpendicular to the conic section, at a point C that is on the evolute.

Example I. (§119) Let the curve *AMD (see Fig. 6.8)* be a Parabola, whose incident rays *PM* are perpendicular to its axis *AP*. We wish to find the points F on the reflected rays *MF* where they touch the caustic *AFK*.

[110] It is clear that if we draw the radius *MC* of the evolute, and drop the perpendicular *CG* to the reflected ray *MF*, we must (see §113) take *MF* equal to half of *MG*. However, this construction can be shortened, considering that if we draw *MN* parallel to the axis *AP*, and draw the straight line *ML* to the focus *L*, then the angles *LMP* and *FMN* are equal, because by the property of the parabola $LMQ = QMN$, and by assumption $PMQ = QMF$. If we now add the same angle *PMF* to both sides, the angle *LMF* is equal to the angle *PMN*, that is to say a right angle. Now, we have just demonstrated (see §118, no. 1) that the perpendicular *LH* on *ML* meets the radius *MC* of the evolute at its midpoint H. Therefore, if we draw *MF* parallel and equal to *LH*, it is one of the reflected rays and it touches the caustic *AFK* at F. This is what we were required to find.

If we suppose that the reflected ray *MF* is parallel to the axis *AP*, it is clear that the point F of the caustic will be the furthest possible from the axis *AP*, because the tangent at this point is parallel to the axis. Thus, in order to determine this point on all caustics, such as *AFK*, formed by incident rays perpendicular to the axis of the given curve, we need to only consider that *MP* is equal to *PQ*. This gives $dy = dx$.

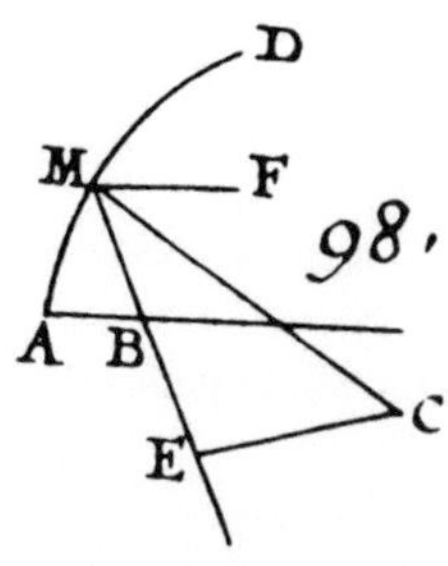

Fig. 6.5 Caustic of the Parabola by Reflection

[5]The transverse axis of an ellipse is the horizontal axis. The transverse of a hyperbola is the line segment joining the vertices. The transverse semi-axis is half the length of the transverse axis.

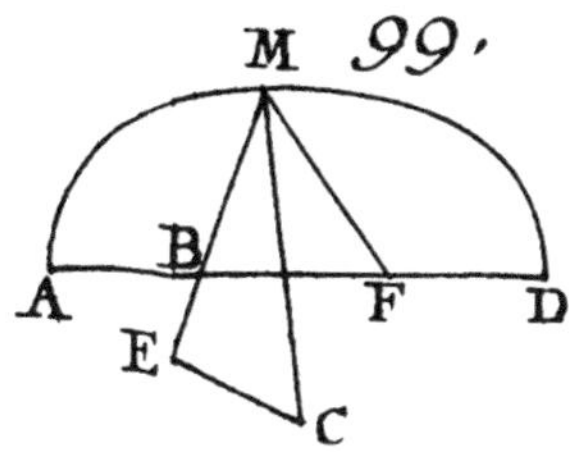

Fig. 6.6 Caustic of the Ellipse by Reflection

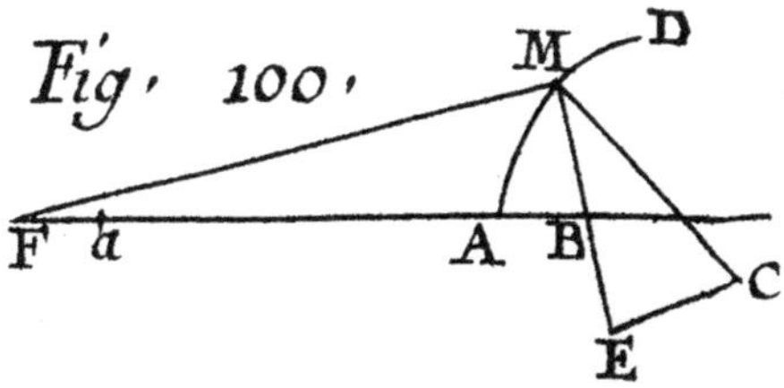

Fig. 6.7 Caustic of the Hyperbola by Reflection

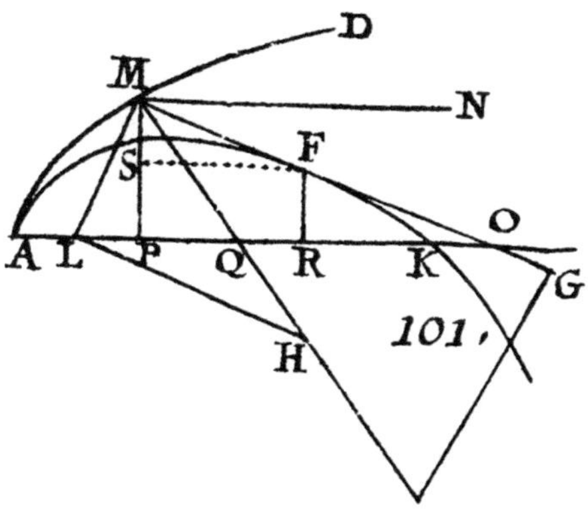

Fig. 6.8 Caustic of the Parabola by Reflection, Incident Rays Perpendicular to the Axis

Let $ax = yy$; we have $dy = \frac{a\,dx}{2\sqrt{ax}} = dx$, from which we conclude $AP(x) = \frac{1}{4}a$, that is to say if the point P falls on the focus L, the reflected ray MF is parallel to the axis. What is also clear is that because in this case MP coincides with LM, MF must also coincide with MN, and LH with LQ. From this we see that MF is therefore equal to ML, and consequently that if we draw FR perpendicular to the axis, we have AR or $AL + MF = \frac{3}{4}a$. We also see that the portion AF of the caustic is equal, in this case, to the parameter, because it is always (see §110) equal to $PM + MF$.

To determine the point K where the caustic AFK meets the axis AP, we must find the value MO, and [111] make it equal to MF, because it is clear that if the point F falls on K, the lines MF and MO become equal to each other. Therefore, denoting the unknown MO by t, the angle PMO cut equally in two by MQ, perpendicular to the curve, gives $MP(y) : MO(t) :: PQ\left(\frac{y\,dy}{dx}\right) : OQ = \frac{t\,dy}{dx}$, and consequently $OP = \frac{t\,dy+y\,dy}{dx} = \sqrt{tt-yy}$, because of the right triangle MPO. Dividing both sides by $t+y$, we find $\frac{dy}{dx} = \sqrt{\frac{t-y}{t+y}}$, from which we conclude $MO(t) = \frac{y\,dx^2+y\,dy^2}{dx^2-dy^2} = MF\left(\frac{1}{2}a\right) = \frac{dx^2+dy^2}{-2ddy}$, because (see §77) $ME(a) = \frac{dx^2+dy^2}{-ddy}$. This gives $dy^2-2y\,ddy = dx^2$, which is used to find the point P such that if we draw the incident ray PM and the reflected ray MF, this latter touches the caustic AFK at the point K where it meets the axis AP.

For the parabola $y = x^{\frac{1}{2}}$, we have $dy = \frac{1}{2}x^{-\frac{1}{2}}\,dx$ and $ddy = -\frac{1}{4}x^{-\frac{3}{2}}\,dx^2$, and substituting these values in the preceding equation, we find $\frac{1}{4}x^{-1}\,dx^2 + \frac{1}{2}x^{-1}\,dx^2 = dx^2$, from which we conclude $AP(x) = \frac{3}{4}$ of the parameter.

To find the nature of the caustic AFK in the manner of *Descartes*,[6] we must find an equation that expresses the relationship of the abscissa $AR(u)$ to the ordinate $RF(z)$, which is done in the following way. Because $MO(t) = \frac{y\,dx^2+y\,dy^2}{dx^2-dy^2}$, we have $PO\left(\frac{t\,dy+y\,dy}{dx}\right) = \frac{2y\,dx\,dy}{dx^2-dy^2}$, and because of the similar triangles MPO and MSF, we form the following proportions: $MO\left(\frac{y\,dx^2+y\,dy^2}{dx^2-dy^2}\right) : MF\left(\frac{dx^2+dy^2}{-2ddy}\right)$ or $-2y\,ddy : dx^2-dy^2 :: MP(y) : MS(y-z) = \frac{dx^2-dy^2}{-2ddy} :: PO\left(\frac{2y\,dx\,dy}{dx^2-dy^2}\right) : SF$ or $PR(u-x) = \frac{dx\,dy}{-ddy}$. We therefore have the following two [112] equations $z = y + \frac{dy^2-dx^2}{-2ddy}$ and $u = x + \frac{dx\,dy}{-ddy}$, which can be used with the equation of the given curve to form a new equation where x and y are no longer present, and which consequently expresses the relationship of $AR(u)$ to $FR(z)$.

When the curve AMD is a parabola, as we have assumed in this example, we find $z = \frac{3}{2}x^{\frac{1}{2}} - 2x^{\frac{3}{2}}$, or (by squaring each side) $\frac{9}{4}x - 6xx + 4x^3 = zz$ and $u = 3x$, from which we derive the equation we wish to find, $azz = \frac{4}{27}u^3 - \frac{2}{3}auu + \frac{3}{4}aau$, which expresses the nature of the caustic AFK. We may remark that PR is always twice AP, because $AR(u) = 3x$. This again gives us a new method for determining the point F that we wish to find on the reflected ray MF.

Example II. (§120) Let the curve AMD *(see Fig. 6.9)* be a semi-circle that has the line AD as its diameter and its center at the point C. Let the incident rays PM be perpendicular to AD.

Because the evolute of the circle is a single point which is its center, it follows (see §113) that if we cut the radius CM equally in two at the point H, and we drop the perpendicular HF to the reflected ray MF, then it cuts this ray at a point F where it touches the caustic AFK. It is clear that the reflected ray MF is equal to half the incident ray PM. From this it follows that:

1. If the point P falls on C, then the point F falls on the midpoint of CB at K.

[6] I.e., to give an equation for the curve.

2. The portion AF is three times MF, and the caustic AFK is three times BK.

We also see that if we make the angle AMC half of a right angle, the reflected ray MF is parallel to AC, and consequently the point F is higher above the diameter AD than any other point of the caustic.

The circle with diameter MH passes through the point F, because the angle HFM is a right angle. If we describe the circle KHG with [113] center C and radius CK or CH, half of CM, then the arc HF is equal to the arc HK. This is because the angle CMF is equal to[7] CMP or HCK, so the arcs $\frac{1}{2}HF$ and HK, which measure these angles in the circles MFH and KHG, are to each other as $\frac{1}{2}MH$ is to HC, the radii of these circles. From this we see that the Caustic AFK is a Roulette formed by the revolution of the mobile circle MFH around the immobile circle KHG, whose origin is at K, and whose vertex is at A.

Example III. (§121) Let the curve AMD *(see Fig. 6.10)* be a circle with the line AD as diameter and the point C as center. Let the radiant point, from which all the incident rays AM emanate, be A, one of the extremities of this diameter.

If we drop the perpendicular CE from the center C to the incident ray AM, it is clear by the property of the circle, that the point E cuts the chord AM into two equal parts, and thus that $ME(a) = \frac{1}{2}y$. We therefore have $MF\left(\frac{ay}{2y-a}\right) = \frac{1}{3}y$, that is to say we must take the reflected ray MF equal to one-third of the incident ray

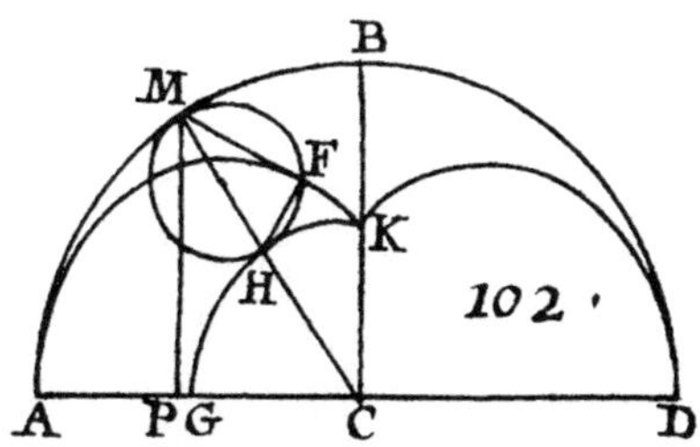

Fig. 6.9 Caustic of the Semi-circle by Reflection, Incident Rays Perpendicular to the Diameter

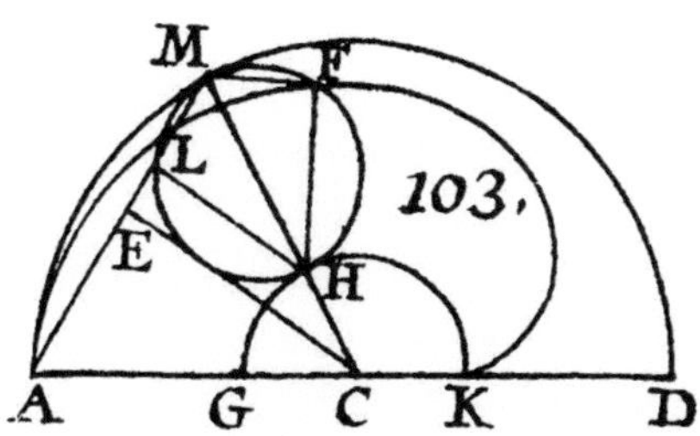

Fig. 6.10 Caustic of the Circle by Reflection, Radiant Point on the Circumference

[7] In L'Hôpital (1696) this was written as CPM, but corrected to CMP in the *Errata*.

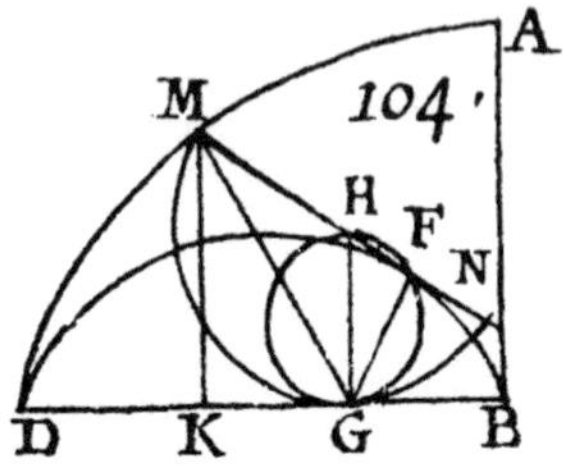

Fig. 6.11 Caustic of the Half-Cycloid by Reflection, Incident Rays Parallel to the Axis

AM. From this we see that $DK = \frac{1}{3}AD$, $CK = \frac{1}{3}CD$, and that (see §110) the caustic $AFK = \frac{4}{3}AD$, as well as that its portion $AF = \frac{4}{3}AM$. If we take $AM = AC$, the reflected ray *MF* is parallel to the diameter *AD*, and consequently the point *F* is the highest possible above this diameter.

If we take $CH = \frac{1}{3}CM$, and we draw *HF* perpendicular to *MF*, the point *F* is on the caustic, for if we draw *HL* perpendicular to *AM*, it is clear that $ML = \frac{2}{3}ME = \frac{1}{3}AM$, because $MH = \frac{2}{3}CM$. The circle with diameter *MH* therefore passes through the point *F* of the caustic, and if we describe another circle *KHG* with center *C* and with radius *CK* or *CH*, it will be equal to it, and the arc *HK* [114] will be equal to the arc *HF*, because in the isosceles triangle *CMA*, the external angle $KCH = 2CMA = AMF$. Consequently the arcs *HK* and *HF*, the measures in the equal circles, are also equal. From this it follows that the Caustic *AFK* is again a Roulette described by the revolution of the mobile circle *MFH* around the immobile circle *KHG*, whose origin is at *K*, and whose vertex is at *A*.

We might also prove this by the following other method. If we describe a roulette by the revolution of a circle equal to the circle *AMD* around that circle, starting at the point *A*, we demonstrated in the second corollary (see §111) that its evolute is the caustic *AFK*. Now (see §100), this evolute is a roulette of the same kind, that is to say that the diameters of the generating circles are equal, and we determine the point *K* by taking *CK* as the third proportional to $CD + DA$ and to *CD*, that is to say equal to $\frac{1}{3}CD$. Therefore, etc.

Example IV. (§122) Let the curve *AMD (see Fig. 6.11)* be an ordinary half-roulette described by the revolution of the semi-circle *NGM* on the straight line *BD*, whose vertex is at *A*, and whose origin at *D*. Let the incident rays *KM* be parallel to the axis *AB*.

Because (see §95) *MG* is equal to half the radius of the evolute, it follows (see §113) that if we draw *GF* perpendicular to the reflected ray *MF*, the point *F* will be on the caustic *DFB*. From this we see that *MF* must be taken equal to *KM*.

If we draw the radii *HG* and *HM* from the center *H* of the generating circle *MGN* to the point of contact *G* and to the describing point *M*, then it is clear that *HG* is perpendicular to *BD*, and that the angle $GMH = MGH = GMK$, from which we see that the reflected ray *MF* passes through the center *H*. Now, the circle with diameter *GH* also passes through the point *F*, because the angle *GFH* is a right

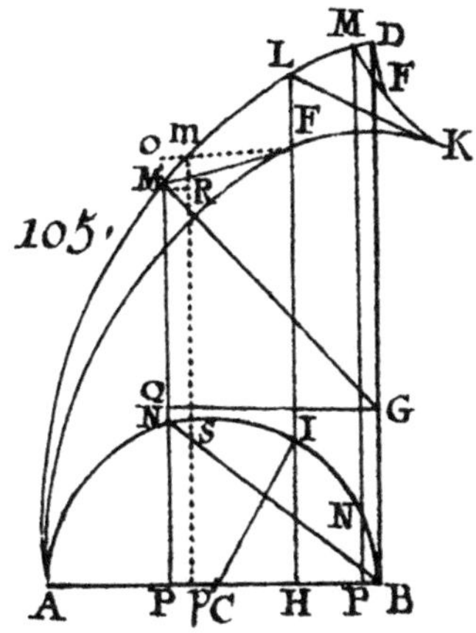

Fig. 6.12 Caustic of the Half-Cycloid by Reflection, Incident Rays Parallel to the Base

angle. Therefore, the arcs *GN* and $\frac{1}{2}GF$, which measure the same angle *GHN*, are to each other as the *MN* is to *GH*, the diameters [115] of their circles, and consequently the arc $GF = GN = GB$. It is therefore clear that the Caustic *DFB* is a Roulette described by an entire revolution of the circle *GFH* on the straight line *BD*.

Example V. (§123) Let the curve *AMD (see Fig. 6.12)* again be an ordinary half-roulette, whose base *BD* is equal to the semi-circumference *ANB* of the generating circle. Additionally, let the incident rays *PM* now be parallel to the base *BD*.

If we draw *GQ* perpendicular to *PM*, the right triangles *GQM* and *BPN* are equal and similar, and therefore $MQ = PN$. From this we see (see §95, 113) that we must take *MF* equal to the corresponding ordinate *PN* in the generating semi-circle *ANB*.

In order that the point *F* is the furthest possible from the axis *AB*, it is necessary that the tangent *MF* at this point be parallel to this axis. The angle *PMF* is therefore a right angle, its half *PMG* or *PNB* is half a right angle, and consequently the point *P* falls on the center of the circle[8] *ANB*.

It is worth remarking, that if the point *P* then continually approaches the extremity *B*, the point *F* also moves closer the axis *AB* up to a certain point *K*, after which it moves away to *D*, so that the caustic *AFKFD* has a cusp at *K* .

To determine this, I remark (see §110, 111) that the portion $AF = PM + MF$, the portion $AFK = HL + LK$, and the portion *KF* of the part *KFD* is $= HL + LK - PM - MF$, from which we see that $HL + LK$ must be the *greatest*. That is why if we denote *AH* by x, *HI* by y, and the arc *AI* by u, we have $HL + LK = u + 2y$, whose differential gives $du + 2\,dy = 0$ and $\frac{a\,dx}{y} + 2\,dy = 0$ by substituting the value $\frac{a\,dx}{y}$

[8] In L'Hôpital (1696) this was written as *AND*.

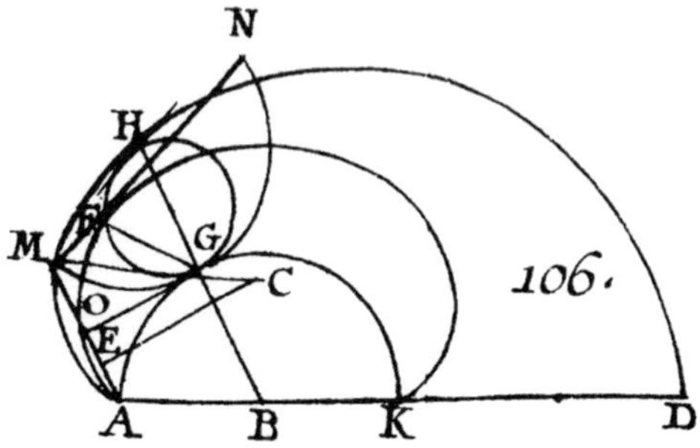

Fig. 6.13 Caustic of an Epicycloid by Reflection

for du. From this we derive $a\,dx = -2y\,dy = 2x\,dx - 2a\,dx$ because of the circle, and consequently $AH(x) = \frac{3}{2}a$.

Corollary. [116] (§124) *The space AFM or AFKFM enclosed by the portions of the curves AF or AFKF, AM, and by the reflected rays MF, is equal to half of the circular space APN. This is because its differential, which is the sector FMO, is equal to half of the rectangle PpSN, the differential of the space APN, because the right triangles MOm and MRm are equal and similar, so MO is equal to MR or NS or Pp, and furthermore MF* $=$ *PN.*

Example VI. (§125) Let the curve *AMD (see Fig. 6.13)* be the half-roulette formed by the revolution of the circle *MGN* around an equal circle *AGK*, whose origin is at *A*, and whose vertex is at *D*. Let the incident rays be *AM*, which all emanate from the point *A*. The line *BH* that joins the centers of these two generating circles continually passes through the point of contact *G*, and the arcs *GM* and *GA*, as well as their chords, are always equal. So the angle $HGM = BGA$ and the angle $GMA = GAM$. Now, the angle $HGM + BGA = GMA + GAM$, because if we add the same angle *AGM* to both sides, we form two right angles. Thus, the angle *HGM* is always equal to the angle *GMA*, and consequently also to the angle of reflection *GMF*. From this it follows that *MF* is always passes through the center *H* of the mobile circle.

If we now drop the perpendiculars *CE* and *GO* on the incident ray *AM*, it is clear that $MO = OA$, and that $OE = \frac{1}{3}OM$, because (see §100) since the point *C* is on the evolute $GC = \frac{1}{3}GM$. We therefore have $ME = \frac{2}{3}AM$, that is to say $a = \frac{2}{3}y$, and consequently $MF\left(\frac{ay}{2y-a}\right) = \frac{1}{2}y$. From this we see that if we draw *GF* perpendicular to *MF*, the point *F* is on the caustic *AFK*.

The circle with diameter *GH* passes through the point *F*, and the arcs *GM* and $\frac{1}{2}GF$, which measure the same angle *GHM*, being [117] to each other as *MH* is to *GH*, the diameters of their circles, the arc *GF* is equal to the arc *GM*, and consequently to the arc *GA*. From this it is clear that the Caustic *AFK* is a Roulette

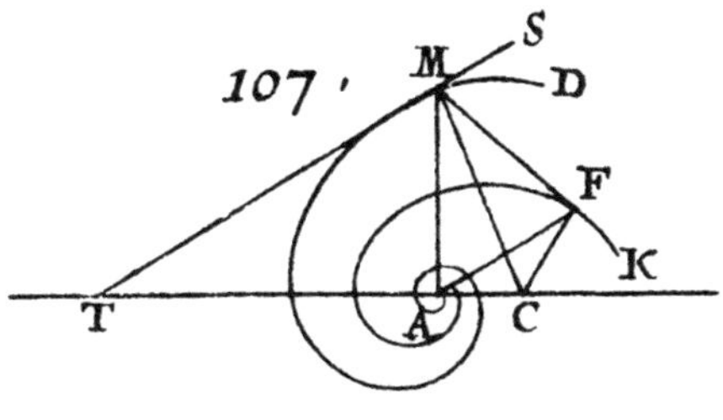

Fig. 6.14 Caustic of the Logarithmic Spiral by Reflection

described by the revolution of the mobile circle *HFG* around the immobile circle *AGK*.

Corollary. (§126) *If we describe a circle with center at the point B, and whose radius is a straight line equal to BH or AK, that has an infinity of straight lines parallel to BD that fall on its circumference, it is clear (see §120) that by reflection the rays form the same caustic AFK.*

Example VII. (§127) Let the curve *AMD (see Fig. 6.14)* be a logarithmic spiral, with incident rays *AM* that all of emanate from the center *A*.

If we draw the straight line *CA* from the extremity *C* of the radius of the evolute perpendicular to the incident ray *AM*, it meets it (see §91) at the center *A*. This is why $AM(y) = a$, and consequently $MF\left(\frac{ay}{2y-a}\right) = y$. The triangle *AMF* is therefore isosceles, and because the angles of incidence and reflection, *AMT* and *FMS*, are equal to each other, it follows that the angle *AFM* is equal to the angle *AMT*. From this it is clear that caustic *AFT* is a logarithmic spiral which differs from the given *AMD* only in its position.

Proposition II.

Problem. (§128) *Given the caustic by reflection HF* (see Fig. 6.15) *with its radiant point B, we wish to find an infinity of curves, such as AM, of which it is the caustic by reflection.*

Take the point *A* at will on any tangent *HA* to be one of the points on the curve *AM* that we wish to find. [118] We describe the circular arc *AP* with center *B* and interval *BA*, and with any other interval *BM*, we describe another circular arc. If we take $AH + HE = BM - BA$ or *PM*, we evolve the caustic *HF* beginning at the point *E*, and we describe by this motion the curved line *EM* that cuts the arc of the circle described from the radius *BM* at a point *M* that is (see §110) on the curve *AM*. This is because by the construction, $PM + MF = AH + HF$.

Alternately, if we attach a thread *BMF* by its extremities at *B* and to *F*, then we make this thread tight by means of a stylus placed at *M*, and we make it move so that we envelop the caustic *HF* with the part *MF* of this thread, it is clear that by this motion the stylus describes the curve *MA* that we wish to find.

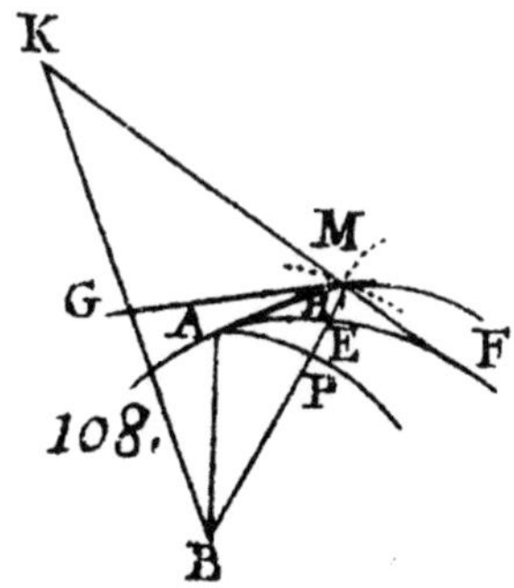

Fig. 6.15 The Inverse Problem: Finding the Original Curve from its Caustic

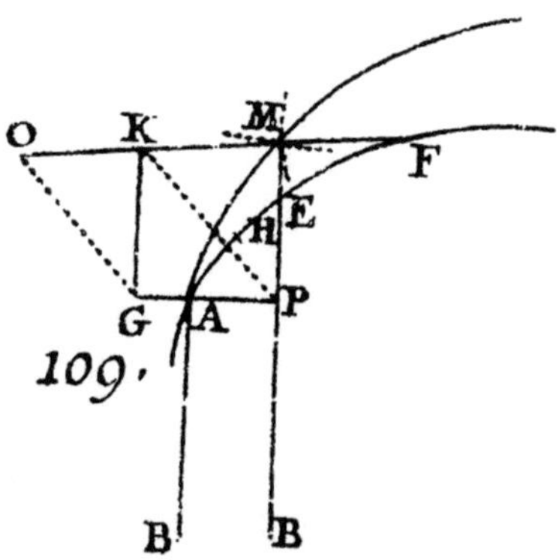

Fig. 6.16 The Inverse Problem: Radiant Point at Infinity

Alternate Solution. (§129) If we draw at will a tangent *FM*, other than *HA*, we wish to find a point *M* on it, such that $BM + MF = BA + AH + HF$. This is done in the following manner.

Let *FK* be taken $= BA + AH + HF$, and if we divide *BK* in the middle at *G*, then let the perpendicular *GM* be drawn; it meets the tangent *FM* at the point *M* that we wish to find. This is because $BM = MK$.

If the point *B* is infinitely far from the curve *AM (see Fig. 6.16)*, that is to say that the incident rays *BA* and *BM* are parallel to a straight line given in position, the first construction will still hold, by considering that the circular arcs described from the center *B* become straight lines perpendicular to the incident rays. However, this latter construction becomes useless, which is why we must substitute it with the following.

Let *FK* be taken $= AH + HF$. If we find the point *M*, so that *MP* is parallel to *AB* and perpendicular to *AP*, and is equal to *MK*, it is clear (see §110) that this point is on the curve *AM* that we wish to find, because $PM + MF = AH + HF$. Now, this is done as follows.

Let *KG* be drawn perpendicular to *AP*. If we take $KO = KG$, let *KP* be drawn parallel to *OG* and *PM* parallel to *GK*, I say that the point *M* is the one we wish

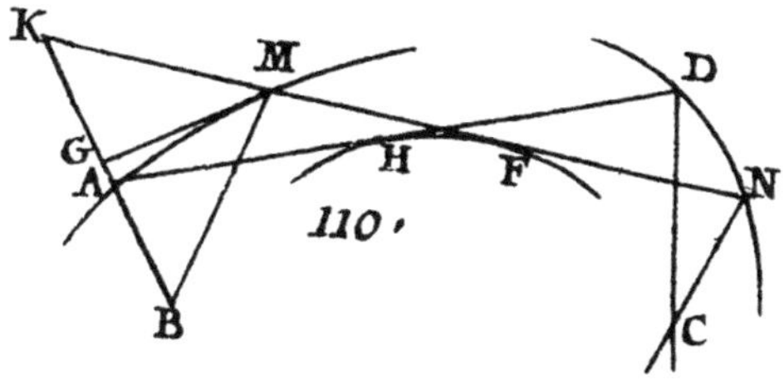

Fig. 6.17 An Application of the Inverse Problem: Constructing a Focus

to find. [119] This is because *GKO* and *PMK* are similar triangles, so that we have $PM = MK$, because $GK = KO$.

If the caustic *HF* is a single point, the curve *AM* becomes a conic section .

Corollary I. (§130) *It is clear that the curve that passes through all the points K is formed by the evolution of the curve HF beginning at A, and that its nature changes as the point A changes its position on the tangent AH. Thus, because the curves AM are all born from these curves by the same construction, which is geometric, it follows (see §108) that they are of a different nature from each other, and that they are geometric only when the caustic HF is geometric and rectifiable.*

Corollary II. (§131) *Given a curved line DN* (see Fig. 6.17) *with a radiant point C, we wish to find an infinity of lines, such as AM, so that the reflected rays DA and NM meet at a given point B, after being reflected again upon meeting any such line AM.*

If we imagine that the curve *HF* is the caustic of the given curve *DN*, formed by the radiant point *C*, it is clear that this line *HF* must also be the caustic of the curve *AM* having the given point *B* as its radiant point, so that $FK = BA + AH + HF$, and $NK = BA + AH + HF + FN = BA + AD + DC - CN$, because (see §118) $HD + DC = HF + FN + NC$. This gives the following construction.

If we take the point *A* at will on any reflected ray to be one of the points on the curve *AM* that we wish to find and, on any other reflected ray *NM* that we wish, we take the part $NK = BA + AD + DC - CN$, then we find the point *M* that we wish as above in §129.

Chapter 7
Use of the Differential Calculus for Finding Caustics by Refraction

Definition. [120] If we imagine that an infinity of rays *BA*, *BM*, and *BD* (see Fig. 7.1), which emanate from the same radiant point B, are refracted[1] when they encounter a curved line *AMD*, by approaching or moving away from its perpendiculars *MC*, so that the sines *CE* of the angles *CME* of incidence are always to the sines *CG* of the angles *CMG* of refraction in the same given ratio as m to n, then the curved line[2] *HFN* (see Fig. 7.2) which touches all the refracted rays or their prolongations *AH*, *MF*, and *DN* is called *the Caustic by refraction*.[3]

Corollary. (§132) *If we envelop*[4] *the caustic HFN beginning at the point A, we describe the curve ALK so that the tangent LF plus the portion FH of the caustic is continually equal to the same straight line AH. Moreover, if we imagine another tangent Fml infinitely close to FML, with another incident ray Bm, and if we describe the little arcs MO and MR with centers F and B, then we form two little right triangles MRm and MOm, which are similar to two others MEC and MGC, pair by pair, because if we remove the same angle EMm from the right angles RME and CMm, then the remaining angles RMm and EMC are equal. Similarly, if we remove the same angle GMm from the right angles GMO and CMm, the remaining angles OMm and GMC are equal. This is why* $Rm : Om :: CE : CG :: m : n$. *Now, because Rm is the differential of BM and Om is the differential of LM, it follows (see §96) that BM − BA, the sum of all the differentials Rm in the portion*

[1] In L'Hôpital (1696) the term *rompre* is used, literally meaning "to break." We consistently translate the term *rayon rompu* as "refracted ray."

[2] In L'Hôpital (1696) the curved line was given as *FHN*, but corrected in the *Errata*.

[3] A caustic by refraction is sometimes called a "Dicaustic."

[4] I.e., describe the involute in reserve order, see §110, Footnote 4.

R.E. Bradley et al., *L'Hopital's Analyse des infiniments petits*, Science Networks. Historical Studies 50, DOI 10.1007/978-3-319-17115-9_7

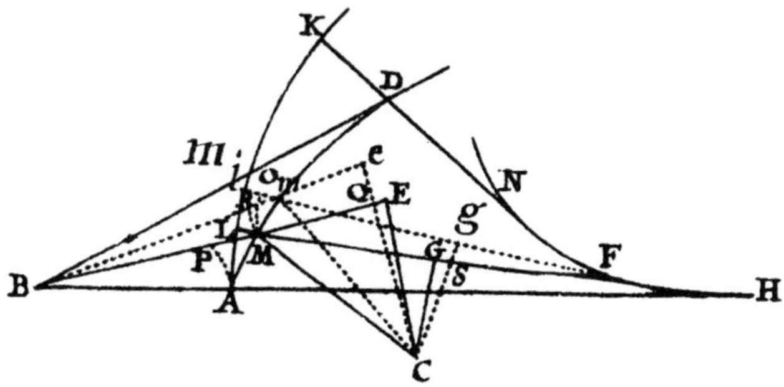

Fig. 7.1 Caustic by Refraction, Convex Case

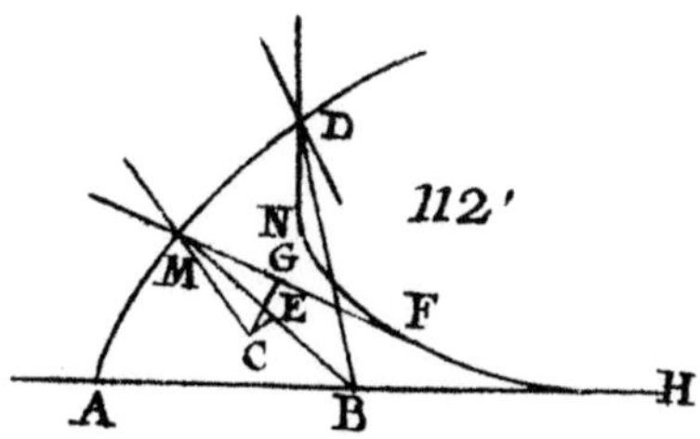

Fig. 7.2 Caustic by Refraction, Concave Case

AM of the curve, is to ML or AH − MF − FH, the sum of all the differentials Om in the same portion [121] AM, as m is to n, and consequently the portion $FH = AH - MF + \frac{n}{m}BA - \frac{n}{m}BM$.

There could be different cases, according to whether the incident ray *BA* is greater or less than *BM*, and whether the refracted ray *AH* envelops or evolves the portion *HF*. However, we will still prove, as we have just done, that the difference of the incident rays is to the difference of the refracted rays (by joining to one of them the portion of the caustic that it evolves before falling on the other) as m is to n. For example (see Fig. 7.2), $BA - BM : AH - MF - FH :: m : n$, from which we conclude that $FH = AH - MF + \frac{n}{m}BM - \frac{n}{m}BA$.

If we describe the circular arc *AP* with center *B* (see Fig. 7.1), then it is clear that *PM* is the difference of the incident rays *BM* and *BA*. Moreover, if we suppose that the radiant point *B* becomes infinitely distant from the curve *AMD*, the incident rays *BA* and *BM* become parallel and the arc *AP* becomes a straight line perpendicular to these rays.

Proposition I.

General Problem. (§133) *Given the nature of the curve AMD* (see Fig. 7.1), *the radiant point B, and the incident ray BM, we wish to find the point F on the refracted ray MF, given in position, where it touches the caustic by refraction.*

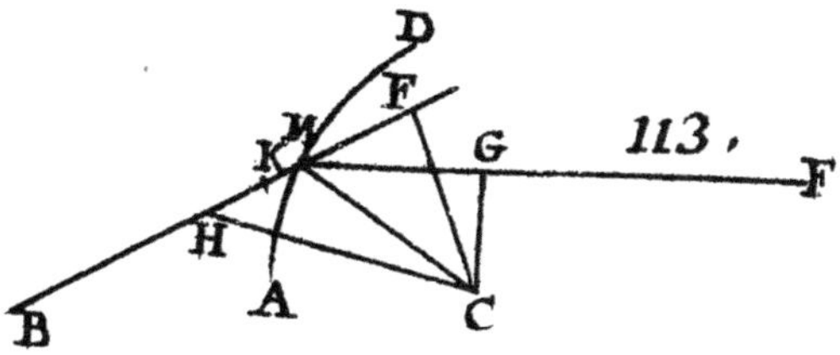

Fig. 7.3 Intersection of the Refracted Ray with the Caustic

We find (see Ch. 5) the length MC of the radius of the evolute at the given point M. We take the infinitely small arc Mm, and draw the straight lines Bm, Cm, and Fm. We describe the little arcs MR and MO with centers B and F, and we drop the perpendiculars CE, Ce, CG, and Cg to the incident and refracted rays. We denote the given quantities BM by y, ME by a, MG by b, and the little arc MR by dx.

Given this, the similar right triangles MEC and MRm, MGC and MOm, and BMR and BQe give $ME(a) : MG(b) :: MR(dx) : MO = \frac{b\,dx}{a}$ and [122] $BM(y) : BQ$ or $BE(y+a) :: MR(dx) : Qe = \frac{a\,dx + y\,dx}{y}$. Now, by the property of the refraction $Ce : Cg :: CE : CG :: m : n$. Consequently, $m : n :: Ce - CE$ or $Qe\left(\frac{a\,dx+y\,dx}{y}\right) : Cg - CG$ or $Sg = \frac{an\,dx + ny\,dx}{my}$. Thus, because of the similar right triangles FMO and FSg, we have $MO - Sg\left(\frac{bmy\,dx - any\,dx - aan\,dx}{amy}\right) : MO\left(\frac{b\,dx}{a}\right) :: MS$ or $MG(b) : MF = \frac{bbmy}{bmy - any - aan}$. This gives the following construction.

Let the angle $ECH = GCM$ (see Fig. 7.3) be constructed towards CM, and let $MK = \frac{aa}{y}$ be taken towards B. I say that if we make $HK : HE :: MG : MF$, then the point F is on the caustic by refraction.

Because of the similar triangles CGM and CEH, we have $CG : CE :: n : m :: MG(b) : EH = \frac{bm}{n}$. From this we conclude that $HE - ME$ or $HM = \frac{bm - an}{n}$, $HM - MK$ or $HK = \frac{bmy - any - aan}{ny}$, and consequently $HK\left(\frac{bmy-any-aan}{ny}\right) : HE\left(\frac{bm}{n}\right) :: MG(b) : MF = \frac{bbmy}{bmy - any - aan}$.

It is clear that if the value of HK is negative, the value of MF is also negative, from which it follows that the point M falls between the points G and F, when the point H is between the points K and E.

If the radiant point B falls on the side of the point E (see Fig. 7.2),[5] or (what is the same thing) if the curve AMD is concave on the side of the radiant point B, then y changes from positive to negative, and consequently we have $MF = \frac{-bbmy}{-bmy + any - aan}$ or $\frac{bbmy}{bmy - any + aan}$. The construction remains the same.

If we suppose that y becomes infinite, that is to say that the radiant point B is infinitely distant from the curve AMD, then the incident rays are parallel to each other, and we have $MF = \frac{bbm}{bm - an}$, because the term aan is null [123] with respect to

[5]In L'Hôpital (1696), the reference here was to figures 7.1 and 7.3.

the other two, *bmy* and *any*, and because $MK\left(\frac{aa}{y}\right)$ therefore vanishes, we need only make $HM : HE :: MG : MF$.

Corollary I. (§134) *We demonstrate, in the same way as for the caustic by reflection (see §114), that a curved line AMD has only one caustic by refraction, given the ratio of m to n. This caustic is always geometric and rectifiable when the given curve AMD is geometric.*

Corollary II. (§135) *If the point E falls on the other side of the perpendicular MC with respect to the point G, and if CE is equal to CG, then it is clear that the caustic by refraction changes to a caustic by reflection. Indeed, we have* $MF\left(\frac{bbmy}{bmy-any\mp aan}\right) = \frac{ay}{2y\mp a}$, *because* $m = n$, *and a changes from negative to positive, and it also becomes equal to b. This agrees with what we proved in the previous chapter.*

If m is infinite with respect to n, then it is clear that the refracted ray MF falls on the perpendicular CM, so that the Caustic by refraction becomes the Evolute. Indeed, we have $MF = b$, *which in this case becomes MC, that is to say that the point F falls on the point C, which is on the evolute.*

Corollary III. (§136) *If the curve AMD is convex with radiant point B, and the value of* $MF\left(\frac{bbmy}{bmy-any-aan}\right)$ *is positive, it is clear that we must take the point F on the same side as the point G with respect to the point M, as we have supposed from making the calculations. On the contrary, if it is negative, we must take it on the opposite side. It is the same when the curve AMD is concave towards the point B, however it should be noted that in this case [124]* $MF = \frac{bbmy}{bmy-any+aan}$. *From this it follows that infinitely close refracted rays are convergent when the value of MF is positive in the first case, and negative in the second case, and, on the contrary, they are divergent when the value of MF is negative in the first case, and positive in the second. Given this, it is clear that:*

1. If the curve *AMD* is convex towards the radiant point B, and m is less than n, or if it is concave towards this point, and m is greater than n, then infinitely close refracted rays are always divergent.
2. If the curve *AMD* is convex towards the radiant point B, and m is greater than n, or if it is concave towards this point and m is less than n, then infinitely close refracted rays are convergent, when $MK\left(\frac{aa}{y}\right)$ is less than $MH\left(\frac{bm}{n} - a \text{ or } a - \frac{bm}{n}\right)$, divergent when *MK* is greater, and parallel when it is equal. Now, because $MK = 0$ when the incident rays are parallel, it follows that in this case infinitely close refracted rays are always convergent.

Corollary IV. (§137) *If the incident ray BM touches the curve AMD at the point M, then we have* $ME(a) = 0$, *and consequently* $MF = b$. *This shows that the point F therefore falls on the point G.*

If the incident ray *BM* is perpendicular to the curve *AMD*, then the straight lines $ME(a)$ and $MG(b)$ each become equal to the radius of the evolute *CM*, because they

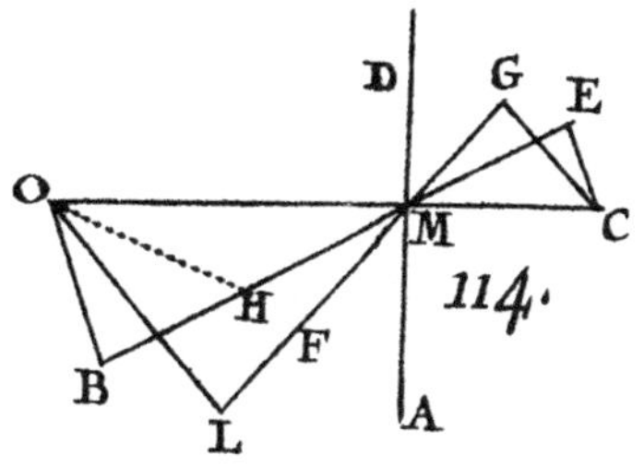

Fig. 7.4 Caustic of a Straight Line by Refraction

coincide with it. We therefore have $MF = \frac{bmy}{my - ny \mp bn}$, which becomes $\frac{bm}{m-n}$ when the incident rays are parallel to each other.

If the refracted ray MF touches the curve AMD at the point M, then we have $MG(b) = 0$. From this we see that the caustic therefore touches the given curve at the point M.

[125] If the radius of the evolute CM is null, then the straight lines $ME(a)$ and $MG(b)$ are also equal to zero. Consequently, the terms aan and $bbmy$ are null with respect to the other terms bmy and any. From this it follows that $MF = 0$, and therefore that the caustic has the point M in common with the given curve.

If the radius of the evolute CM is infinite, then the straight lines $ME(a)$ and $MG(b)$ are also infinite. Consequently, the terms bmy and any are null with respect to other terms aan and $bbmy$, so that we have $MF = \frac{bbmy}{\mp aan}$. Now (see §133), because this quantity is negative when we suppose that the point F falls on the other side of the point B, with respect to the line AMD, and on the contrary it is positive when we suppose that it falls on the same side, it follows (see §136) that we must take the point F on the same side of the point B, that is to say that infinitely close refracted rays are divergent. It is clear that the little arc Mm thus becomes a straight line, and that the preceding construction no longer holds. We may substitute the following one for it, which can be used to determine the points of caustics by refraction when the line AMD is straight.

Draw BO perpendicular to the incident ray BM (see Fig. 7.4), meeting the straight line MC perpendicular to AD at O. If we draw OL perpendicular to the refracted ray MG, and make the angle BOH equal to the angle LOM, then we have $BM : BH :: ML : MF$. I say that the point F is on the caustic by refraction.

Because the right triangles MEC and MBO are similar and the right triangles MGC and MLO are also similar, no matter what magnitude we suppose CM to have, and consequently when it becomes infinite, we still have[6] $ME(a) : MG(b) :: BM(y) : ML = \frac{by}{a}$. Additionally, because the triangles OLM and OBH are similar,

[6]In L'Hôpital (1696) the connective between $BM(y)$ and ML was missing.

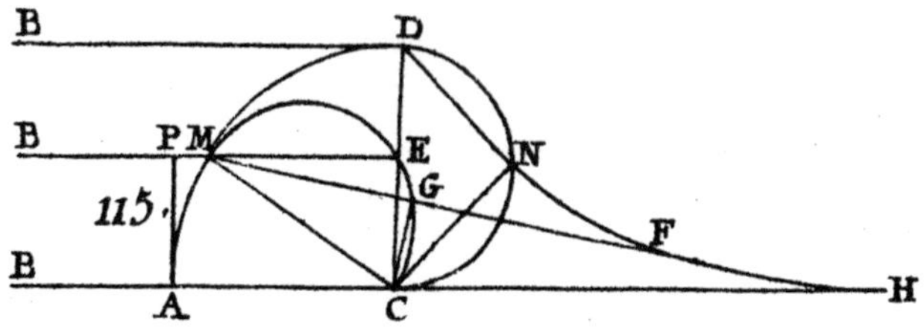

Fig. 7.5 Caustic of a Quarter Circle by Refraction, Convex Case

we also have[7] $OL : OB(n : m) :: ML\left(\frac{by}{a}\right) : BH = \frac{bmy}{an}$. From this we see that $BM(y) : BH\left(\frac{bmy}{an}\right) :: ML\left(\frac{by}{a}\right) : MF\left(\frac{bbmy}{aan}\right)$.

Corollary V. [126] (§138) *It is clear that given any two of the three points B, C, and F, we can easily find the third one.*

Example I. (§139) Let the curve *AMD* (see Fig. 7.5) be a quarter of a circle that has the point C as its center. Let the incident rays *BA*, *BM*, and *BD* be parallel to each other and perpendicular to *CD*. Finally, let the ratio of m to n be as 3 is to 2, which is the ratio for rays of light passing from air into glass. Because the evolute of the circle *AMD* is the point C, which is its center, it follows that if we describe a semi-circumference *MEC*, which has the radius *CM* as its diameter, and if we take the chord $CG = \frac{2}{3}CE$, then the line *MG* is the refracted ray, on which we determine the point F, as we demonstrated above (see §133).

To find the point H where the incident ray *BA*, perpendicular to *AMD*, touches the caustic by refraction, we have (see §137) $AH\left(\frac{bm}{m-n}\right) = 3b = 3CA$.[8] Moreover, if we describe a semi-circumference *CND* with the radius *CD* as its diameter, and if we take the chord $CN = \frac{2}{3}CD$, then it is clear (see §137) that the point N is on the caustic by refraction because the incident ray *BD* touches the circle *AMD* at the point D.

If we draw *AP* parallel to *CD*, then it is clear (see §132) that the portion $FH = AH - MF - \frac{2}{3}PM$, so that the entire caustic $HFN = \frac{7}{3}CA - DN = \frac{7-\sqrt{5}}{3}CA$.

If the quarter circle *AMD* (see Fig. 7.6) is concave towards the incident rays *BM*, and the ratio of m to n, is as 2 is to 3, we take the chord $CG = \frac{3}{2}CE$ on the semi-circumference *CEM* that has the radius *CM* as its diameter, and we draw the refracted ray *MG*, on which we determine the point F by the general construction of §133.

[127] We have (see §137) $AH\left(\frac{bm}{m-n}\right) = -2b$, that is to say that *AH* is on the side (see §136) of the convexity of the quarter circle *AMD*, and twice the radius *AC*. If we

[7]In L'Hôpital (1696) the parentheses around $\frac{by}{a}$ were missing.

[8]In L'Hôpital (1696) the parentheses following *AH* were omitted.

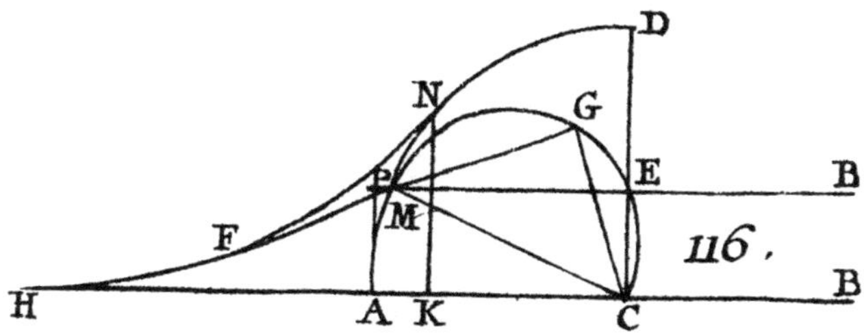

Fig. 7.6 Caustic of a Quarter Circle by Refraction, Concave Case

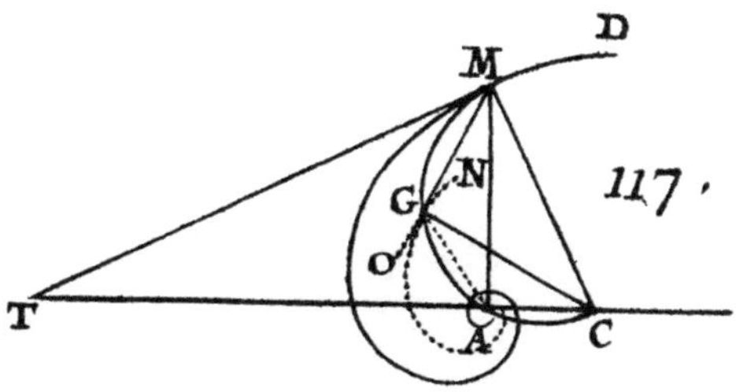

Fig. 7.7 Caustic of the Logarithmic Spiral by Refraction Spiral

suppose that CG or $\frac{3}{2}CE$ is equal to CM, then it is manifest that the refracted ray MF touches the circle AMD at M, because then the point G coincides with the point M. From this it follows that if we take $CE = \frac{2}{3}CD$, the point M falls on the point N, where the caustic HFN (see §137) touches the quarter circle AMD. However, when CE is greater than $\frac{2}{3}CD$, the incident rays BM can no longer be refracted, that is to say pass from glass into the air, because it is impossible for CG, perpendicular to the refracted ray MG, to be greater than CM, so that all the rays that fall on the part ND are reflected.

If we draw AP parallel to CD, then it is clear (see §132) that the portion $FH = AH - MF + \frac{3}{2}PM$, so that if we draw NK parallel to CD, the entire caustic $HFN = 2CA + \frac{3}{2}AK = \frac{7-\sqrt{5}}{2}CA$.

Example II. (§140) Let the curve AMD (see Fig. 7.7) be a logarithmic spiral, which has the point A as its center, from which all the incident rays AM emanate.

It is clear (see §91) that the point E falls on the point A, that is to say that $a = y$. Thus, if we substitute y in the place of a in $\frac{bbmy}{bmy-any+aan}$, the value (see §133) of MF when the curve is concave on the side of the radiant point, then we have $MF = b$. From this we see that the point F falls on the point G.

If we draw the straight line AG and the tangent MT, the angle AGO, supplementary to the angle AGM, is equal the angle AMT. This is because in the circle whose diameter is the line CM, that passes through the points A and G, the angles AGO and AMT each has as measure of half of the same arc AM. Therefore, it is clear that

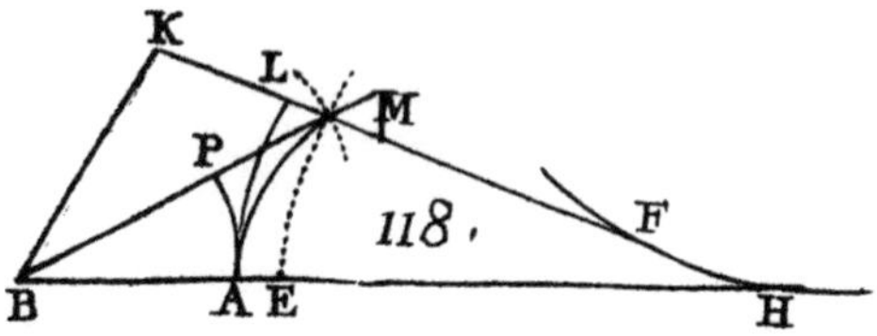

Fig. 7.8 Caustic by Refraction, Inverse Problem

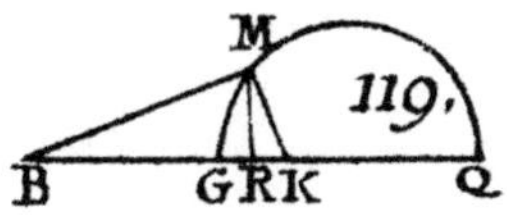

Fig. 7.9 Caustic by Refraction, Inverse Problem, Alternate Solution

the caustic *AGN* is the same [128] logarithmic spiral as the given *AMD*, and that it only differs in its position.

Proposition II.

Problem. (§141) *Given the caustic by refraction HF* (see Fig. 7.8) *with its radiant point B and the ratio of m to n, we wish to find an infinity of curves, such as AM, for which HF is the caustic by refraction.*

Take the point *A* at will on any tangent *HA* as one of the points on the curve *AM*. Describe the circular arc *AP* with center *B* and interval *BA*, and another circular arc with any other interval *BM*. Taking $AE = \frac{n}{m}PM$, we describe a curved line *EM* by enveloping the caustic *HF*, that cuts the circular arc described on the interval *BM* in a point M, which is on the curve we wish to find. This is because (see §132), $PM : AE$ or $ML :: m : n$.

Alternate Solution. (§142) On any tangent *FM*, other than *HA*, we wish to find the point M such that $HF + FM + \frac{n}{m}BM = HA + \frac{n}{m}BA$. This is why if we take $FK = \frac{n}{m}BA + AH - FH$, and we find a point M on *FK*, such that $MK = \frac{n}{m}BM$, this (see §132) will be the point that we wish to find. Now, this can be done by describing a curved line *GM* (see Fig. 7.9) such that when we draw the straight lines *MB* and *MK* from any of its point M to the given points B and K, they are always to each other in the same ratio as m is to n. It is therefore only a matter of finding the nature of this place.[9]

To this end, let *MR* be drawn perpendicular to *BK* and denote the given *BK* by a, and the indeterminates *BR* by x and *RM* by y. The right triangles *BRM* and *KRM* give $BM = \sqrt{xx + yy}$ and $KM = \sqrt{aa - 2ax + xx + yy}$, [129] so that to satisfy the condition of the problem, we must have $\sqrt{xx + yy} : \sqrt{aa - 2ax + xx + yy} :: m : n$.

[9]I.e., the place of the general point M of the curved line *GM*.

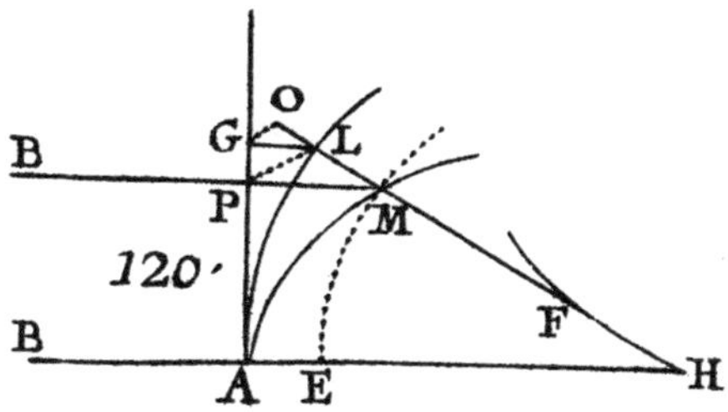

Fig. 7.10 Caustic by Refraction, Inverse Problem, Radiant Point at Infinity

From this we conclude $yy = \frac{2ammx-aamm}{mm-nn} - xx$, which is a place on the circle that we construct as follows.

Let us take $BG = \frac{am}{m+n}$ and $BQ = \frac{am}{m-n}$, and let the semi-circumference *GMQ* be described with diameter *GQ*; I say this is the required place. Because we have *QR* or $BQ - BR = \frac{am}{m-n} - x$ and *RG* or $BR - BG = x - \frac{am}{m+n}$, the property of the circle, which gives $QR \times RG = RM^2$, yields $yy = \frac{2ammx-aamm}{mm-nn} - xx$ in analytic terms.

If the incident rays *BA* and *BM* (see Fig. 7.10) are parallel to a straight line given in position, the first solution will still hold, but this latter becomes useless, and we may substitute it with the following.

Let us take $FL = AH - HF$, and draw *LG* parallel to *AB* and perpendicular to *AP*. We take $LO = \frac{n}{m}LG$, and draw *LP* parallel to *GO*, and *PM* parallel to *GL*. It is clear (see §132) that the point *M* is the one that we wish to find; because $LO = \frac{n}{m}LG$, so it follows that $ML = \frac{n}{m}PM$.

If the caustic by refraction *FH* meets in a point, then the curves *AM* become the Ovals of *Descartes*, which have caused such a stir among Geometers.[10]

Corollary I. (§143) *We prove as we did for the caustics by reflection (see §130) that the curves AM have different natures, and they are not geometric except when the caustic by refraction HF is geometric and rectifiable.*

Corollary II. (§144) *Given a curved line AM* (see Fig. 7.11) *with the radiant point B, and the ratio of m to n, we wish to find an [130] infinity of lines such as DN, so that the refracted rays MN break again when they encounter these lines DN to meet at a given point C.*

If we imagine that the curved line *HF* is the caustic by refraction of the given curve *AM*, formed by the radiant point *B*, then it is clear that this same line *HF* must also be the caustic by refraction of the curve *DN* that we wish to find, having the given point *C* as its radiant point. This is why (see §132)

[10]The Ovals of Descartes, or Cartesian Ovals, are a quartic curve (Lockwood 1971, p. 188).

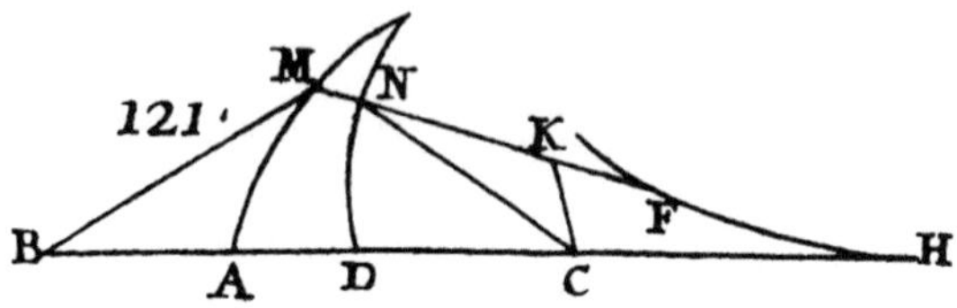

Fig. 7.11 Using the Inverse Construction to Focus Refracted Rays

$$\frac{n}{m}BA + AH = \frac{n}{m}BM + MF + FH \quad \text{and}$$

$$NF + FH - \frac{n}{m}NC = HD - \frac{n}{m}DC.$$

Consequently

$$\frac{n}{m}BA + AH = \frac{n}{m}BM + MN + HD - \frac{n}{m}NC,$$

and transposing as usual,

$$\frac{n}{m}BA - \frac{n}{m}BM + \frac{n}{m}DC + AD = MN + \frac{n}{m}NC.$$

This gives the following construction.

Take the point D at will on any refracted ray AH as one of the points of the curve DN that we wish to find. On any other refracted ray MF we take the part $MK = \frac{n}{m}BA - \frac{n}{m}BM + \frac{n}{m}DC + AD$, and find the point N, as above (see §142), such that $NK = \frac{n}{m}NC$. It is then clear (see §132) that the point M will be on the curve DN.

General Corollary.

For the Three Precceding Chapters. (§145) *It is manifest (see §80, 85, 107, 108, 114, 115, 128, 129, 134, 143) that a curved line can have only one evolute, only one caustic by reflection, and only one caustic by refraction, given the radiant point and the ratio of sines. These lines are always geometric and rectifiable when the given curve is geometric. On the other hand, the same curved line may be the evolute, or one or the other caustic in the same ratio of sines, with the same position of the radiant point, of an*[11] *infinity of very different lines, which are only geometric when the given curve is geometric and rectifiable.*

[11]In L'Hôpital (1696), the indefinite article *une* was omitted, but this was corrected in the *Errata*.

Chapter 8
Use of the Differential Calculus for Finding the Points of Curved Lines That Touch An Infinity of Lines Given in Position, Whether Straight or Curved

Proposition I.

Problem. [131] (§146) *Let any line AMB* (see Fig. 8.1) *be given, which has the straight line AP as its axis. Let it be understood that there is an infinity of parabolas AMC and AmC, which all pass through the point A, and which have as their axes the ordinates PM and pm. We wish to find the curved line that touches all of these parabolas.*[1]

It is clear that the point of contact of each parabola *AMC* is the intersection point *C* where the infinitely close parabola *AmC* cuts it. Given this, we draw *CK* parallel to *MP*, and denote the given quantities *AP* by x and *PM* by y, and the unknowns *AK* by u and *KC* by z. By the property of the parabola, we have $\overline{AP}^2(xx) : \overline{PK}^2(uu - 2ux + xx) :: MP(y) : MP - CK(y - z)$. This gives $zxx = 2uxy - uuy$, which is the equation common to all the parabolas such as *AMC*. Now, I remark that the unknowns $AK(u)$ and $KC(z)$ remain the same, while the given quantities $AP(x)$ and $PM(y)$ vary by becoming *Ap* and *pm*, and $KC(z)$ remains the same only when the point *C* is the intersection point. This is because it is clear that in every other place the straight line *KC* cuts the two parabolas *AMC* and *AmC* at two different points, and consequently there would be two values that would correspond to the same *AK*. This is why if we treat u and z as constants, and take the differential of the equation that we have just found, we determine the point *C* to be the intersection point. Therefore, we have $2zx\,dx = 2ux\,dy + 2uy\,dx - uu\,dy$. From this we find the unknown [132] $AK(u) = \frac{2xx\,dy - 2yx\,dx}{x\,dy - 2y\,dx}$ by substituting the value $\frac{2uxy - uuy}{xx}$ for z. If the nature of the curve *AMB* is given, then we find a value for dy in terms of dx, which is substituted into the value of *AK*, so that this unknown will finally be expressed in entirely known terms and freed of differentials. This is what was proposed.

[1]The modern term for such a curve is the *envelope* of this family of parabolas.

R.E. Bradley et al., *L'Hopital's Analyse des infiniments petits*, Science Networks. Historical Studies 50, DOI 10.1007/978-3-319-17115-9_8

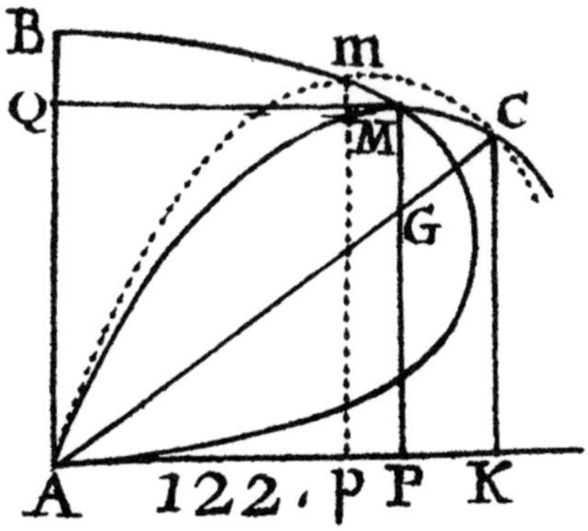

Fig. 8.1 The Envelope of a Family of Parabolas

If instead of parabolas *AMC*, we were to propose other straight or curved lines, the positions of which are determined, we would always solve the Problem in much the same way, as we will see in the following Propositions.

Example. (§147) Let the equation $xx = 4ay - 4yy$ express the nature of the curve *AMB*, which is a half-ellipse that has the straight line $AB = a$ as its minor axis, perpendicular to *AP*, the major axis of which is double the minor axis.

We find that $x\,dx = 2a\,dy - 4y\,dy$ and consequently $AK\left(\frac{2xx\,dy - 2xy\,dx}{x\,dy - 2y\,dx}\right) = \frac{ax}{y} = u$. From this it follows that if we take *AK* to be the fourth proportional to *MP*, *PA*, and *AB*, and if we draw *KC* perpendicular to *AK*, it will cut the parabola *AMC* at the point *C* that we wish to find.

To find the nature of the curve that touches all the parabolas, or that passes through all the points *C* found in this way, we find the equation that expresses the relationship of $AK(u)$ to $KC(z)$ in the following way. Substituting the value $\frac{ax}{y}$ in place of u in $zxx = 2uxy - uuy$, we conclude that $y = \frac{aa}{2a-z}$ and consequently x or $\frac{uy}{a} = \frac{au}{2a-z}$. Therefore, if we substitute these values in place of x and y in $xx = 4ay - 4yy$, we form the equation $uu = 4aa - 4az$, in which x and y are no longer found and which expresses the relationship of *AK* to *KC*. From this we see that the curve that we wish to find is a parabola that has the line *BA* as its axis, the point *B* as its vertex, the point *A* as its focus, and whose parameter is consequently four times *AB*.

[133] We have just found that $y = \frac{aa}{2a-z}$, from which we conclude that $KC(z) = \frac{2ay-aa}{y}$. Now, because this value is positive when $2y$ is greater than a, negative when it is less than a, and null when it is equal to a, it follows that the point of contact *C* falls above *AP* in the first case, as we had assumed in doing the calculation, below it in the second case, and finally on *AP* in the third case.

If we draw the straight line *AC*, that cuts *MP* at *G*, then I say that $MG = BQ$, and that the point *G* is the focus of the parabola *AMC*. This is because

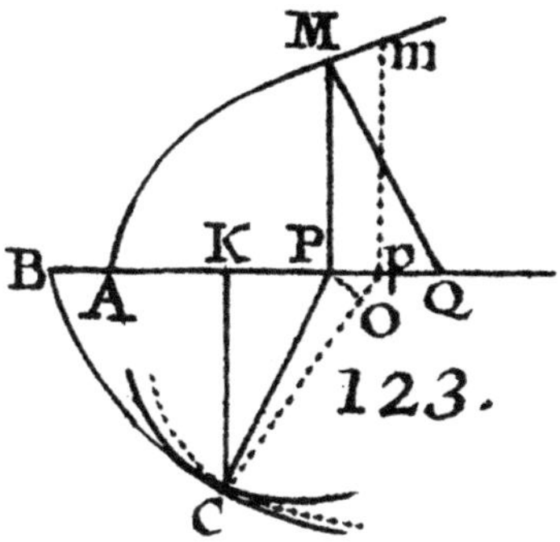

Fig. 8.2 The Envelope of a Family of Circles

1. $AK\left(\frac{ax}{y}\right) : KC\left(\frac{2ay-aa}{y}\right) :: AP(x) : PG = 2y - a$ and consequently $MG = a - y = BQ$.
2. The parameter of the parabola AMC is $= 4a - 4y$, by substituting the value $4ay - 4yy$ for xx. Consequently, $MG(a - y)$ is one-fourth of the parameter, from which we see that the point G is the focus of the parabola, and therefore that the angle BAC must be divided equally into two by the tangent at A.

It follows from the fact that the parameter of the parabola AMC is four times BQ, that when the vertex M falls on A the parameter is four times AB, and therefore the parabola that has the point A as its vertex is asymptotic to the one that passes through all the points C.

Because the parabola BC touches all parabolas such as AMC, it is clear that all these parabolas cut the determined line AC at points that are closer to the point A than the point C. Now, we demonstrate in Ballistics (supposing that AK is horizontal) that all parabolas such as AMC mark the path described in the air by Bombs which are launched by a Mortar placed at A, at all possible elevations with the same force. From this it follows that if we draw a straight line that divides the angle BAC in the middle, it will mark the position that the mortar should have so that the bomb that it launches falls in the plane AC, given in position, at a point C more distant from the mortar than with any other elevation.

Proposition II.

Problem. [134] (§148) *Let any curve AM be given* (see Fig. 8.2), *which has the straight line AP as its axis. We wish to find another curve BC such that if we draw the ordinate PM at will, and the perpendicular PC to this curve, the two lines PM and PC are always equal to each other.*

If we imagine an infinity of circles described with centers P and p, and radii PC and pC, equal to PM and pm, then it is clear that the curve BC that we wish to find must touch all of these circle, and that the point of contact C of each circle is the intersection point where the circle that is infinitely close to it touches it. Given this, let CK be drawn perpendicular to AP, and denote the given and variable quantities

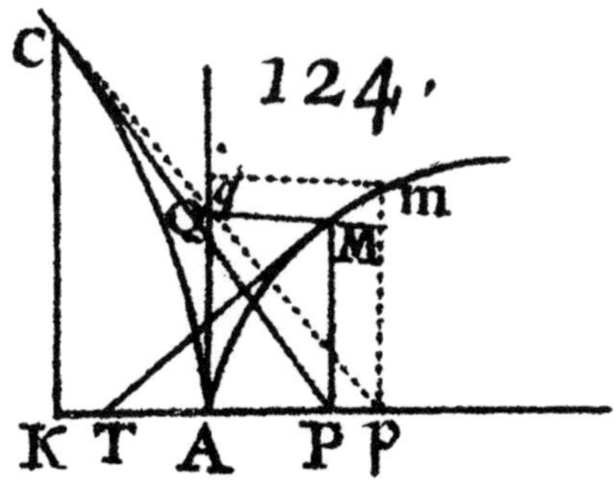

Fig. 8.3 Envelope of a Family of Straight Lines

AP by x and *PM* or *PC* by y, and the unknown and constant quantities *AK* by u and *KC* by z. By the property of the circle we have $\overline{PC}^2 = \overline{PK}^2 + \overline{KC}^2$, in analytic terms that is to say $yy = xx - 2ux + uu + zz$, which is the equation common to all of these circles, the differential of which is $2y\,dy = 2x\,dx - 2u\,dx$. From this we conclude[2] $PK(x-u) = \frac{y\,dy}{dx}$, which gives the following general construction.

Let *MQ* be drawn perpendicular to the curve *AM*, and taking $PK = PQ$, let *KC* be drawn parallel to *PM*. I say that it will meet the circle described with center point *P* and radius $PC = PM$ at the point C, where it touches the curve *BC* that we wish to find. This is clear, because $PQ = \frac{y\,dy}{dx}$.

We may also find the value of *PK* in this alternate manner.

If we draw *PO* perpendicular to *Cp*, the right triangles *pOP* and *PKC* are similar and consequently $Pp(dx) : Op(dy) :: PC(y) : PK = \frac{y\,dy}{dx}$.

When $PQ = PM$, it is clear that the circle described with radius *PC* touches *KC* at the point K, so that the point [135] of contact C will coincide with the point K, and consequently will fall on the axis.

However, when *PQ* is greater than *PM*, the circle described with radius *PC* cannot touch the curve *BC*, because it can never meet the straight line *KC* at any point.

Example. (§149) Let the given curve *AM* (see Fig. 8.2) be a parabola that has its equation as $ax = yy$. We have *PQ* or $PK(x-u) = \frac{1}{2}a$, and consequently $x = \frac{1}{2}a + u$ and $yy = \frac{1}{4}aa + zz$, because of the right triangle *PKC*. Now, if we substitute these values in $ax = yy$, we form the equation $\frac{1}{2}aa + au = \frac{1}{4}aa + zz$ or $\frac{1}{4}aa + au = zz$, which expresses the nature of the curve *BC*. From this it is clear that this curve is the same parabola as *AM*, because they both have the same parameter a, and that its vertex B is at a distance of $BA = \frac{1}{4}a$ from the vertex A.

[2] In L'Hôpital (1696), the right parenthesis was missing.

Proposition III.

Problem. (§150) *Let any curved line AM* (see Fig. 8.3) *be given that has the line AP as its diameter and whose ordinates PM and pm are parallel to the straight line AQ given in position. We draw MQ and mq parallel to AP, and then the straight lines PQC and pqC. We wish to find the curve AC that has all these straight lines as tangents or, what amounts to the same thing, to determine the point of contact C on each straight line PQC.*

If we imagine another tangent pqC infinitely close to PQC, and draw CK parallel to AQ, we denote the given and variable quantities AP by x and PM or AQ by y, and the unknown and constant quantities AK by u and KC by z. The similar triangles PAQ and PKC then give $AP(x) : AQ(y) :: PK(x+u) : KC(z) = y + \frac{uy}{x}$, which is the equation [136] common to all the straight lines such as[3] KC. Its differential is $dy + \frac{ux\,dy - uy\,dx}{xx} = 0$, from which we conclude $AK(u) = \frac{xx\,dy}{y\,dx - x\,dy}$. This gives the following general construction.

Let the tangent MT be drawn, and let AK be taken to be the third proportional to AT and AP. I say that if we draw KC parallel to AQ it will cut the straight line PQC at the point C that we wish to find.

This is because[4] $AT\left(\frac{y\,dx - x\,dy}{dy}\right) : AP(x) :: AP(x) : AK = \frac{xx\,dy}{y\,dx - x\,dy}$.

Example I. (§151) Let the given curve AM (see Fig. 8.3) be a parabola whose equation is $ax = yy$. We have $AT = AP$, from which it follows that $AK(u) = x$, that is to say that the point K falls on the point T. If we now wish to have an equation which expresses the relationship of $AK(u)$ to $KC(z)$, then we find that $KC(z) = 2y$, because we have just found that PK is twice AP. Thus, substituting for x and y their values u and $\frac{1}{2}z$ in $ax = yy$, we have $4au = zz$. From this, we see that the curve AC is a parabola whose vertex is the point A and whose parameter is a line which is four times the parameter of the parabola AM.

Example II. (§152) Let the given curve AM (see Fig. 8.4) be a quarter of the circle BMD which has the point A as its center, and the line AB or AD as its radius, which I call a. It is clear that PQ is always equal to the radius AM or AB, that is to say that they are everywhere the same, so that we may imagine that its extremities P and Q slide along the sides BA and AD of the right angle BAD. We have $AK(u) = \frac{x^3}{aa}$, because $AT = \frac{aa}{x}$, and the parallels KC and AQ give $AP(x) : PQ(a) :: AK\left(\frac{x^3}{aa}\right) : QC = \frac{xx}{a}$. From this we see that in order to find the point of contact C, we need to

[3]In L'Hôpital (1696), PC was written in place of KC, but corrected in the *Errata*.

[4]In L'Hôpital (1696), AK was preceded by the symbol :: instead of a colon.

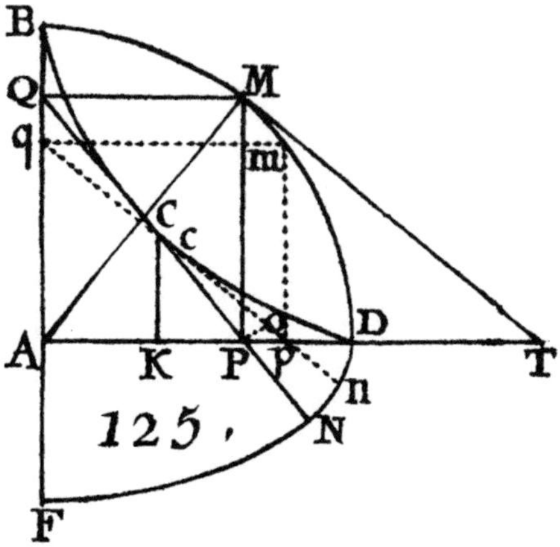

Fig. 8.4 Envelope of a Family of Straight Lines Generated by a Quarter Circle

only take QC to be the third [137] proportional to PQ and AP. If wish to find the equation which expresses the nature of the curve BCD, then we find the following:[5]

$$\begin{array}{llllllll} u^6 & - 3aau^4 & + 3a^4uu & - & a^6 = 0. \\ & + & 3zz & + 21aazz & + 3a^4zz \\ & & + & 3z^4 & - 3aaz^4 \\ & & & + & z^6 \end{array}$$

Corollary I. (§153) *If we wish to find the ratio of the portion DC of the curve BCD to its tangent CP, then we imagine another tangent cp infinitely close to CP, and describe the small arc PO with center C. We will then have $cp - CP$ or $Op - Cc = -\frac{2x\,dx}{a}$, as the differential of $CP = \frac{aa-xx}{a}$, from which we conclude that $Cc = Op + \frac{2x\,dx}{a}$. Now, because of the similar right triangles QPA and PpO, we have $PQ(a) : AP(x) :: Pp(dx) : Op = \frac{x\,dx}{a}$ and consequently $Cc = \frac{3x\,dx}{a} = DC - Dc$. It is then manifest that in whatever location we take the point C, we always have $DC - Dc\left(\frac{3x\,dx}{a}\right) : CP - cp\left(\frac{2x\,dx}{a}\right) :: 3 : 2$. From this it follows that the sum of all the differentials $DC - Dc$ which correspond to the straight line PD, that is to say (see §96) the portion DC of the curve BCD, is to the sum of all the differentials $CP - cp$ which correspond to the same straight line PD, that is to say (see §96) to the tangent $CP :: 3 : 2$. Similarly, the entire curve BCD is to its tangent $BA :: 3 : 2$.*

Corollary II. (§154) *If we evolve the curve BCD, starting at the point D, then we form the curved line DNF such that $CN : CP :: 3 : 2$, because CN is always equal to the portion DC of the curve BCD. From this it follows that the similar sectors*

[5] In the equation that follows, the factor u^4 is understood in the first term of the second line and the factor uu is understood in the next term, as well as the first term of the third line.

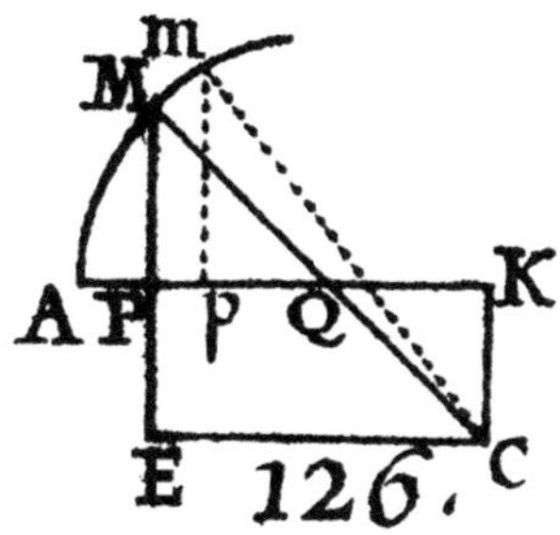

Fig. 8.5 The Envelope of a Family of Perpendiculars

CNn and CPO are to each other as :: 9 : 4, and thus that the space DCN contained by the curves DC and DN, and by the straight line CN which is tangent at C and perpendicular at [138] N, is to the space DCP contained by the curve DC and the two tangents DP and CP, as 9 is to 4.

Corollary III. (§155) *The center of gravity of the sector CNn must be situated on the arc PO, because* $CP = \frac{2}{3}CN$. *In addition, because this arc is infinitely small, it follows that this center must be on the straight line AD and consequently that the center of gravity of the spaces DCN and BDF, which are composed all these sectors, must be on this straight line AD, so that if we describe a figure exactly like BDF on the other side of BF, then the center of gravity of the whole figure would be at the point A.*

Corollary IV. (§156) *Because of the similar right triangles PQA and pPO, we have* $PQ(a) : AQ$ *or* $PM\left(\sqrt{aa-xx}\right) :: Pp(dx) : PO = \frac{dx\sqrt{aa-xx}}{a}$. *In addition, because of the similar sectors CPO and CNn we also have* $CP : CN$ *or* $2 : 3 :: PO\left(\frac{dx\sqrt{aa-xx}}{a}\right) : Nn = \frac{3\,dx\sqrt{aa-xx}}{2a}$. *Now, the rectangle* $MP \times Pp$, *that is to say (see §2) the little circular space* $MPpm = dx\sqrt{aa-xx}$. *Therefore, we have* $AB \times Nn = \frac{3}{2}MPpm$, *from which it follows that the portion ND of the curve DNF, when multiplied by the radius AB, is three-halves of the circular segment DMP, and that the entire curve DNF is equal to three quarters of BMD, which is one fourth of the circumference of the circle.*

Proposition IV.

Problem. (§157) *Let any curve AM* (see Fig. 8.5) *be given that has the straight line AP as its axis, and let an infinity of perpendiculars MC and mC to this curve be understood. We wish to find the curve [139] that has all of these perpendiculars as tangents or, what amounts to the same thing, we wish to find the point of contact C on every perpendicular MC.*

We imagine another perpendicular *mC* infinitely close to *MC* with ordinate *MP*, and from the point of intersection *C*, we draw the straight line *CK* perpendicular to the axis and the straight line *CE* parallel to the axis. If we then denote the given and

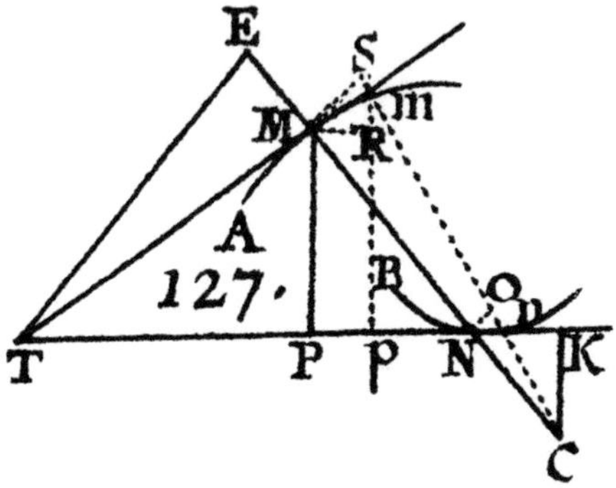

Fig. 8.6 The Envelope of a Sliding Line Segment, or Glissette

variable quantities AP by x and PM by y and the unknown and constant quantities AK by u and KC by z, then we have $PQ = \frac{y\,dy}{dx}$, PK or $CE = u-x$, and $ME = y+z$. The similar right triangles MPQ and MEC give $MP(y) : PQ\left(\frac{y\,dy}{dx}\right) :: ME(y+z) : EC(u-x) = \frac{y\,dy+z\,dy}{dx}$, which is the equation common to all the perpendiculars such as MC, and whose differential (supposing dx to be constant) gives $-dx = \frac{y\,ddy+dy^2+z\,ddy}{dx}$, from which we conclude that $ME(z+y) = \frac{dx^2+dy^2}{-ddy}$. Now, if the nature of the curve AM is given, we have the values of dy^2 and ddy in terms of dx^2, which being substituted in $\frac{dx^2+dy^2}{-ddy}$, gives a value for ME entirely known and free of differentials. This is what was proposed.

It is clear that the curve that goes through all the points C is the Evolute of the curve AM, and because we have specifically treated them in Chapter 5, it is unnecessary to give new examples here.

Proposition V.

Problem. (§158) *Let any two lines AM and BN be given* (see Fig. 8.6) *with a straight line MN that always remains the same. We suppose that the extremities M and N of this line slide continually along the two others. We wish to find the curve that it always touches in this motion.*

Draw the tangents MT and NT, and imagine [140] another straight line mn infinitely close to MN, which consequently cuts it at the point C where it touches the curve whose points we wish to determine. It is clear that, in order to reach mn, the extremities of the straight line MN traveled along the little portions Mm and Nn of the lines AM and BN, which coincide with the tangents TM and TN, on account of their infinite smallness. In this way, we may imagine that in order to arrive at the infinitely close position mn, the line MN slides along the straight lines TM and TN given in position.

With this understood, let the perpendiculars MP and CK be dropped to NT. We denote the given and variable quantities TP by x and PM by y, and the unknown and constant quantities TK by u and KC by z, and the given MN that

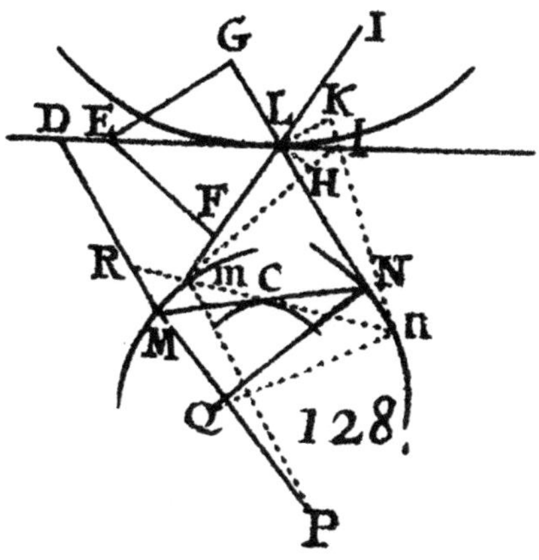

Fig. 8.7 The Envelope to a Family of Cords

always remains the same by a. The right triangle MPN gives $PN = \sqrt{aa-yy}$ and because of the similar triangles NPM and NKC we have $NP\left(\sqrt{aa-yy}\right) : PM(y) :: NK\left(u - x - \sqrt{aa-yy}\right) : KC(z) = \frac{uy-xy}{\sqrt{aa-yy}} - y$, whose differential gives $aau\,dy - aax\,dy - aay\,dx + y^3\,dx = \overline{aa\,dy - yy\,dy}\sqrt{aa-yy}$. Using $\sqrt{aa-yy} = m$ to abbreviate, we conclude from this that $PK(u - x) = \frac{m^3\,dy + mmy\,dx}{aa\,dy} = \frac{m^3 + mmx}{aa}$, substituting the value $x\,dy$ for $y\,dx$, because of the similar triangles mRM and MPT. Consequently, $MC = \frac{mm+mx}{a}$, which gives the following construction.

Let TE be drawn perpendicular to MN, and take $MC = NE$. I say that the point C is the one we wish to find. Because of the similar right triangles MNP and TNE, we have $MN(a) : NP(m) :: NT(m + x) : NE$ or $MC = \frac{mm+mx}{a}$.

Alternate method. Draw TE perpendicular to MN, and describe the little arcs MS and NO with center C. Denote the given quantities NE by r, ET by s, and MN by a, and the unknown CM by t. We have Sm or $On = dt$, and the similar right triangles [141] MET and mSM give $ME(r - a) : ET(s) :: mS(dt) : SM = \frac{s\,dt}{r-a}$. Additionally, the similar right triangles NET and nON give $NE(r) : ET(s) :: nO(dt) : ON = \frac{s\,dt}{r}$. Finally, the similar right triangles CMS and CNO give $MS{-}NO\left(\frac{as\,dt}{rr-ar}\right) : MS\left(\frac{s\,dt}{r-a}\right) :: MN(a) : MC(t) = r$. This gives the same construction as above.

If we suppose that the lines AM and BN are straight lines that form a right angle, then it is clear that the curve that we wish to find is the same as that of §152.

Proposition VI.

Problem. (§159) *Let any three lines L, M, and N be given* (see Fig. 8.7), *and let it be understood that from each of the points L and l of the line L there are two tangents LM and LN, and lm and ln, to the two curves M and N, one to each. We wish to find the fourth curve C, which has as its tangents all the straight lines MN and mn, that join the points of contact M and N of the curves.*

We draw the tangent LE and, from any of its points E, we drop the perpendiculars EF and EG to the other two tangents ML and NL. We imagine that the point l is infinitely close to the point L and draw the little straight lines LH and LK perpendicular to ml and nl. We also erect the perpendiculars MP, mP, NQ, and nQ on the tangents ML, ml, NL, and nl; these perpendiculars intersect each other at the points P and Q. All of this forms the similar right triangles EFL and LHl, and EGL and LKl. We also form the right triangles LMH and MPm, and LnK and NQn, with right angles at H and M, and at K and N, which are similar to each other, because the angles LMH and MPm make right angles when we add the angle PMm to each of them. Similarly, we prove that the angles LnK and NQn are equal to each other.

Given this, we denote the little side Mm of the polygon which composes the curve M by du, and the given quantities EF by m, EG by n, MN or mn by a, ML or ml by b, NL or nl by c, MP or [142] mP by f, NQ or nQ by g (here I take the straight lines MP and NQ as given, because we may always find them (see §78) from the nature of the curves M and N, which are given by hypothesis). We have:

1. $MP(f) : ML(b) :: Mm(du) : LH = \frac{b\,du}{f}$.
2. $EF(m) : EG(n) :: LH\left(\frac{b\,du}{f}\right) : LK = \frac{bn\,du}{mf}$.
3. LN or $Ln(c) : nQ(g) :: LK\left(\frac{bn\,du}{mf}\right) : nN = \frac{bgn\,du}{cfm}$.
4. (Drawing MR parallel to NL or nl) $ml(b) : ln(c) :: mM(du) : MR = \frac{c\,du}{b}$.
5. $MR + Nn\left(\frac{c\,du}{b} + \frac{bgn\,du}{cfm}\right) : MR\left(\frac{c\,du}{b}\right) :: MN(a) : MC = \frac{accfm}{ccfm+bbgn}$.

This is what we were required to find.

If the tangent EL falls on the tangent ML, it is clear that EF (m) becomes null or zero, and consequently the point C that we wish to find falls on the point M. Similarly, if the tangent EL coincides with the tangent LN, then EG (n) becomes null, and consequently we have $MC = a$, from which we see that the point C that we wish to find also falls on the point N. Finally, in the case that the tangent EL falls inside the angle GLI, EG (n) becomes negative, which therefore gives $MC = \frac{accfm}{ccfm-bbgn}$ and the point C that we wish to find does not fall between the points M and N, but on one side or the other.

Example I. (§160) Suppose that the curves M and N (see Fig. 8.8) are the same circle. It is clear that in this case $b = c$ and $f = g$, which gives $MC = \frac{am}{m+n}$, from which we see that we need to only cut the straight line MN in the given ratio of m to n to find the point C that we wish to find, that is to say so that $MC : NC :: m : n$.

Example II. (§161) Suppose that the curves M and N form [143] any conic section. The general construction can be changed to this other one, which is much simpler, if we pay attention to a property of conic sections, which we find proven in books that treat them. That is, if from each of the points L and l on the straight line EL we draw two tangents LM and LN, and lm and ln, to the conic section, then all the straight lines MN and mn, which join the points of contact, cut each other at the same point C, through which the diameter AC passes, the ordinates to which are parallel to the

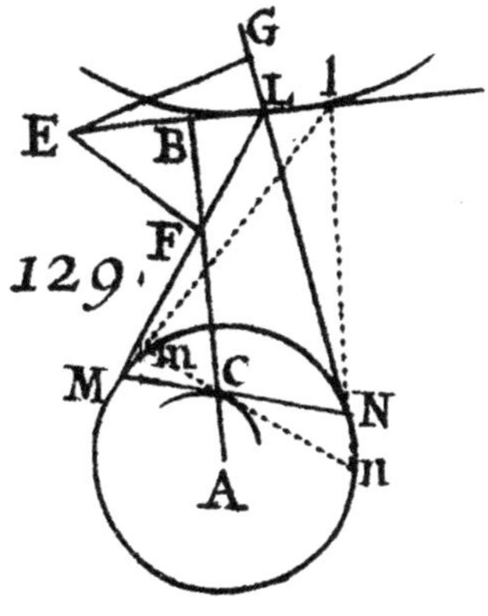

Fig. 8.8 Envelope of a Family of Chords in a Conic Section

straight line *EL*. Hence, it follows from this that to find the point C we need to only draw a diameter that has its ordinates parallel to the tangent *EL*.

It is clear that in the circle, the diameter should be perpendicular to the tangent *EL*, that is to say that if we drop a perpendicular *AB* from the center *A* to this tangent, it will cut through the straight line *MN* at the point C that we wish to find.

Remark. (§162) By means of this Problem (see Fig. 8.7) we may solve the following problem, which depends on the Method of Tangents.

Given three curves C, M, and N, we make a straight line *MN* roll around the curve C, so that it touches it continually. From the points M and N, where it cuts the curves M and N, we draw the tangents *ML* and *NL*, which intersect at the point L, which describes by this motion a fourth curve *Ll*. We wish to draw the tangent *LE* to this curve, given the position of the straight lines *MN*, *ML*, and *NL*, along with the point of contact C.

It is clear that this problem is the inverse of the previous one, and that here *MC* is given. What we wish to find is the ratio of *EF* to *EG*, which determines the position of the tangent *EL*. This is why, if we denote the given *MC* by h, then we have $\frac{accfm}{ccfm+bbgn} = h$. From this we conclude that $m = \frac{bbghn}{accf-ccfh}$, and consequently that the tangent *LE* must be so situated in the given angle *MLG*, that if we drop [144] the perpendiculars *EF* and *EG* from any of its points *E* to the sides of this angle, they always have the given ratio of *bbgh* to $accf - ccfh$ to each other. Now, this is done by drawing *MD* parallel to *NL*, and equal to $\frac{b^3gh}{accf-ccfh}$.

It is clear (see §161) that if the two curves M and N (see Fig. 8.8) form a conic section, then we need to only draw the tangent *LE* parallel to the ordinates of the diameter that passes through the point C.

Chapter 9
The Solution of Several Problems That Depend upon the Previous Methods

Proposition I.

Problem. [145] (§163) *Let AMD* (see Fig. 9.1) *be a curved line* ($AP = x$, $PM = y$, *and* $AB = a$) *such that the value of the ordinate* y *is expressed by a fraction, in which the numerator and the denominator each becomes zero when* $x = a$, *that is to say, when the point* P *falls on the given point* B. *We ask what the value of the ordinate BD ought to be.*[1]

Let it be understood that there are two curved lines *ANB* and *COB* that have the line *AB* as a common axis, and which are such that the ordinate *PN* expresses the numerator, and the ordinate *PO* the denominator of the general fraction that corresponds to all of the ordinates *PM*, so that $PM = \frac{AB \times PN}{PO}$. It is clear that these two curves meet at the point B because, by the assumption, *PN* and *PO* each becomes zero when the point P falls on B. Given this, if we imagine an ordinate *bd* infinitely close to *BD*, which meets the curved lines *ANB* and *COB* at f and g, then we will have $bd = \frac{AB \times bf}{bg}$, which (see §2) does not differ from *BD*. It is therefore only a question of finding the ratio of *bg* to *bf*. Now, it is clear that as the abscissa *AP* becomes *AB*, the ordinates *PN* and *PO* become null, and that as *AP* becomes *Ab*, they become *bf* and *bg*. From this, it follows that these ordinates themselves, *bf* and *bg*, are the differentials of the ordinates at B and b with respect to the curves *ANB* and *COB*. Consequently, if we take the differential of the numerator and we divide it by the differential of the denominator, after [146] having let $x = a = Ab$ or *AB*, we will have the value that we wish to find for the ordinate *bd* or *BD*. This is what we were required to find.

[1]This is the rule known as l'Hôpital 's Rule. Compare this to Bernoulli's Letter 28 on p. 267.

R.E. Bradley et al., *L'Hopital's Analyse des infiniments petits*, Science Networks. Historical Studies 50, DOI 10.1007/978-3-319-17115-9_9

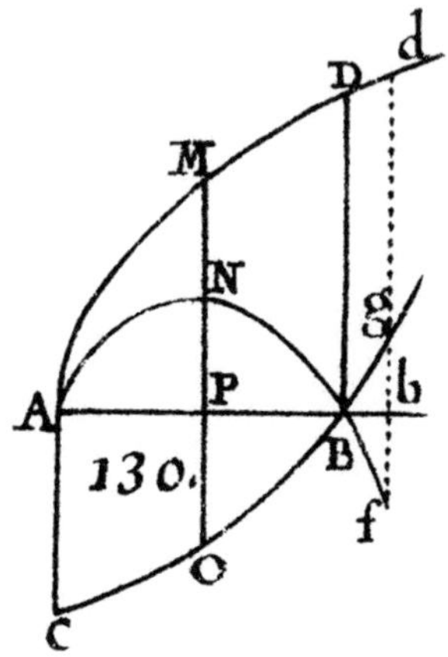

Fig. 9.1 L'Hôpital 's Rule

Example I. (§164) Let[2]

$$y = \frac{\sqrt{2a^3x - x^4} - a\sqrt[3]{aax}}{a - \sqrt[4]{ax^3}}.$$

It is clear that when $x = a$, then the numerator and denominator of the fraction both become equal to zero. This is why we take the differential of the numerator

$$\frac{a^3\,dx - 2x^3\,dx}{\sqrt{2a^3x - x^4}} - \frac{aa\,dx}{3\sqrt[3]{axx}}$$

and we divide it by the differential of the denominator

$$-\frac{3a\,dx}{4\sqrt[4]{a^3x}},$$

after having let $x = a$. That is to say, we divide $-\frac{4}{3}a\,dx$ by $-\frac{3}{4}\,dx$, which gives $\frac{16}{9}a$ as the value of BD that we wish to find.

Example II. (§165) Let[3]

$$y = \frac{aa - ax}{a - \sqrt{ax}}.$$

---

[2]This example is Bernoulli's $\frac{0}{0}$ Challenge Problem, which appears frequently in l'Hôpital 's correspondences with Bernoulli. It first appeared in Bernoulli's Letter 11 on p. 239.

[3]In his letter 28, of July 22, 1694, Bernoulli gave the similar example $y = \frac{a\sqrt{ax}-xx}{a-\sqrt{ax}}$, and found that $y = 3a$ when $x = a$.

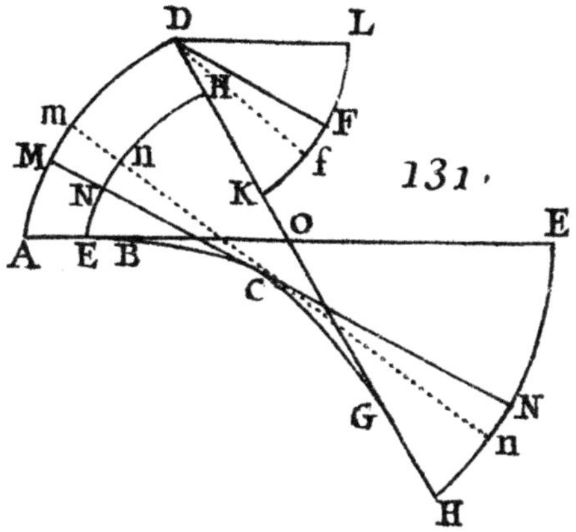

Fig. 9.2 The Rolling Tangent Lemma

We find $y = 2a$ when $x = a$.

We might have solved this example without the need of the calculus of differentials in the following way.

Having removed the incommensurables, we will have $aaxx + 2aaxy - axyy - 2a^3x + a^4 + aayy - 2a^3y = 0$, which being divided by $x - a$, reduces to $aax - a^3 + 2aay - ayy = 0$, and substituting a for x, it follows as before that[4] $y = 2a$.

Lemma I. (§166) *Let BCG be any curved line* (see Fig. 9.2)*, with a straight line AE that touches it at the point B, and on which two fixed points A and E are marked at will. If we make this straight line roll along the curve, so that it continually touches it, it is clear that the fixed points A and E describe two curves AMD and ENH by this motion. We now draw DL parallel to AB. The angle KDL, which DL makes with DK (on which I suppose the straight line AE to be when it* [147] *touches the curve BCG at G) is equal to the angle AOD made by the tangents at B and G. We also describe at will the arc KFL with center D.*

I say that DK : KFL :: AE : AMD ± ENH, namely + when the point of contact always falls between the describing points and − when they always remain on the same side.[5]

Suppose that the straight line *AE*, while rolling along the curve *BCG*, is brought to the positions *MCN* and *mCn*, infinitely close to each other, and that we draw the radii *DF* and *Df* parallel to *CM* and *Cm*. It is clear that the sectors *DFf*, *CMm*, and *CNn* are similar, and thus that *DF* : *Ff* :: *CM* : *Mm* :: *CN* : *Nn* :: *CM* ± *CN* or *AE* : *Mm* ± *Nn*. Now, because this always holds no matter where the point of contact *C* is found, it follows that the radius *DK* is to the arc *KFL*, which is the sum of all the

[4]There is no mention here that $y = 0$ is also a root of the equation $2aay - ayy = 0$, which is the result of substituting a for x. Bernoulli's example in letter 28 has the spurious root $y = -a$.

[5]The point E appears twice in Fig. 9.2. In the case of the point E on the right, we use $AMD + ENH$ in the proportion. In the case of the point E on the left, we use $AMD - ENH$ in the proportion.

little arcs Ff :: AE : $AMD \pm ENH$, which is the sum of all the little arcs $Mm \pm Nn$. This is what we were required to show.

Corollary I. (§167) *It is clear that the curves AMD and ENH are formed by the involution of the same curve BCG and that therefore the straight line AE is everywhere perpendicular to these two curves in any position where it is found, so that their distance is always the same, which is the property of parallel lines. From this we see that when the curved line AMD is given, we can find an infinity of points of the curve ENH without needing its evolute BCG, by drawing as many perpendiculars as we wish to this curve, and taking them all equal to the straight line AE.*

Corollary II. (§168) *If the curve BCG has its two halves BC and CG entirely similar and equal, and if we take the straight lines BA and GH equal to each other, then it is clear that the curves AMD and ENH are similar and equal, so that [148] they differ only in their position. From this it follows that the curve AMD is to the circular arc KFL ::* $\frac{1}{2}AE$: DK*, that is to say, in the given ratio.*

Proposition II.

Problem. (§169) *Let AEV and BCG be any two curves* (see Fig. 9.3) *with a third one AMD, such that when we describe a portion of the curve EM by the evolution of the curve BCG, the relationship among the curved portions AE and EM, and the*

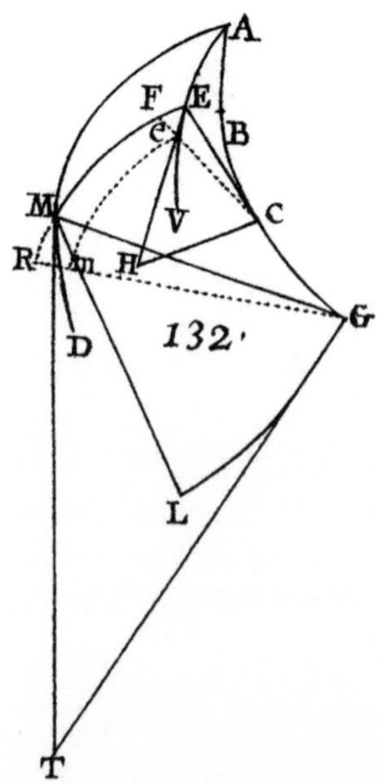

Fig. 9.3 Tangent to a Curve Defined from Two Other Curves by Evolution

radii of the evolutes EC and MG are expressed by any given equation. We wish to draw the tangent MT to the curve AMD at the given point M.

We imagine another portion of a curve *em* infinitely close to *EM* and the radii of the evolutes *CeF* and *GmR*.

1. Let *CH* be perpendicular to *CE*, which meets the tangent *EH* to the curve *AEV* at *H*.
2. Let *ML* be parallel to *CE*, which meets the arc *GL* described by the center *M* and radius *MG* at *L*.
3. Let *GT* be perpendicular to *MG*, which meets the tangent *MT* that we wish to find at *T*.

We denote the given quantities *AE* by x, *EM* by y, *CE* by u, *GM* by z, *CH* by s, *EH* by t, and the arc *GL* by r. From this we have $Ee = dx$, Fe or $Rm = du = dz$, and the similar right triangles *eFE* and *ECH* give $CE(u) : CH(s) :: Fe(dz) : FE = \frac{s\,dz}{u}$. Also, $CE(u) : EH(t) :: Fe(dz) : Ee(dx) = \frac{t\,dz}{u}$. Now, by the Lemma (see §166) $RF - me = \frac{r\,dz}{z}$, and consequently $RM\left(\overline{RF - me} + \overline{me - ME} + \overline{ME - MF}\right) = \frac{r\,dz}{z} + dy + \frac{s\,dz}{u}$. Thus, because of the similar right triangles *mRM* and *MGT*, we have $mR(dz) : RM\left(\frac{r\,dz}{z} + \frac{s\,dz}{u} + dy\right) :: MG(z) : GT = r + \frac{sz}{u} + \frac{z\,dy}{dz}$. However, if we substitute the values dz and $\frac{t\,dz}{u}$ in place of du and dx in the differential of the given equation, we then find the value of dy in terms of dz which being [149] substituted in $\frac{z\,dy}{dz}$ gives a value entirely known and free of differentials for the subtangent *GT* that we wish to find. This is what was proposed.

If we suppose that the curve *BCG* is reduced to a point *O* (see Fig. 9.4), it is clear that the portion of the curve $ME(y)$ becomes an arc of a circle equal to the arc $GL(r)$, and that the radii $CE(u)$ and $GM(z)$ of the evolute become equal to each other, so that *GT*, which in this case becomes *OT*, is found to be equal to $y + s + \frac{z\,dy}{dz}$.

Example. (§170) Let $y = \frac{xz}{a}$ (see Fig. 9.4), the differential of which gives $dy = \frac{z\,dx - x\,dy}{a}$ (we take $-x\,dz$ in place of $+x\,dz$ (see §8), because as x and y increase, z decreases) $= \frac{t\,dz - x\,dz}{a}$, substituting the value $\frac{t\,dz}{z}$ for dx. Consequently, $OT\left(y + s + \frac{z\,dy}{dz}\right) = y + s + \frac{tz - xz}{a} = \frac{as + tz}{a}$, substituting the value $\frac{xz}{a}$ for y.

Remark. (§171) If the point *O* falls on the axis *AB* (see Fig. 9.5), and the curve *AEV* is a semi-circle, then the curve *AMD* will be a half-roulette, formed by the revolution of a semi-circle *BSN* around an equal arc *BGN* of a circle described with center *O* such that the generating point *A* falls outside, inside, or on the circumference of the mobile semi-circle *BSN*, depending on whether the given a is greater than, less than, or equal to *OV*. We wish to prove this and at the same time determine the point *B*.

I assume that which is in question, namely that the curve *AMD* is a half-roulette, formed by the revolution of the semi-circle *BSN* – which has as its center the point *K*, which is the center of the semi-circle *AEV* – around the arc *BGN* described from the center *O*. Imagining that this semi-circle *BSN* stops at the position *BGN* such

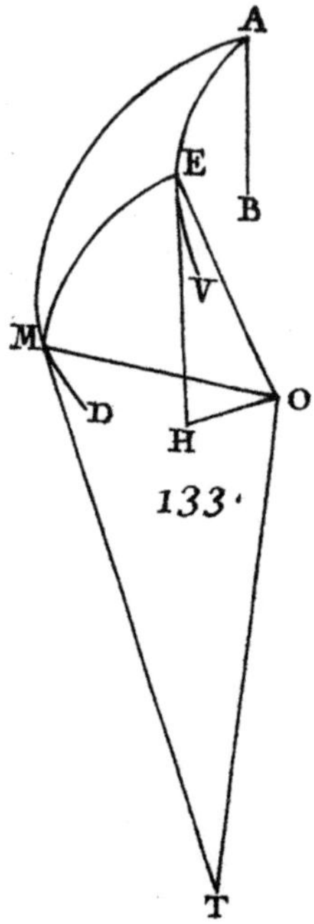

Fig. 9.4 Degenerate Case of the Construction in Figure 9.3

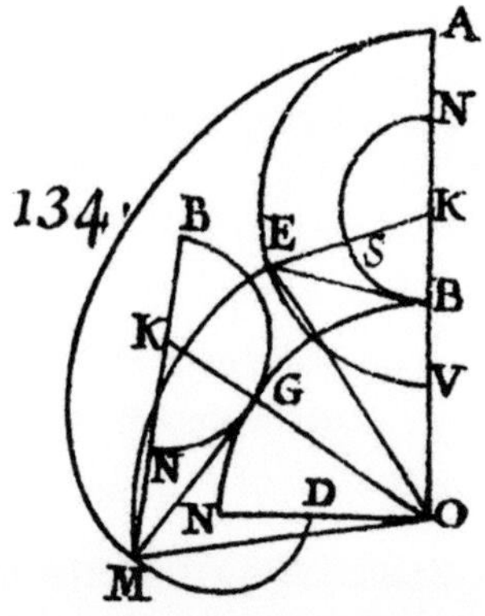

Fig. 9.5 Special Case of Figure 9.3: The Half-Roulette

that the describing point A falls [150] on the point M, I draw the straight line OK through the centers of the generating circles, which consequently passes through the point of contact G. Drawing KSE, I observe that the triangles OKE and OKM are equal and similar, because their three sides are equal one to another. From this it follows that:

1. The extreme angles MOK and EOK are equal, and therefore that the angles EOM and GOB are also equal, which gives $GB : ME :: OB : OE$.

2. The angles MKO and EKO are also equal, and thus that the arcs GN and BS, which measure them, are also equal. The same thing can be said of their supplements GB and SN, because they belong to equal circles. Now, by the generation of the roulette, the arc GB of the mobile circle is equal to the arc GB of the immobile circle. Therefore, I have $SN : ME :: OB : OE$.

Given this, I denote the given quantities OV by b, KV or KA by c, and the unknown KB by u. I have $OB = b + c - u$, and the similar sectors KEA and KSN give me $KE(c) : KS(u) :: AE(x) : SN = \frac{ux}{c}$. Consequently,

$$OB(b + c - u) : OE(z) :: SN\left(\frac{ux}{c}\right) : EM(y) = \frac{uxz}{bc + cc - cu} = \frac{xz}{a}.$$

From this I conclude $KB(u) = \frac{bc+cc}{a+c}$. It is therefore clear that if we take $KB = \frac{bc+cc}{a+c}$, and we describe the semi-circle BSN and the arc BGN from the centers K and O, the curve AMD is a half-roulette described by the revolution of the semi-circle BSN around the arc BGN, for which the describing point A falls outside, inside, or on the circumference of this circle depending on whether $KV(c)$ is greater than, less than, or equal to $KB\left(\frac{bc+cc}{a+c}\right)$, that is to say depending on whether a is greater than, less than, or equal to $OV(b)$.

Corollary I. (§172) *It is clear that* $EM(y) : AE(x) :: KB \times OE(uz) : OB \times KV\,(bc + cc - uc)$. *Now, if we suppose that OB becomes infinite, then the straight line OE will also be infinite and will become parallel to OB, because it will never meet it. The [151] concentric arcs BGN and EM become straight lines parallel to each other, and perpendicular to OB and OE. Therefore, the straight line EM is to the arc AE :: KB : KV, because the infinite straight lines OE and OB, which differ from each other only by a finite magnitude, must be regarded as equal.*

Corollary II. (§173) *Because the angles MKO and EKO are equal, it follows that the triangles MKG and EKB are equal and similar. Therefore, the straight lines MG and EB are equal to each other. From this we see (see §43) that in order to draw the perpendicular MG from any given point M on the roulette, we need to only describe the arc ME with center O, and from the center M with radius EB, we describe a circular arc that cuts the base BGN at a point G. Through this and the given point M we draw the perpendicular that we wish to find.*

Corollary III. (§174) *Given a point G on the circumference of the mobile semi-circle BGN, if we wish to find the point M on the roulette on which the describing point A falls when the given point G touches the base, we need to only take the arc SN equal to the arc BG and draw the radius KS to meet the circumference AEV at E. We then describe the arc EM with center O. It is clear that this arc cuts the roulette at the point M that we wish to find.*

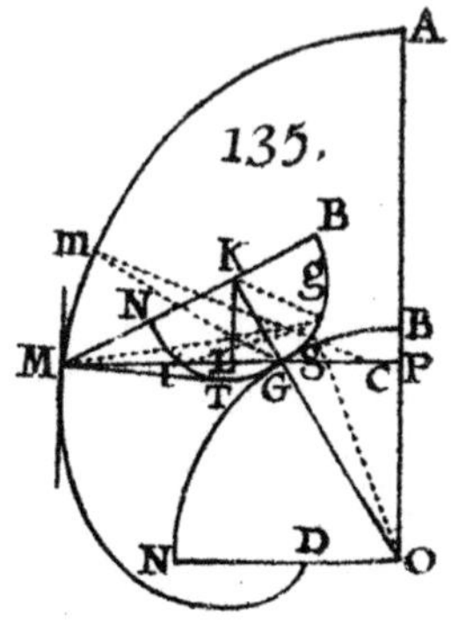

Fig. 9.6 Greatest Ordinate of a Roulette with respect to an Axis – External Describing Point

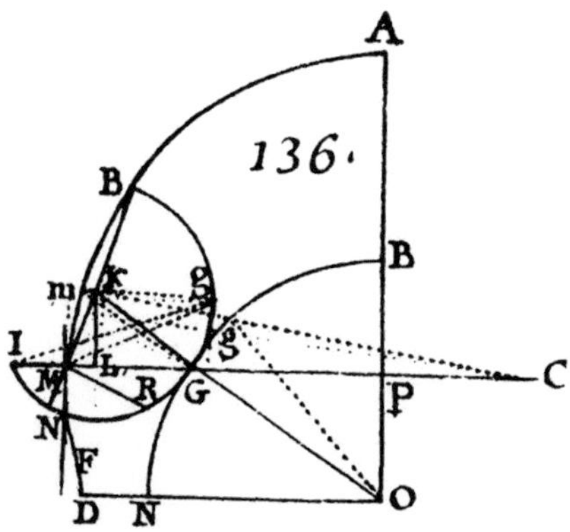

Fig. 9.7 Greatest Ordinate of a Roulette with respect to an Axis – Internal Describing Point

Proposition III.

Problem. (§175) *Let AMD be a half-roulette* (see Figs. 9.6, 9.7) *described by the revolution of the semi-circle BGN around an equal arc BGN of another circle, so that the revolved parts BG and BG are always equal to each other. Let the describing point M be taken on the diameter BN either outside, inside, or on the circumference of the mobile circle BGN. We wish to find the point M on the half-roulette with the greatest magnitude with respect to its axis OA.*

Assuming that the point M is the one that we wish to find, it [152] is clear (see §47) that the tangent at M must be parallel to the axis OA, and thus the perpendicular MG to the roulette must also be perpendicular to the axis, which it meets at P. Given this, if we draw OK through the centers of the generating circles, then it passes through the point of contact G, and if we draw KL perpendicular to MG, we make the equal angles GKL and GOB. Consequently, the arc IG, which is twice the measure of the angle GKL, is to the arc GB, which is the measure of the angle GOB, as the diameter BN is to the radius OB. From this it follows that to determine

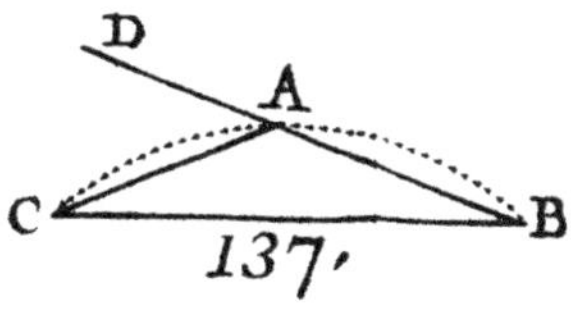

Fig. 9.8 Triangle with Infinitely Small Angles

the point G on the semi-circle BGN, where it touches the arc that serves as its base when the describing point M falls on the point of greatest magnitude, we must cut the semi-circle BGN at a point G such that when we draw the chord IG through the given point M, then the arc IG is to the arc BG in the given ratio of BN to OB. The question therefore reduces to a problem of common geometry, which may always be solved geometrically when the given ratio is of a number to a number. However, when the ratio is composed to some other power,[6] it maybe solved with the assistance of lines whose equation is raised to the same power.

If we suppose that the radius OB becomes infinite, as happens when the base BGN becomes a straight line, it follows that the arc IG is infinitely small with respect to the arc GB. From this we see that the secant MIG therefore becomes the tangent MT, when the describing point M falls outside of the mobile circle and that there cannot be a point of the greatest magnitude when it falls inside.

When the point M falls on the circumference at N, we need to only divide the semi-circumference BGN in the given ratio of BN to OB at the point G. This is because the point G thereby found is the one where the mobile circle BGN touches the base, when the describing point falls on the point that we wish to find.

Lemma II. [153] (§176) *In any triangle BAC* (see Fig. 9.8)*, such that the angles ABC, ACB, and CAD, the supplement to the obtuse angle BAC, are infinitely small, I say that these angles have the same ratios to one another as the sides AC, AB, and BC, to which they are opposite.*

If we circumscribe a circle around the triangle BAC, then the arcs AC, AB, and BAC, which measure the doubles of these angles, are infinitely small and consequently do not differ (see §3) from their chords or subtenses.[7]

If the sides AC, AB, and BC of the triangle BAC are not infinitely small, but have a finite magnitude, then it follows that the circumscribed circle must be infinitely large, because the arcs AC, AB, and BAC which have a finite magnitude must be infinitely small with respect to this circle, being the measures of infinitely small angles.

[6]I.e., involving 3rd, 4th, etc., proportionals.

[7]Some seventeenth century authors used the term *subtense* synonymously with the term chord, e.g., Cohen (1999, p. 130).

Proposition IV.

Problem. (§177) *The same things being given* (see Figs. 9.6, 9.7)*, we wish to determine the point C on each perpendicular MG where it touches the evolute of the roulette.*

We imagine another perpendicular *mg* infinitely close to *MG*, which consequently cuts it at the point *C* that we wish to find, and we draw the straight line *Gm*. Taking the little arc *Gg* on the circumference of the mobile circle equal to the arc *Gg* on the immobile circle, we draw the straight lines *Mg*, *Ig*, *Kg*, and *Og*. Given this, if we consider the little arcs *Gg* and *Gg* as little straight lines perpendicular to the radii *Kg* and *Og*, then it is clear that when the little arc *Gg* of the mobile circle falls on *Gg* of the immobile circle, then the describing point *M* falls on *m*, so that the triangle *GMg* coincides with the triangle *Gmg*. From this we see that the angle *MGm* is equal to the angle $gGg = GKg + GOg$, because when adding to both sides the same angles *KGg* and *OGg*, we make up two right angles.

Now, denoting the given quantities *OG* by *b*, *KG* by *a*, *GM* or *Gm* by *m*, and [154] *GI* or *Ig* by *n*, we find that:

1. $OG : GK :: GKg : GOg$ and $OG(b) : OG + GK$ or $OK(b + a) :: GKg : GKg + GOg$ or $MGm = \frac{a+b}{a}GKg$.
2. (see §176) $Ig : MI :: GMg : MgI$ and $Ig \pm MI$ or $MG(m) : Ig(n) :: GMg \pm MgI$ or GIg or $\frac{1}{2}GKg : GMg$ or $Gmg = \frac{n}{2m}GKg$.
3. (see §176) The angle MCm or $MGm - Gmg \left(\frac{a+b}{b} - \frac{n}{2m}GKg\right) : Gmg \left(\frac{n}{2m}Gkg\right) :: Gm(m) : GC = \frac{bmn}{2am+2bm-bn}$.

Consequently, the radius *MC* of the evolute that we wish to find is $= \frac{2amm+2bmm}{2am+2bm-bn}$.

If we suppose that the radius *OG*(*b*) of the immobile circle becomes infinite, then its circumference becomes a straight line. Thus, erasing the terms $2amm$ and $2am$, because they are null with respect to the others $2bmm$ and $2bm - bn$, we have $MC = \frac{2mm}{2m-n}$.

Corollary I. (§178) *Because the angle* $MGm = \frac{a+b}{b}GKg$, *and because the arcs of the different circles are to each other in the ratio composed of the radii and the angles that they measure, it follows that* $Gg : Mm :: KG \times GKg : MG \times \frac{a+b}{b}GKg$. *Consequently, we also have that* $KG \times Mm = \frac{a+b}{b}MG \times Gg$, *or (what amounts to the same thing) that* $KG \times Mm : MG \times Gg :: OK(a + b) : OG(b)$, *which is a constant ratio. From this we see that the magnitude of the portion AM of the half-roulette AMD depends on the sum of the MG × Gg in the arc GB. This is what Mr.* Pascal *has shown with respect to roulettes whose bases are straight lines.*

Mr. *Varignon*[8] came across the same property by a very different path from this.

---

[8]Pierre Varignon (1654–1722).

Corollary II. (§179) *When the describing point M* (see Fig. 9.6) *falls outside of [155] the circumference of the mobile circle, then necessarily one of the following three cases holds. If we draw the tangent MT, the point of contact G falls:*

1. *On the arc TB, as we have supposed in the figure when doing the calculation, and hence $MC\left(\frac{2amm+2bmm}{2am+2bm-bn}\right)$ is always greater than $MG(m)$.*
2. *On the point T of tangency, and we therefore have $MC\left(\frac{2amm+2bmm}{2am+2bm-bn}\right) = m$, because $IG(n)$ disappears.*
3. *On the arc TN, and hence the value of $GI(n)$ goes from positive to negative, and we have $MC = \frac{2amm+2bmm}{2am+2bm-bn}$, so that MC is less than $MG(m)$, and still positive.*

From this it is clear that in all of these cases, the value of the radius MC of the evolute is always positive.

Corollary III. (§180) *When the describing point M falls inside the circumference of the mobile circle* (see Fig. 9.7), *we still have $MC = \frac{2amm+2bmm}{2am+2bm-bn}$ and it may happen that bn is greater than $2am+2bm$, and thus that the value of the radius of MC of the evolute is negative. From this we see that when it ceases to be positive in order to become negative, as occurs (see §81) when the point M becomes an inflection point, it must therefore necessarily be the case that $bn = 2am + 2bm$ and consequently that $MI \times MG(mn - mm) = \frac{2amm+bmm}{b}$. Now, if we denote the given KM by c, then by the property of the circle we have $MI \times MG\left(\frac{2amm+bmm}{b}\right) = BM \times MN(aa - cc)$, which gives the unknown $MG(m) = \sqrt{\frac{aab-bcc}{2a+b}}$. Therefore, if we describe a circle with the given point M as center, and radius $MG = \sqrt{\frac{aab-bcc}{2a+b}}$, it cuts the mobile circle at the point G, where it touches the immobile circle that serves as its base, when the describing point M falls on the inflection point F.*

[156] If we draw MR perpendicular to BN, then it is clear that this $MG\left(\sqrt{\frac{aab-bcc}{2a+b}}\right)$ is less than $MR\left(\sqrt{aa-cc}\right)$ and that it must be equal to it if b becomes infinite, that is to say when the base of the roulette becomes a straight line.

It should be noted that, in order that the circle described with radius MG cuts the mobile circle, it is necessary that MG is greater than MN, that is to say that $\sqrt{\frac{aab-bcc}{2a+b}}$ is greater than $a-c$, and thus $KM(c)$ is greater than $\frac{aa}{a+b}$. From this it is clear that in order that there be an inflection point in the roulette AMD, it is necessary that KM be less than KN and greater than $\frac{aa}{a+b}$.

Lemma III. (§181) *Let ABb and CDd* (see Fig. 9.9) *be two triangles, each of which has one of their sides Bb and Dd infinitely small with respect to the others. I say that the triangle ABb is to the triangle CDd in a ratio composed of the of the angle BAb to the angle DCd, and of the square of the side AB or Ab to the square of the side CD or Cd.*

If we describe the circular arcs BE and DF with centers A and C, and radii AB and CD, then it is clear (see §2) that the triangles ABb and CDd do not differ from the sectors ABE and CDF. Therefore, etc.

If the sides AB and CD are equal, then the triangles ABb and CDd are to each other as their angles BAb and DCd.

Proposition V.

Problem. (§182) *The same things still being given, we wish to find the quadrature of the space MGBA* (see Fig. 9.6)*, enclosed by the perpendiculars MG and BA to the roulette, by the arc GB, and by the portion AM of the half-roulette AMD, assuming the quadrature of the circle.*

The angle $GMg\left(\frac{n}{2m}GKg\right)$ is to the angle $MGm\left(\frac{a+b}{b}GKg\right)$, [157] as (see §181) the little triangle MGg, which has as its base the arc Gg of the mobile circle, to the little triangle or sector GMm. Consequently, the sector

$$GMm = \frac{2m}{n}MGg \times \frac{a+b}{b} = \frac{2a+2b}{b}MGg + \frac{2ap+2bp}{bn}MGg,$$

denoting MI by p and substituting the value $p+n$ for m. Now (see §181), the little triangle or sector KGg is to the little triangle MGg, in a ratio composed of the square of KG to the square MG, and of the angle GKg to the angle GMg, that is to say :: $aa \times GKg : mm \times \frac{n}{2m}GKg$, and consequently the little triangle $MGg = \frac{mn}{2aa}KGg$. Thus, substituting this value in place of the triangle MGg in $\frac{2ap+2bp}{bn}MGg$, we have that the sector

$$GMm = \frac{2a+2b}{b}MGg + \frac{\overline{a+b} \times pm}{aab}KGg.$$

However, because of the circle, $GM \times MI(pm) = BM \times MN(cc-aa)$, which is a constant quantity, and always remains the same in whatever place we find the describing point M. Consequently, $GMm + MGg$ or mGg, that is to say the little space of the roulette

$$GMmg = \frac{2a+3b}{b}MGg + \frac{\overline{a+b} \times \overline{cc-aa}}{aab}KGg.$$

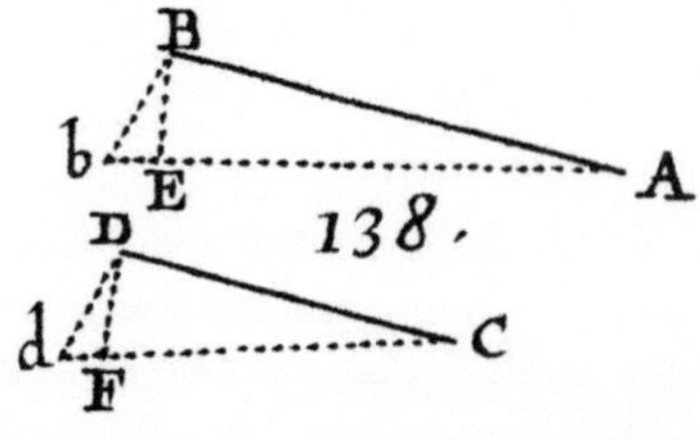

Fig. 9.9 Triangles with an Infinitely Small Side

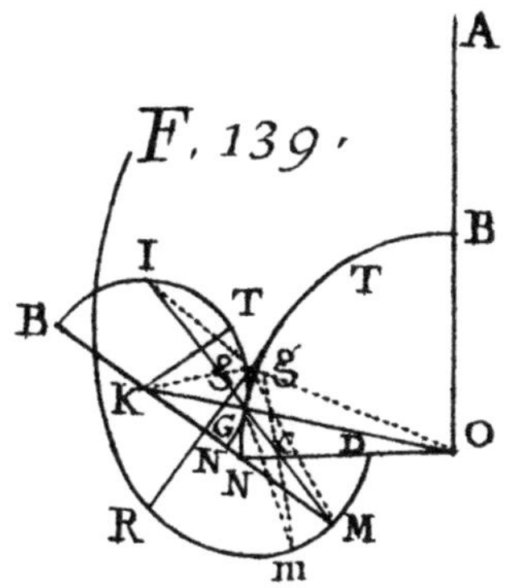

Fig. 9.10 Quadrature of the Roulette

Thus, because *GMmg* is the differential of the space of the roulette *MGBA* and *MGg* is the differential of the circular space *MGB*, enclosed by the straight lines *MG* and *MB*, and by the arc *GB*, and because also the little sector *KGg* is the differential of the sector *KGB*, it follows (see §96) that the space of the roulette

$$MGBA = \frac{2a+3b}{b}MGB + \frac{\overline{a+b} \times \overline{cc-aa}}{aab}KGB.$$

This is what we were required to find.

When the describing point *M* (see Fig. 9.10) falls outside the circumference of the mobile circle *BGN*, and when the point of contact *G* falls on the arc *NT*, then it is clear (see §180) that the perpendiculars *MG* and *mg* intersect each other at a point *C*, and that we therefore have $m = p - n$. From this it follows that from the little sector

$$\begin{aligned} GMm &= -\frac{2a-2b}{b}MGg + \frac{2ap+2bp}{bn}MGg \\ &= -\frac{2a-2b}{b}MGg + \frac{amp+bmp}{aab}KGg, \end{aligned}$$

substituting as before [158] the value $\frac{mn}{2aa}KGg$ for the little triangle *MGg*. Consequently, $GMm - MGg$ or *mGg*, that is to say

$$MCm - GCg = -\frac{2a-3b}{b}MGg + \frac{\overline{a+b} \times \overline{cc-aa}}{aab}KGg,$$

substituting the value $cc - aa$ for *pm*. Now, supposing that *TH* is the position of the tangent *TM* on the mobile circle, when the point *T* touches the base at the point *T*, it is clear that $MCm - GCg = MGTH - mgTH$, that is to say the differential of the space *MGTH*, and that *MGg* is the differential of *MGT*, and similarly that *KGg* is the differential of *KGT*. Therefore (see §96), the space

$$MGTH = -\frac{2a-3b}{b}MGT + \frac{\overline{a+b}\times\overline{cc-aa}}{aab}KGT.$$

However, as we have just shown the space

$$HTBA = \frac{2a+3b}{b}MTB + \frac{\overline{a+b}\times\overline{cc-aa}}{aab}KTB.$$

Consequently, in all cases, we will always have that the space

$$MGBA\ (MGTH + HTBA) = \frac{2a+3b}{b}\overline{MTB - MGT} \text{ or } MGB$$
$$+\frac{\overline{a+b}\times\overline{cc-aa}}{aab}\overline{KGT + KTB} \text{ or } KGB.$$

Therefore, the entire space *DNBA* (see Fig. 9.6) enclosed by the two perpendiculars *DN* and *BA* to the roulette, by the arc of the circle *BGN*, and by the half-roulette *AMD* is

$$= \overline{\frac{2a+3b}{b} + \frac{\overline{a+b}\times\overline{cc-aa}}{aab}} \times KNGB,$$

because the sector *KGB* and the circular space *MGB* each become the semi-circle *KNGB*, when the point of contact *G* falls on *N*.

When the describing point *M* falls inside the mobile circle (see Fig. 9.7), we must put $aa - cc$ in the place of $cc - aa$ in the preceding formulas, because then $BM \times MN = aa - cc$.

If we let $c = a$, then we will have the quadrature of roulettes that have their describing point on the circumference of the mobile circle and if we suppose b to be infinite, we will have the quadrature of roulettes that have straight lines as their bases.

Alternate Solution. (§183) We describe the arc *DV* (see Fig. 9.11) with radius *OD* and the semi-circles *AEV* and *BSN* with diameters *AV* and *BN*. We describe [159] at will the arc *EM* with center *O* contained between the semi-circle *AEV* and the half-roulette *AMD*, and we draw the ordinate *EP*. We wish to find the quadrature of the space *AEM* contained between the arcs *AE* and *EM* and the portion *AM* of the half-roulette *AMD*.

To do this, let *em* be another arc concentric with and infinitely close to *EM*, and let *ep* be another ordinate, and let *Oe* meet the arc *ME* prolonged (if necessary) to the point *F*. Denote the variables *OE* by z, *VP* by u, and the arc *AE* by x, and as before, the constants *OB* by b, *KB* or *KN* by a, and *KV* or *KA* by c. We have $Fe = dz$, $Pp = du$, $OP = a + b - c + u$, $\overline{PE}^2 = 2cu - uu$, and the arc (see §172)

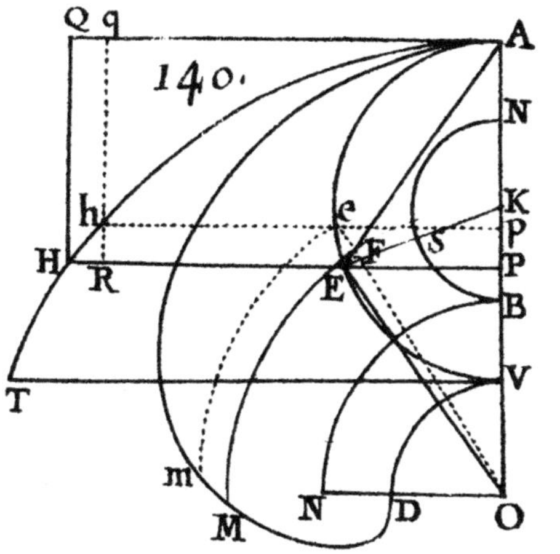

Fig. 9.11 Quadrature of the Roulette, Alternate Solution

$EM = \frac{axz}{bc}$. Consequently, the rectangle made on the arc *EM* by the little straight line *Fe*, that is to say (see §2) the little space $EMme = \frac{ax\,dz}{bc}$. Now, because *OPE* is a right triangle, $zz = aa + 2ab + bb - 2ac - 2bc + cc + 2au + 2bu$, the differential of which gives $z\,dz = a\,du + b\,du$. Therefore, substituting the value $\frac{axz\,dz}{bc}$ in place of $z\,dz$, we have the little space

$$EMme = \frac{aax\,du + abx\,du}{bc}.$$

If we now describe the half-roulette *AHT* by the revolution of the semi-circle *AEV* on the straight line *VT* perpendicular to *VA*, and if we prolong the ordinates *PE* and *pe* until they meet *AHT* in the points *H* and *h*, then it is clear (see §172) that $EH \times Pp$, that is to say that the little space $EHhe = x\,du$, and thus that

$$EMme\left(\frac{aax\,du + abx\,du}{bc}\right) : EHhe(x\,du) :: aa + ab : bc,$$

which is a constant ratio. Now, because this always happens no matter where the arc *EM* is found, it follows that the sum of all the little spaces *EMme*, that is to say the space *AEM*, is to the sum of all the little spaces *EHhe*, that is to say the space $AEH :: aa + ab : bc$. However (see §99), we have the quadrature space *AEH* depending on that of the circle, and consequently also the quadrature of the space *AEM*, which we wish to find.

This may also be shown without any calculus, as I have shown in the *Acts* of Leipzig for the month of August of the year 1695.

[160] We may also find the quadrature of the space *AEH* without recourse to §99. If we complete the rectangles *PQ* and *pq*, then we will have *Qq* or *HR* : *Pp* or *Rh* :: *EP* : *PA* or *HQ*, because (see §18) the tangent at *H* is parallel to the chord *AE*. Thus, $HQ \times Qq = EP \times Pp$, that is to say that the little spaces *HQqh* and *EPpe* are

always equal to each other. From this it follows that the space AHQ contained by the perpendiculars AQ and QH and by the portion AH of the half-roulette AHT is equal to the space APE contained by the perpendiculars AP and PE and by the arc AE. The space AEH is therefore equal to the rectangle PQ minus twice the circular space APE, that is to say the rectangle made on PE by KA plus or minus the rectangle made on KP by the arc AE, depending on whether the point P falls below or above the center. Consequently, the required space

$$AEM = \frac{aa + ab}{bc}\overline{PE \times KA \pm KP \times AE}.$$

Corollary I. (§184) *When the point P falls on K, the rectangle $KP \times AE$ vanishes, and the rectangle $PE \times KA$ becomes equal to the square on KA. From this we see that the space AEM is therefore* $= \frac{aac+abc}{b}$, *and consequently is absolutely quadrable and independent of the quadrature of the circle.*

Corollary II. (§185) *If we add the sector AKE to the space AEM, then the space $AKEM$ contained by the radii AK and KE, by the arc EM, and by the portion AM of the half-roulette AMD is found (when the point P falls above the center K)*

$$= \frac{bcc + 2aac + 2abc - 2aau - 2abu}{abc}AE + \frac{aa + ab}{bc}PE \times KA.$$

Consequently, if we take[9]

$$VP(u) = \frac{2aac + 2abc + bcc}{2aa + 2ab}$$

(which makes the value of

$$\frac{bcc + 2aac + 2abc - 2aau - 2abu}{2bc}AE$$

null), we have [161] the space $AKEM$ $= \frac{aa+ab}{bc}PE \times KA$. *From this we see that its quadrature is also independent of that of the circle.*

It is clear that among all the spaces AEM and $AKEM$, the only ones whose quadratures are absolute are the two that we have just considered.

Note. Everything that has just been shown with regard to exterior roulettes should also be understood for interior roulettes, that is to say of those whose mobile circles roll inside the immobile circle, observing that the radii $KB(a)$ and $KV(c)$ change from positive to negative. For this reason, we must change the signs of the terms in the preceding formulas, where a and c are found with an odd power.

[9] In L'Hôpital (1696), the equal sign after $VP(u)$ was omitted.

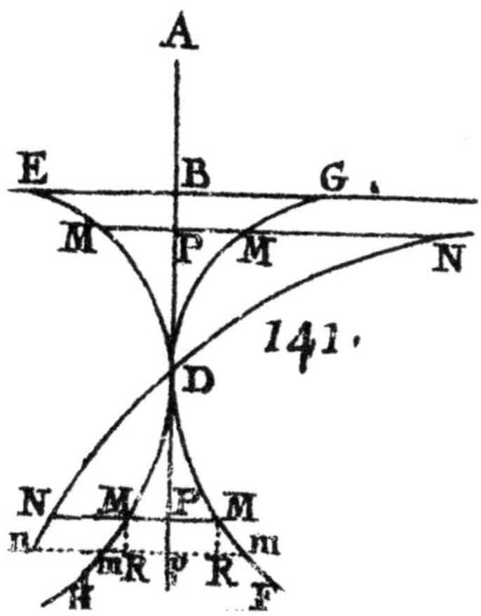

Fig. 9.12 Curve Described by $z = \frac{xx-aa}{\sqrt{2xx-aa}}$

Remark. (§186) There are certain curves that appear to have an inflection point, but nevertheless do not have one. I believe it is appropriate to explain this with an example, because this matter could cause some difficulty.

Let the curve *NDN* (see Fig. 9.12) be geometric, whose nature is expressed by the equation $z = \frac{xx-aa}{\sqrt{2xx-aa}}$ ($AP = x$ and $PN = z$), in which it is clear:

1. That when x is equal to a, $PN(z)$ vanishes.
2. That when x is greater than a the value of z is positive and on the contrary when x is less than a, z is negative.
3. When $x = \sqrt{\frac{1}{2}aa}$, the value of *PN* is infinite.

From this we see that the curve *NDN* passes from one side to the other of its axis in cutting it at a point D, such that $AD = a$ and that it has the perpendicular *BG* drawn from the point B such that $AB = \sqrt{\frac{1}{2}aa}$ as its asymptote.

We now describe another curve *EDF*, such that when we draw at will the perpendicular *MPN*, the rectangle made from the ordinate *PM* by the constant *AD* [162] is always equal to the corresponding space *DPN*. It is clear that if we denote *PM* by y and take differentials, we have $AD \times Rm\,(a\,dy) = NPpn$ or $NP \times Pp\left(\dfrac{xx\,dx - aa\,dx}{\sqrt{2xx-aa}}\right)$ and consequently $Rm\,(dy) : Pp$ or $RM\,(dx) :: PN : AD$. From this it follows that the curve *EDF* touches the asymptote *BG* prolonged on the other side of B at a point E and touches the axis *AP* at the point D, and thus it ought to have an inflection point at D. Nevertheless, we find (see §78) $-\frac{x^3}{2aa}$ for the value of the radius of its evolute, which is always negative and becomes equal to $-\frac{1}{2}a$ when the point M falls on D. From this we ought to conclude (see §81) that the curve that passes through all the points M is always convex towards the axis *AP*

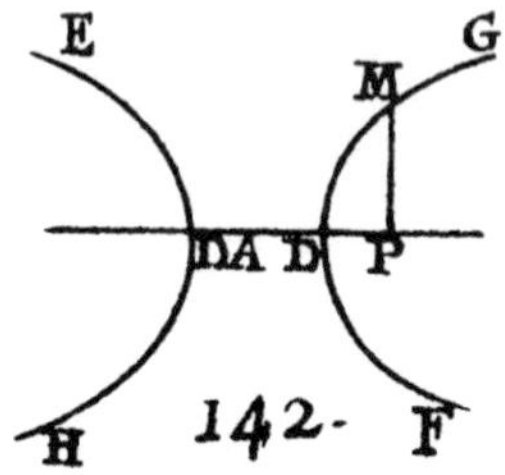

Fig. 9.13 Curve Described by $y^4 = x^4 + aaxx - b^4$

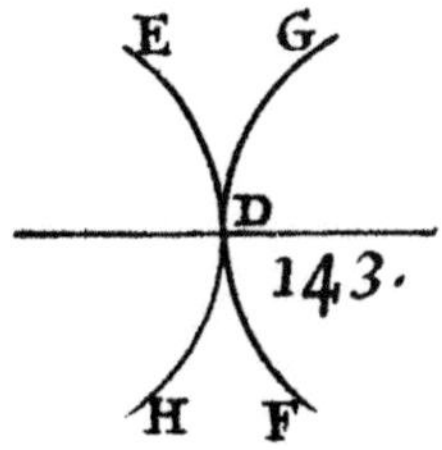

Fig. 9.14 Curve Described by $y^4 = x^4 + aaxx$

and that it does not have an inflection point at D. Then how are we to reconcile all of this? Here is the dénouëment.

If we take *PM* on the same side as *PN*, then we form another curve *GDH*, which is exactly the same as *EDF* and which should be a part of it because its construction is the same. That being so, we should consider the parts that compose the whole curve are not *EDF* and *GDH* as we had imagined, but rather *EDH* and *GDF*, which touch each other at the point D, because everything reconciles perfectly under this last assumption. This is further confirmed by the following example.

Let *DMG* be the curve (see Fig. 9.13), which has as its equation $y^4 = x^4 + aaxx - b^4$ ($AP = x$ and $PM = y$). It follows from this equation that the entire curve has two parts *EDH* and *GDF*, opposed to one other as in the usual hyperbola, so that their distance *DD* or $2AD = \sqrt{-2aa + 2\sqrt{a^4 + 4b^4}}$.

If we suppose that b vanishes, the distance *DD* (see Fig. 9.14) also vanishes, and consequently the two parts *EDH* and *GDF* touch each other at the point D, so that we might now think at first that this curve has an inflection point or cusp at D, depending on whether we imagine that its parts [163] are *EDF* and *GHD* or *EDG* and *HDF*. However, we easily disabuse ourselves of this by finding the radius of the evolute, because we would find that it is always positive and that it becomes equal to $\frac{1}{2}a$ at the point D.

We may remark in passing that the quadrature space *DPN* (see Fig. 9.12) depends on that of the hyperbola or (what amounts to the same thing) on the rectification of the parabola.

Chapter 10
A New Method for Using the Differential Calculus with Geometric Curves, from Which We Deduce the Method of Messrs. Descartes and Hudde

Definition I. [164] Let *ADB* (see Figs. 10.1, 10.2, 10.3) be a curved line, such that the parallels *KMN* to its diameter *AB* meet it in two points *M* and *N*, and let it be understood that the intersected part *MN* or *PQ* becomes infinitely small. It is therefore called the *Differential* of the abscissa *AP* or *KM*.

Corollary I. (§187) *When the part MN or PQ becomes infinitely small, it is clear that the abscissas AP and AQ both become equal to AE, and that the points M and N meet at the point D, so that the ordinate ED is the greatest or least of all the similar ordinates PM and NQ.*

Corollary II. (§188) *It is clear that among all the abscissas AP it is only AE that has a differential, because it is only in this case that PQ becomes infinitely small.*

Corollary III. (§189) *If we denote the indeterminates AP or KM by* x*, and PM or AK by* y*, it is clear that if AK*(y) *remains the same, then it must have two different values of* x*, namely KM and KN, or AP and AQ. This is why the equation that expresses the nature of the curve ADB must be cleared of incommensurables, so that the same unknown* x *that denotes the roots (because we consider* y *as known) may have different values. This is what we must observe in the following.*

Proposition I.

Problem. [165] (§190) *If the nature of the geometric curve ADB is given, we wish to determine the greatest or least of its ordinates ED.*

If we take the differential of the equation that expresses the nature of the curve, treating y as constant, and x as variable, then it is clear (see §188) that we form a new equation that will have a value *AE* for one of its roots x, such that the ordinate *ED* is the greatest or the least of all the similar ordinates.

R.E. Bradley et al., *L'Hopital's Analyse des infiniments petits*, Science Networks. Historical Studies 50, DOI 10.1007/978-3-319-17115-9_10

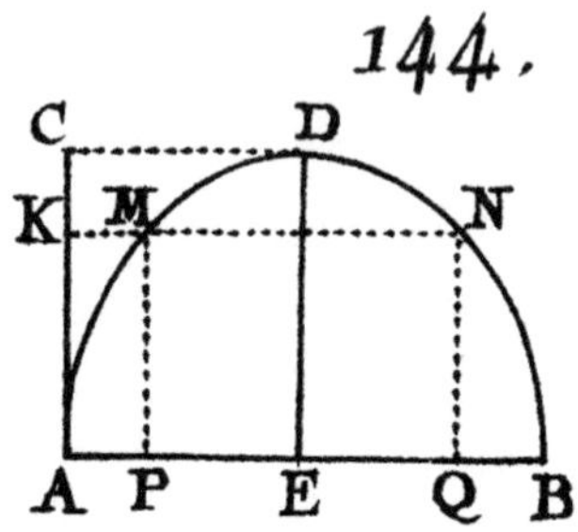

Fig. 10.1 Greatest Ordinate – Perpendicular Case

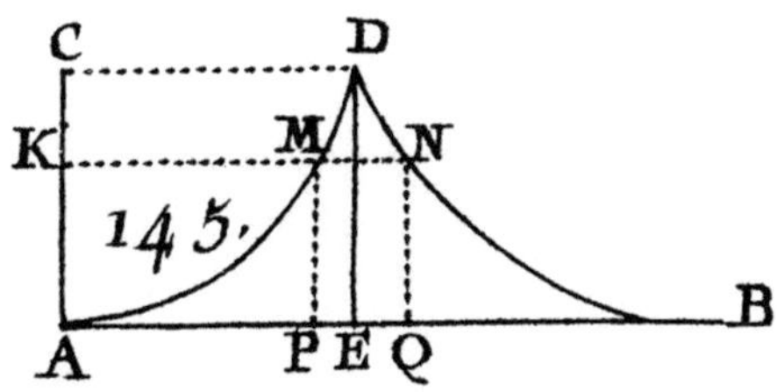

Fig. 10.2 Greatest Ordinate – Tangential Case

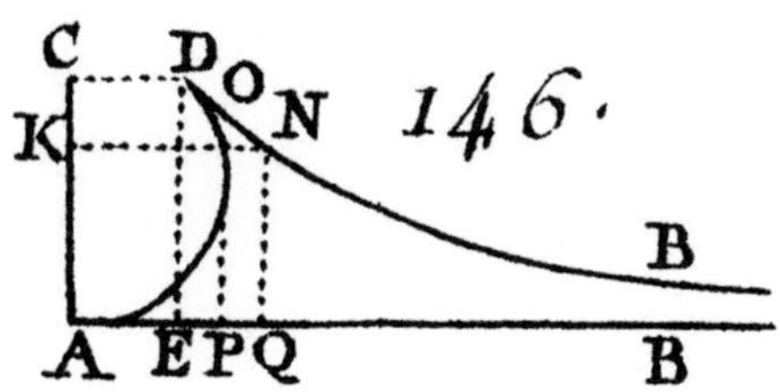

Fig. 10.3 Greatest Ordinate – Oblique Cusp

For example, let $x^3 + y^3 = axy$, whose differential, treating x as variable and y as constant, gives $3xx\,dx = ay\,dx$, and consequently $y = \frac{3xx}{a}$. If we substitute this value in place of y in the equation of the curve $x^3 + y^3 = axy$, then we have a value of $AE = \frac{1}{3}a\sqrt[3]{2}$ for x, such that the ordinate ED is the greatest of all the similar ordinates, as we have already found in §48.

It is clear that in a similar way we determine not only the points D, where the ordinates ED are perpendicular or tangent to the curve ADB, but also when it is oblique to the curve, that is to say when the points D are cusps of the first or second kind. From this we see that this new method of considering differentials in geometric curves is simpler and less troubling in many instances than the first method (see Ch. 3).

Remark. (§191) We may remark that in curves with cusps (see Fig. 10.3), the ordinates PM parallel to AK meet them in two points M and O, in the same way that the parallel lines KM to AP meet them M and N, so that if AP (x) remains the same, then y has two [166] different values PM and PO. This is why we may treat x as constant and y as variable, in taking the differential of the equation that expresses the nature of this curve. From this we see that if we treat x and y as variables, and take this differential, it must be the case that all the terms multiplied by dx, on the one hand, and all those multiplied by dy, on the other hand, are equal to zero. However, we must take great care that dx and dy here denote the differentials of two ordinates emanating from the same point, and not (as above in Chapter 3) the differential of two infinitely close ordinates.[1]

Corollary. (§192) *If, after ordering the equation that expresses the nature of the curve, in which only the unknown x is considered variable, we take the differential, then it is clear that:*

1. We do nothing other than to multiply each term by the exponent of the power of x and by the differential dx, and then divide it by x.
2. This division by x, as well as the multiplication by dx, can be neglected, because it is the same for all the terms.
3. The exponents of the powers of x make an arithmetic progression, the first term of which is the exponent of highest power, and the last of which is zero, because we assume that we have marked with an asterisk the terms that may be missing from the equation.[2]

For example, let $x^3 * -ayx + y^3 = 0$. If we multiply each term by those of the arithmetic progression 3, 2, 1, 0, then we form the new equation $3x^3 - ayx = 0$.

$$\begin{array}{cccc} x^3 & * & -ayx & +y^3 = 0. \\ 3, & 2, & 1, & 0. \\ \hline 3x^3 & * & -ayx & * = 0. \end{array}$$

From this we conclude $y = \frac{3xx}{a}$, the same as we would have found by taking the differential in the accustomed manner.

Given this, I say that instead of the arithmetic progression [167] 3, 2, 1, 0, we may use whatever other arithmetic progression that we may wish: $m+3, m+2, m+1, m+0$ or m (we denote by m any number, whole or fractional, positive or negative). Because when we multiply $x^3 * -ayx + y^3 = 0$ by x^m, we have $x^{m+3}*$, etc. $= 0$, whose terms must be multiplied by the progression $m+3, m+2, m+1, m$, each one by its corresponding term to find the differential.

[1]See the discussion in Letters 25 and 26 on p. 258 and p. 263.

[2]Descartes used an asterisk to denote the absence of a term in a complete polynomial, see, e.g., Descartes (1954, pp. 162ff).

$$\begin{array}{llll} x^{m+3} & * & -ayx^{m+1} & +y^3x^m = 0. \\ m+3,\ m+2, & & m+1, & m. \\ \hline \overline{m+3}x^{m+3} & * & -\overline{m+1}ayx^{m+1} & +my^3x^m = 0. \end{array}$$

This gives $\overline{m+3}\,x^{m+3} - \overline{m+1}\,ayx^{m+1} + my^3x^m = 0$, and dividing by x^m, it becomes $\overline{m+3}\,x^3 - \overline{m+1}\,ayx + my^3 = 0$, as we would have found right away by simply multiplying the given equation by the progression $m+3, m+2, m+1, m$.

If $m = -3$, the progression is $0, -1, -2, -3$, and the equation would be $2ayx - 3y^3 = 0$. If $m = -1$, the progression is $2, 1, 0, -1$, and the equation would be $2x^3 - y^3 = 0$.

We may change the signs of all the terms of the progression, that is to say instead of $0, -1, -2, -3$, and $2, 1, 0, -1$, we may take $0, 1, 2, 3$, and $-2, -1, 0, 1$, because this does nothing but to change the signs of all the terms in the new equation, which must be equal to zero. Indeed, instead of $2ayx - 3y^3 = 0$ and $2x^3 - y^3 = 0$, we would have $-2ayx + 3y^3 = 0$ and $-2x^3 + y^3 = 0$, which is the same thing.

Now, it is clear that what we have just shown with regard to this example can be applied in the same way to all others. From this it follows that if, after ordering an equation that must have two roots equal to each other, we multiply the terms by those of an arbitrary arithmetic progression, we form a new equation which contains among its roots one of the two equal roots of the first equation. By the same reasoning, if this new equation must still have two equal roots, and if we multiply it by an arithmetic progression [168], we form a third equation from it, which has among its roots one of the two equal roots of the second equation, and so on. In this way, if we multiply an equation which must have three equal roots by the product of two arithmetic progressions, then we form a new equation from it, which has among its roots one of the three equal roots from the first equation. Similarly, if the equation must have four equal roots, we must multiply it by the product of three arithmetic progressions. If five, the product of four, etc.

This is precisely what constitutes the Method of Mr. *Hudde*.[3]

Proposition II.

Problem. (§193) *Let the given point T be on the diameter AB* (see Fig. 10.4)*, or the given point H be on AH parallel to the ordinates. We wish to draw the tangent THM.*

Draw the ordinate MP from the point of contact M, and denote AT by s, and AH by t, of which one or the other is given, and the unknowns AP by x, and PM by y. The similar triangles TAH and TPM give $y = \frac{st+tx}{s}$ and $x = \frac{sy-st}{t}$. Substituting these values in place of y or x in the given equation, which expresses the nature of the curve AMD, we form a new equation in which y or x is no longer present.

[3]For a recent exposition on the work of Hudde, see Suzuki (2005).

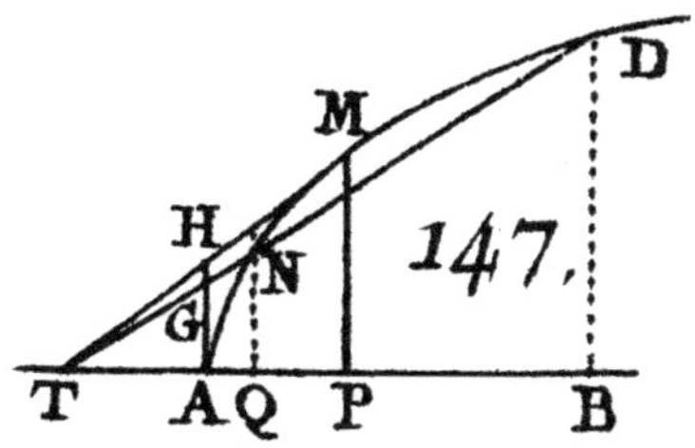

Fig. 10.4 Finding a Tangent Through a Given Point

If we now draw a straight line *TD* that cuts the straight line *AH* at G, and the curve *AMD* at two points N and D, from which we drop the ordinates *NQ* and *DB*, then it is clear that if t expresses *AG* in the preceding equation, x or y will have the two values *AQ* and *AB*, or *NQ* and *DB*. These become equal to each other, namely to the *AP* or *PM* that we wish to find, when t expresses *AH*, that is to say when the secant *TDN* becomes the tangent *TM*. From this it follows that this equation must have two equal roots. This is why we multiply it by an arbitrary arithmetic progression, which is repeated, if it [169] is necessary, by multiplying this very equation again by any other arithmetic progression, so that through a comparison of the equations that result, we may find one that contains only the unknown x or y, with the given s or t. The example that follows gives sufficient clarification of this Method.

Example.[4] (§194) Let $ax = yy$ be the equation that expresses the nature of the curve *AMD*. If we substitute the value $\frac{sy-st}{t}$ in the place of x, we have tyy, etc., which must have two equal roots.

$$\begin{array}{lll} tyy & -asy & +ast = 0. \\ 1, & 0, & -1. \\ \hline tyy & * & -ast = 0. \end{array}$$

This is why if we multiply these terms in order by those of the arithmetic progression $1, 0, -1$, we find that $as = yy = ax$, and consequently $AP\ (x) = s$. From this we see by taking $AP = AT$, and drawing the ordinate *PM*, the line *TM* is tangent at M. However, if $AH\ (t)$ is given instead of $AT\ (s)$, then we multiply the same equation tyy, etc., by this other progression 0, 1, 2, and we have the desired $PM\ (y) = 2t$.

We would have found the same construction by substituting the value $\frac{st+tx}{s}$ for y in $ax = yy$. Because it becomes $ttxx$, etc., whose terms, when multiplied by 1, 0, -1, give $xx = ss$, and consequently $AP\ (x) = s$.

Corollary. (§195) *If we now wish that the point of contact M be given, and that we wish to find the point T or H, at which the tangent MT meets the diameter AB or the parallel AH to the ordinates, we need only look at the previous equation, which*

[4]In chapter 10 of L'Hôpital (1696) examples were not given numbers.

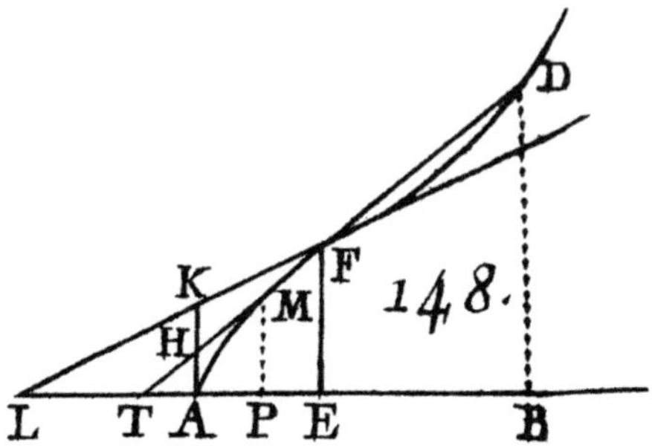

Fig. 10.5 Using the Method of Hudde to Find an Inflection Point

expresses the value of the unknown x or y with respect to the given s or t, and consider the latter as the unknown, and x or y as known.

Proposition III.

Problem. [170] (§196) *Let the nature of the geometric curve AFD* (see Fig. 10.5) *be given. We wish to determine its inflection point F.*

We draw the ordinate *FE* with the tangent *FL* from the point *F* that we wish to find, and *AK* parallel to the ordinates from the point *A* (the origin of the x's). We denote the unknowns *LA* by s, *AK* by t, *AE* by x, and *EF* by y. The similar triangles *LAK* and *LEF* once again give $y = \frac{st+tx}{s}$ and $x = \frac{sy-st}{t}$, so that if we substitute these values in place of y or x in the equation of the curve, we form a new equation in which y or x is no longer present, as in the previous proposition.

If we now draw a straight line *TD* that cuts the straight line *AK* at H, which touches the curve *AFD* at M, and cuts it at D, from which we drop the ordinates *MP* and *DB*, it is clear that:

1. If s expresses *AT* and t expresses *AH*, the equation that we have just found must have two equal roots, that is to say, namely (see §193) one at either *AP* or *PM*, depending on whether we make y or x vanish, and the other at *AB* or *BD*.
2. If s expresses *AL* and t expresses *AK*, the point of contact M coincides with the point of intersection D at the point F that we wish to find, because (see §67) the tangent *LF* must touch and cut the curve at the inflection point F, and thus the values of x, *AP* and *AB*, or the values of y, *PM* and *BD*, become equal to each other, namely one or the other of the values *AE* or *EF* that we wish to find.

From this it follows that this equation must have three equal roots. That is why we multiply it by the product of two arbitrary arithmetic progressions, which we repeat, if necessary, in multiplying it in the same way by the product of any two arithmetic progressions, so that by comparing the equations that result, we can make the unknowns s and t vanish.

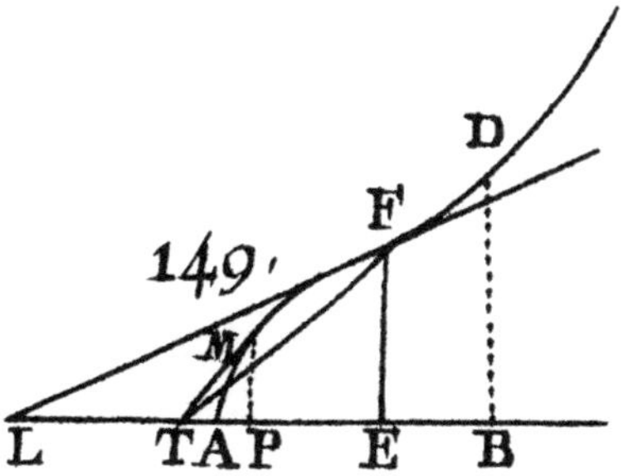

Fig. 10.6 Inflection Point as a Meeting of Two Tangents – First Case

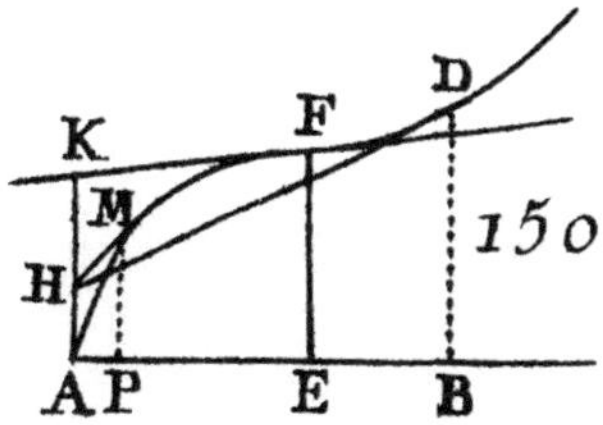

Fig. 10.7 Inflection Point as a Meeting of Two Tangents – Second Case

Example. [171] (§197) Let the equation $ayy = xyy + aax$ express the nature of the curve AFD. If we substitute the value $\frac{sy-st}{t}$ for x, we form the equation $sy^3 - styy - atyy$, etc.,[5]

$$\begin{array}{rrrr}
sy^3 & -styy & +aasy & -aast = 0. \\
 & -at & & \\
1, & 0, & -1, & -2. \\
3, & 2, & 1, & 0. \\
\hline
3sy^3 & * & -aasy & * = 0.
\end{array}$$

which when multiplied by 3, 0, −1, 0, the product of the two arithmetic progressions 1, 0, −1, −2, and 3, 2, 1, 0, gives $yy = \frac{1}{3}aa$. Substituting this value into the equation of the curve, we find the unknown $AE(x) = \frac{1}{4}a$. This is what we found in §68.

Alternate Solution. (§198) We may also solve this problem (see Figs. 10.6, 10.7) by noting that from the same point L or K we may only draw a single tangent LF or KF, because it touches outside the concave part AF, and inside the convex part FD, unlike any other point T or H, taken on AL or AK between A and L, or A and

[5]In the calculation that follows, the term on the second row represents $-atyy$, a second term of the second order, even though the yy is suppressed and only the coefficient $-at$ is written. Similar conventions are used in the remainder of this chapter.

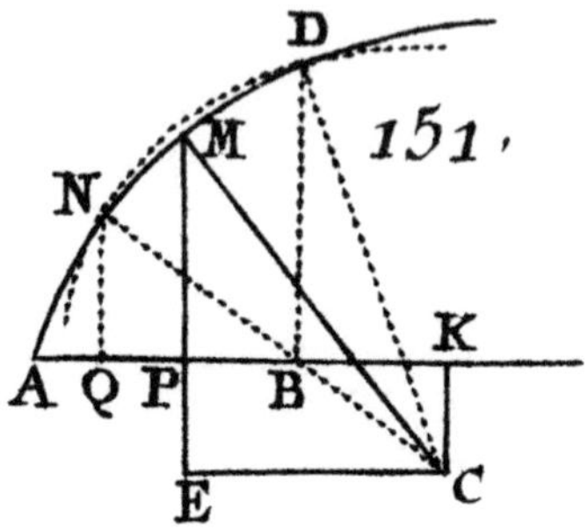

Fig. 10.8 Finding the Normal to a Curve Using Descartes' Construction

K, from which we can draw two tangents TM and TD,[6] or HM and HD,[7] one from the concave part, and the other from the convex part. In this way we may consider the inflection point F as the meeting of the two points of contact M and D. If we therefore suppose that $AT(s)$ or $AH(t)$ is given, and we wish to find (see §194) the value of x or y with respect to s or t, we would have an equation that has two roots AP and AB, or PM and BD, which each become equal to the AE or EF that we wish find, when s expresses AL and t expresses AK. This is why we multiply this equation by an arbitrary arithmetic progression, etc.

Example. [172] (§199) As above, let $ayy = xyy + aax$. We still have $sy^3 - styy - atyy + aasy - aast = 0$, which when multiplied by the arithmetic progression 1, 0, −1, −2, gives $y^3 * -aay - 2aat = 0$, in which s is no longer present, and which has two unequal roots, namely PM and BD, when t expresses AH, and two equal roots, each equal to the EF that we wish to find, when t expresses AK. This is why when we multiply this latter equation again by the arithmetic progression 3, 2, 1, 0, we have $3yy - aa = 0$, and consequently $EF\ (y) = \sqrt{\frac{1}{3}aa}$. This is what we were required to find.

Proposition IV.

Problem. (§200) *From a point C not on a curved line AMD* (see Fig. 10.8), *we wish to draw the perpendicular CM to this curve.*[8]

We draw the perpendiculars MP and CK on the diameter AB, and describe a circle with center C and interval CM. It is clear that it touches the curve AMD at the point M. Next, we denote the unknowns AP by x, PM by y, and CM by r, and the known quantities AK by s and KC by t. We have PK or $CE = s - x$, $ME = y + t$ and, because of the right triangle MEC, $y = -t + \sqrt{rr - ss + 2sx - xx}$,

[6] As in Fig. 10.6.

[7] As in Fig. 10.7.

[8] See Descartes (1954, pp. 94ff) for this construction when the point C lies on the axis.

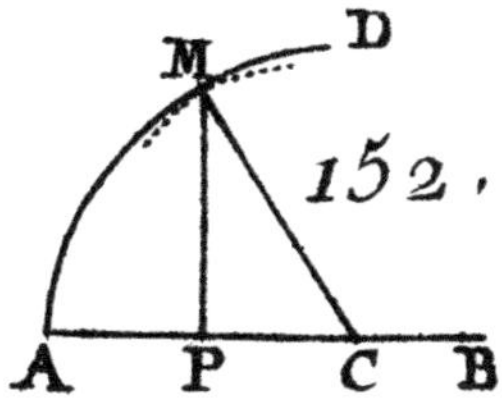

Fig. 10.9 Normal to the Parabola from a Point on the Axis

and $x = s - \sqrt{rr - tt - 2ty - yy}$, so that when we substitute these values in place of y or x in the equation of the curve, we form a new equation where y or x is no longer present.

If we now describe another circle from the same center C that cuts the curve at two points N and D, from which we drop the perpendiculars NQ and DB, it is clear that if r expresses the radius CN or CD in the preceding equation, then x or y has the two values AQ and AB, or NQ and DB, which become equal to each other, namely to the AP or PM that we wish to find, when r expresses the radius CM. From this it follows that this equation must have two equal roots. That is why we multiply it, etc.

Example. [173] (§201) Let the equation $ax = yy$ express the nature of the curve AMD, in which when we substitute the value $s - \sqrt{rr - tt - 2ty - yy}$ for x, we have $as - yy = a\sqrt{rr - tt - 2ty - yy}$, so that when we square both sides, and then order the equation, we have y^4, etc., which must have two equal roots when y expresses the PM that we wish to find.

$$\begin{array}{ccccc}
y^4 & * \ -2asyy & +2aaty & +aass & = 0. \\
 & +aa & & -aarr & \\
 & & & +aatt & \\
4,\ 3, & 2, & 1, & 0. & \\
\hline
4y^4 & * \ -4asyy & +2aaty & * & = 0. \\
 & +2aa & & &
\end{array}$$

This is why we multiply it by the arithmetic progression 4, 3, 2, 1, 0, which gives $4y^3 - 4asy + 2aay + 2aat = 0$, whose solution provides for y the value PM that we wish to find.

If the given point C falls on the diameter AB (see Fig. 10.9), we therefore have $t = 0$, and consequently it is necessary to erase all the terms in which t is found. This gives $4as - 2aa = 4yy = 4ax$, where we substitute the value ax for yy. From this we conclude $x = s - \frac{1}{2}a$, that is to say that if we take CP equal to half of the parameter, and if we draw the ordinate PM perpendicular to AB, the straight line CM is drawn perpendicular to the curve AMD.

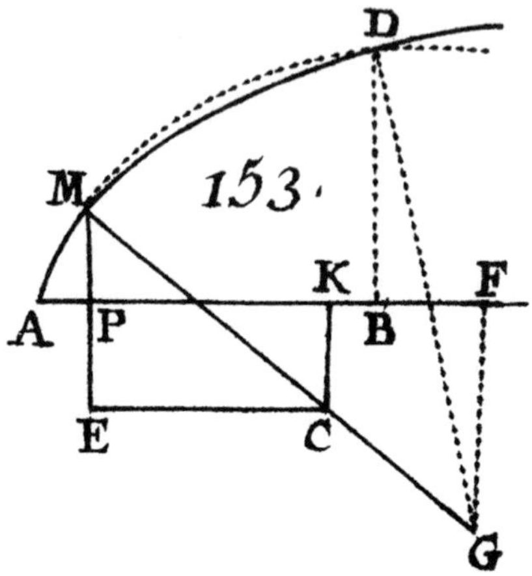

Fig. 10.10 The Osculating Circle

Corollary. (§202) *If we now want the point* M *to be given* (see Fig. 10.9), *and the point* C *is the one we wish to find, then in the previous equation that expresses the value of AC* (s) *with respect to AP* (x) *or PM* (y), *we must consider the latter as known quantities, and former as the unknown.*

Definition II. [174] If we describe a circle on any radius of the evolute, it is called the *osculating circle*.[9]

The point where this circle touches or kisses the curve is called the *osculating point*.

Proposition V.

Problem. (§203)[10] *Given the nature of the curve AMD* (see Fig. 10.10) *and one of its points* M, *we wish to find the center* C *of the circle that kisses it at this point* M.

We draw the perpendiculars *MP* and *CK* on the axis, and denote the lines by the same letters as in the previous Problem. We arrive at the same equation, in which it must be noted that the letter x or y, which we regarded there as the unknown, here denotes a given magnitude. On the contrary, s and t, which we regarded there as known, are in fact here the unknowns, as well as r.

Given this, it is clear that:

1. The point C that we wish to find is situated on the perpendicular *MG* to the curve.
2. We may always describe a circle that touches the curve at M, and that cuts it in at least two points (of which I assume that the closest is D, from which we

[9]In L'Hôpital (1696), the French term *baisant*, literally "kissing," is used for both the circle and the point. We have translated it as "osculating" because this Latin term is the standard one in English.

[10]Descartes (1954, pp. 100ff).

drop the perpendicular *DB*), because we can always find a circle that cuts any curved line, other than a circle, in at least four points, and the point of contact M is equivalent to only two intersections.[11]

3. The more the center G of this circle approaches the point C that we wish to find, the more the point of intersection D approaches the point of contact M, so that when the point G falls on the point C, the point D meets the point M, because (see §76) the circle described with radius *CM* must touch and cut the curve at the same point M.

From this we see that if s expresses *AF* and t expresses *FG*, the equation must have two equal roots, namely (see §200) one at *AP* or *PM*, depending on whether [175] we made y or x vanish, and another at *AB* or *BD*, which also becomes equal to *AP* or *PM* when s and t express the *AK* and *KC* that we wish to find. Thus, this equation must have three equal roots.

Example. (§204) Let the equation $ax = yy$ express the nature of the curve *AMD*, and we find (see §201) y^4, etc., which being multiplied by 8, 3, 0, −1, 0, the product of two arithmetic progressions 4, 3, 2, 1, 0 and 2, 1, 0, −1, −2, gives $8y^4 = 2aaty$.

$$\begin{array}{ccccc} y^4 & * & -2asyy & +2aaty & +aass = 0. \\ & & +aa & & -aarr \\ & & & & +aatt \\ 4, & 3, & 2, & 1, & 0. \\ 2, & 1, & 0, & -1, & -2. \\ \hline 8y^4 & * & * & -2aaty & * = 0. \end{array}$$

From this we conclude that the desired *KC* or *PE* $(t) = \frac{4y^3}{aa}$.

If we want to have the equation that expresses the nature of the curve that passes through all the points C, we again multiply y^4, etc., by 0, 3, 4, 3, 0, the product of two progressions 4, 3, 2, 1, 0, and 0, 1, 2, 3, 4, and we find that $8asy - 4aay = 6aat$. Hence, letting $s - \frac{1}{2}a = u$ to abbreviate, we conclude that $y = \frac{3at}{4u}$ and $4y^3 = \frac{27a^3t^3}{16u^3} = aat$, and consequently $16u^3 = 27att$. From this it follows that the curve that passes through all the points C is a semi-cubical parabola, of which the parameter is $\frac{27a}{16}$, and whose vertex is at a distance of $\frac{1}{2}a$ from the vertex of the given parabola, because $u = s - \frac{1}{2}a$.

When the position of the parts of the curve, neighboring the given point M, is entirely similar on both sides of this point, as happens when the curvature is greatest or least, it follows that one of the intersections of the touching circle cannot join with the point of contact, except when the other one joins it [176] at the

[11] The result known as Bézout's Theorem implies that a circle (of degree 2) and a curve of degree $n \geq 2$, generically intersect in $2n$ points. Although Etienne Bézout (1730–1783) lived much later, the result was widely accepted in the late seventeenth century.

same time, so that the equation must therefore have four equal roots. Indeed, if we multiply y^4, etc., by 24, 6, 0, 0, 0, the product of the three arithmetic progressions $4, 3, 2, 1, 0$, and $3, 2, 1, 0, -1$, and $2, 1, 0, -1, -2$, we have $24y^4 = 0$. This shows us that the point M must fall on the vertex A of the parabola, in order that the position of the neighboring parts of the curve be similar on both sides.

Alternate Solution. (§205) We can also solve this Problem (see Fig. 10.11) by remembering that we have shown in §76 that we can only draw one perpendicular CM from the point C that we wish to find to the curve AMD, whereas there is an infinity of other points G on this perpendicular MC, from which we may draw two perpendiculars MG and GD to the curve. Therefore, if we assume that the point G is given, and that we wish to find (see §200) the value of x or y in terms of the given quantities s and t, then it is clear that this equation must have two unequal roots, namely AP and AB, or PM and BD, which become equal to each other when the point G falls on the point C that we wish to find. This is why we multiply the equation by any arithmetic progression, etc.

Example. (§206) Let $ax = yy$ as above, and we have (see §201) $4y^3$, etc.

$$\begin{array}{cccc} 4y^3 & * & \begin{array}{c}-4asy\\+2aa\end{array} & +2aat = 0. \\ 2, & 1, & 0, & -1. \\ \hline 8y^3 & * & * & -2aat = 0. \end{array}$$

When multiplied by the arithmetic progression $2, 1, 0, -1$, this gives $t = \frac{4y^3}{aa}$, as before (see §204).

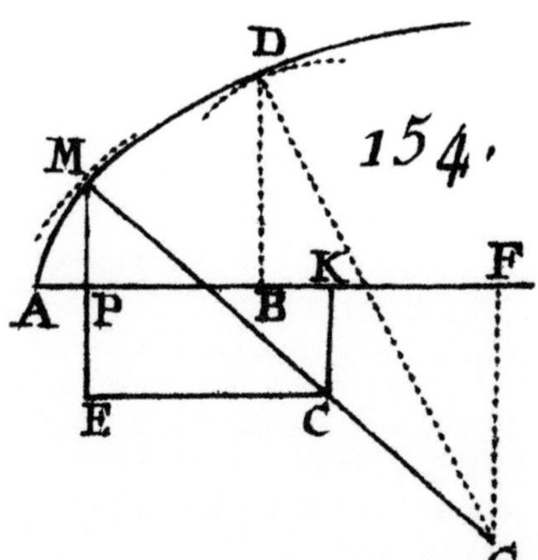

Fig. 10.11 Radius of the Osculating Circle from a Property of the Evolute

Corollary. [177] (§207) *It is clear that we may consider the osculating point* (see Figs. 10.10, 10.11) *as (see §203) the uniting of a point of contact with a point of intersection of the same circle, or else as (see §205) the uniting of two points of contact of two different and concentric circles. Similarly, the inflection point may be regarded as (see §196) the uniting of a point of contact with a point of intersection on the same straight line, or (see §198) as the uniting of two points of contact on two different straight lines that emanate from the same point.*

Proposition VI.

Problem. (§208) *We wish to find an equation that expresses the nature of the caustic AFGK* (see Fig. 10.12), *formed by the quarter circle CAMNB, by reflected rays MH, NL, etc., whose incident rays PM, QN, etc., are parallel to CB.*

I remark that:

1. If we extend the reflected rays *MF* and *NG*, which touch the caustic at F and G, until they meet the radius *CB* at the points H and L, then we will have *MH* equal to *CH*, and *NL* equal to *CL*. This is because the angle $CMH = CMP = MCH$, and similarly the angle $CNL = CNQ = NCL$.
2. From a given point F on the caustic *AFK*, we can only draw one straight line *MH* that is equal to *CH*, whereas from a given point D between the quarter circle *AMB* and the caustic *AFK*, we may draw two lines *MH* and *NL*, such that $MH = CH$ and $NL = CL$. This is because we may draw only a single tangent *MH* at the point F, whereas from the point D, we may draw two tangents *ML* and *NL*.

Given all this, we propose to draw the straight line *MH* through the given point D, such that it is equal to the part *CH*, which it determines on the radius *CB*.

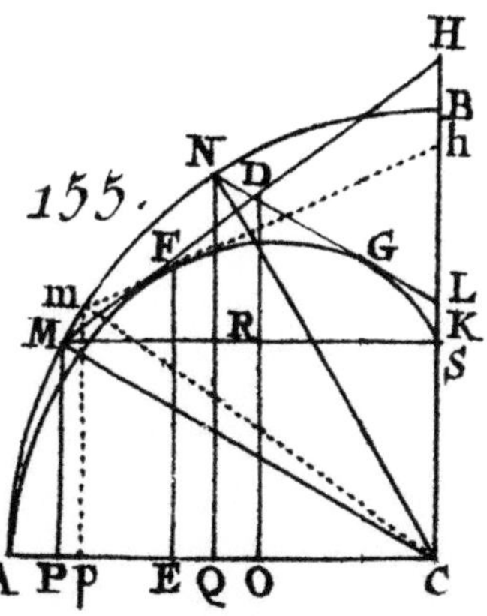

Fig. 10.12 Caustic by Reflection – Parallel Incident Rays

We draw *MP* and *DO* parallel to *CB*, and *MS* parallel to *CA*. We denote the given quantities *CO* or *RS* by u, *OD* by z, and *AC* [178] or *CB* by a, and the unknowns *CP* or *MS* by x, *PM* or *CS* by y, and *CH* or *MH* by r. The right triangle *MSH* gives $rr = rr - 2ry + yy + xx$, from which we conclude $CH(r) = \frac{xx+yy}{2y}$. Furthermore, the similar triangles *MRD* and *MSH* give $MR(x-u) : MS(x) :: RD(z-y) : SH = \frac{zx-xy}{x-u}$, and consequently $CS + SH$ or $CH = \frac{zx-uy}{x-u} = \frac{xx+yy}{2y} = \frac{aa}{2y}$, substituting the value aa for $xx + yy$. From this we form (by cross multiplication) the equation $aax - aau = 2zxy - 2uyy$, and substituting the value $aa - xx$ for yy, it becomes $2zxy = aax + aau - 2uxx$. We then square both sides to remove the incommensurables, and again substitute the value $aa - xx$ for yy, and we finally have:

$$\begin{array}{lllll} 4uux^4 & -4aaux^3 & -4aauuxx & +2a^4ux & +a^4uu = 0. \\ 4zz & & -4aazz & & \\ & & +a^4 & & \end{array}$$

Now, it is clear that if u expresses *CO* and z expresses *OD*, this equation must have two unequal roots, namely *CP* and *CQ*, and on the contrary if u expresses *CE* and z expresses *EF*, *CQ* becomes equal to *CP*, so that this time it has two equal roots. This is why if we multiply the terms by those of two arithmetic progressions 4, 3, 2, 1, 0, and 0, 1, 2, 3, 4, we form two new equalities by means of which we find, after having eliminated the unknown x, this equation

$$\begin{array}{lllll} 64z^6 & -48aaz^4 & +12a^4zz & -a^6 & = 0, \\ & +192uu & -96aauu & -15a^4uu & \\ & & +192u^4 & -48aau^4 & \\ & & & +64u^6 & \end{array}$$

which expresses the relationship of the abscissa $CE(u)$ to the ordinate $EF(z)$. This is what we were required to find.

We may determine the point of contact F by using the Method explained in Chapter 8. For if we imagine another incident ray *pm* infinitely close to *PM*, it is clear that the reflected ray *mh* cuts *MH* at the point F that we wish to find, from which we draw *FE* parallel [179] to *PM*. We denote *CE* by u, *EF* by z, *CP* by x, *PM* by y, and *CM* by a, and we find as above

$$\frac{aax + aau - 2uxx}{xy} = 2z.$$

Now, it is clear that *CM*, *CE*, and *EF* remain the same while *CP* and *PM* vary. That is why we take the differential of this equation by treating a, u, and z as constants, and x and y as variables, which gives $2uyxx\,dx + aauy\,dx - aaxx\,dy + 2ux^3\,dy = 0$. In this we substitute the value $-\frac{y\,dy}{x}$ for dx (which we find by taking the differential of $yy = aa - xx$), then we substitute the value $aa - xx$ for yy, and we finally have $CE(u) = \frac{x^3}{aa}$.

If we suppose that the curve *AMB* is no longer a quarter circle, but any other curve that has the straight line *MC* as the radius of the evolute at the point M, then it is clear (see §76) that its little part *Mm* may be considered as an arc of the circle described with center C. From this it follows that if we draw the perpendicular *CP* from this center to the incident ray *PM*, and take $CE = \frac{x^3}{aa}$ ($CP = x$ and $CM = a$), and we draw *EF* parallel to *PM*, it will cut the reflected ray *MH* at the point F, where it touches the caustic *AFK*.

Suppose that from all the points M and m of any curved line *AMB*, we draw the straight lines *MC* and *mC* to a fixed point C on its axis *AC*, and other straight lines *MH* and *mh* terminated by the perpendicular *CB* to the axis, so that the angles $CMH = MCH$ and $Cmh = mCh$. We wish to find the point F on each *MH* where it touches the curve *AFK* formed by the continual intersection of the straight lines *MH* and *mh*. We find as before

$$CH = \frac{xx + yy}{2y} = \frac{zx - uy}{x - u},$$

from which we conclude

$$\frac{x^3 + uyy + xyy - uxx}{xy} = 2z,$$

whose differential (treating u and z as constants and x and y as variables) gives $2x^3y\,dx - uxxy\,dx - x^4\,dy + ux^3\,dy + xxyy\,dy + uxyy\,dy - uy^3\,dx = 0$, and consequently the desired [180]

$$CE(u) = \frac{2x^3y\,dx - x^4\,dy + xxyy\,dy}{xxy\,dx - x^3\,dy + y^3\,dx - xyy\,dy}.$$

Now, given the nature of the line *AMB*, we have the value for *dy* in terms of *dx*, which when substituted in the expression of *CE*, gives an expression freed of differentials and entirely known.

Proposition VII.

Problem. (§209) *Let AO be an indefinite straight line* (see Fig. 10.13), *which has a fixed origin at A. Let an infinity of parabolas BFD and CDG be understood which have the straight line AO as common axis, and the straight lines AB and AC intercepted between the fixed point A and their vertices B and C as their parameters. We wish to find the nature of the line AFG that touches all of these parabolas.*

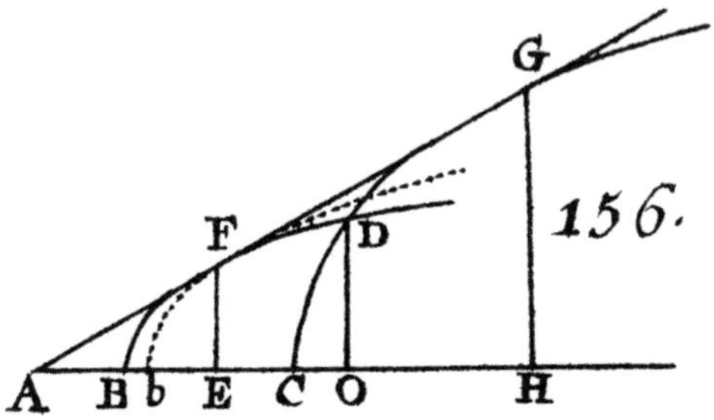

Fig. 10.13 Envelope of a Family of Parabolas with a Common Axis

First, I remark that any two of these parabolas *BFD* and *CDG* cut each other at a point *D* situated between the line *AFG* and the axis *AO*, and that as *AC* becomes equal to *AB*, the point of intersection *D* falls on the point of contact *F*.

Given all this, we propose to draw from the given point *D* a parabola that has the property we have noted. If we draw the ordinate *DO*, and denote the given quantities *AO* by u and *OD* by z, and the unknown *AB* by x, the property of the parabola gives $AB \times BO(ux - xx) = \overline{DO}^2(zz)$, and ordering the equation, we have $xx - ux + zz = 0$. Now, it is clear that if u expresses *AO* and z expresses *OD*, this equality has two unequal roots, namely *AB* and *CA*, and on the contrary, if u expresses *AE* and z expresses *EF*, then *AC* becomes equal to *AB*, that is to say it now has two equal roots. This is why we multiply it by the arithmetic progression $1, 0, -1$, which gives $x = z$, and substituting this value in place of x, it becomes the equation $u = 2z$, which should express the nature of the line *AFG*. From this we see that *AFG* is a straight line making the angle *FAO* with *AO* so that *AE* is twice *EF*.

[181] If we want to solve this question in general, where the parabolas *BFD* and *CDG* may be of any degree, we make use of the Method explained in Chapter 8, in the following manner. Denoting *AE* by u, *EF* by z, and *AB* by x, we have $\overline{u-x}^m \times x^n = z^{m+n}$, which expresses the general nature of the parabola *BF*, whose differential gives (treating u and z as constants, and x as variable) $-m \times \overline{u-x}^{m-1} dx \times x^n + nx^{n-1} dx \times \overline{u-x}^m = 0$. Dividing by $\overline{u-x}^{m-1} dx \times x^{n-1}$, it becomes $-mx + nu - nx = 0$, from which we conclude that $x = \frac{n}{m+n}u$, and consequently $u - x = \frac{m}{m+n}u$. Therefore, substituting these values in place of $u - x$ and x in the general equation, and making (to abbreviate) $\frac{m}{m+n} = p$, $\frac{n}{m+n} = q$, and $m + n = r$, we have $z = u\sqrt[r]{p^m q^n}$. From this we see that the line *AFG* is still straight, no matter how the parabolas might be composed, where only the ratio of *AE* to *EF* changes.

We clearly see from what we have just explained in this chapter, the way in which one should use the Method of Messrs. Descartes *and* Hudde *to solve these kinds of questions when the Curves are Geometric. However, we also see at the same time that it is not comparable to that of Mr.* Leibniz, *which I have tried to explain thoroughly in this Treatise, because this latter gives general solutions, where the other gives only particular ones, that it extends to Transcendental lines, and that it is not necessary to remove incommensurables, which is very often impractical.*

THE END.

Chapter 11
Bernoulli's *Lectiones de Calculo Differentialis*

On the Differential Calculus

[1][1] Postulates[2]

1. Quantities that decrease or increase by an infinitely small quantity neither decrease nor increase.
2. Any Curved line consists of infinitely many straight lines, each of which is infinitely small.
3. The figure contained within two ordinates applied to different abscissas and the infinitely small portion of the Curve is considered to be a Parallelogram.

On the Addition and Subtraction of Differential[3]

Rule 1. The differential of the addition of quantities is the sum of the differentials of each of the quantities, taken separately and then added together.

E.g. The differential of the quantity $x + y$ is $dx + dy$. Indeed, let $e = dx =$ the differential of the indeterminate x and $f = dy =$ the differential of the

[1]Numbers in [square brackets] indicate page numbers in the original manuscript, which were given in the same form in Schafheitlin (1922).

[2]Compare Postulates I and II to §2 and §3 on p. 3. Postulate 3 is used for the integral calculus and is not needed in the *Lectiones de Calculo Differentialis*.

[3]Compare to §4 on p. 3.

R.E. Bradley et al., *L'Hopital's Analyse des infiniments petits*, Science Networks. Historical Studies 50, DOI 10.1007/978-3-319-17115-9_11

indeterminate y. Add the larger $x+e$ and the larger $y+f$. The sum is $x+y+e+f$, from which, if we subtract the smaller sum $x+y$, the result is the differential $e+f=dx+dy$. Q.E.D.

The differential of the quantity $a+x$ is dx if a denotes a certain and determined quantity, as we will suppose it to be here and in what follows.[4] Indeed, adding $a+0$ and $x+e$, the sum is $a+x+e$. Of course, subtract the smaller $a+x$. The remainder is $+e=dx$. Q.E.D.

As it is with added Quantities, the necessary changes having been made, so in turn with subtracted quantities Rule 1 can be applied.

[2] On the Differentials of Composed Quantities

The differential of the quantity ax is $a\,dx$, which is proved as follows[5]:

multipl.	$x+e$	supposing that $e=dx$
with	$a+0$	it is a plus nothing, because a is
prod.	$\overline{ax+ae}$	a determined quantity, which has
from which subtr.	ax	no differential
leaving	$\overline{ae=a\,dx}$	q.e.d.

The differential of the quantity xx is $2x\,dx$, which is demonstrated as follows: Multiply $x+e$ by $x+e$. The product is $xx+2ex+ee$. Subtract xx. The remainder is $2ex+ee$, which by the first postulate is $=2ex=2x\,dx$. Q.E.D.

The different. of x^3 is $=3xx\,dx$. Multiply $x+e$, $x+e$, $x+e$. The product is $x^3+3exx+3eex+e^3$. Subtr. x^3. The remainder is $3exx+3eex+e^3$ = by the first postul. $3exx=3xx\,dx$. By the same reasoning we prove that of the quant. x^4 the different. $=4x^3\,dx$ and of x^5 the different. $=5x^4\,dx$ and of x^6 the different. is $6x^5\,dx$ and similarly for the others.

From which the following General Rule can be stated:

Rule 2.[6] The differential of an indeterminate quantity, whatever dimension the quantity has, is the product of that same quantity raised to that same dimension diminished by unity, in its own differential, taken as many times as the dimension that the indeterminate quantity has. Or if the character of the rule is expressed in a more helpful way: the differential of x^p is $=px^{p-1}dx$.

[4] The manuscript employs the abbreviation "et in seqq." here. Where possible, we will translate abbreviated Latin words with similarly abbreviated English words.

[5] The proof that follows is formatted like a sequence of calculations, with justifications in the right-hand column. It is perhaps easier to read the proof in this form: "multipl. $x+e$, supposing that $e=dx$, with $a+0$, which is a plus nothing, because a is a determined quantity, which has no differential. The prod. is $ax+ae$, from which subtr. ax, leaving $ae=a\,dx$ q.e.d."

[6] This case of the Power Rule is given on p. 6, using the Product Rule.

[3] The differential of the quantity[7] xy is $x\,dy + y\,dx$. Multipl. $x + e$ by $y + f$ (supposing $e = dx$ and $f = dy$). Prod. $xy + ey + fx + ef$. Subtract xy. What remains $ey + fx + ef$ = by Postul. 1 $ey + fx = y\,dx + x\,dy$ q. e. d. Different. of xyz is $= xy\,dz + zx\,dy + zy\,dx$. Multipl. $x + e$, $y + f$, $z + g$ supposing $g = dz$. Product $xyz + zye + zxf + xyg + zef + yeg + xfg + gef$. subtr. zxy what remains $zye + zxf + xyg + zef + yeg + xfg + gef$ = by postulate 1 $zye + zxf + xyg = zy\,dx + zx\,dy + xy\,dz$. In a similar way it is demonstrated that the differential of $xyzu$ is $= xyz\,du + xyu\,dz + xzu\,dy + yzu\,dx$. And so it is for the other cases, which gives the following.

Rule 3. The differential of the product of many quantities with each other is equal to the sum of the products of each of these differentials into the product of the others.

On the Differentials of Divided Quantities[8]

Of the quantity $\frac{1}{x}$ the different. $= \frac{-dx}{xx}$. It is proved as followed: Subtract $\frac{1}{x}$ from $\frac{1+0}{x+e}$. The remainder will be $\frac{-e}{xx+ex}$ = by postul. 1 $\frac{-e}{xx} = \frac{-dx}{xx}$. Q. E. D.

Or otherwise. Suppose $\frac{1}{x} = z$ then $1 = xz$ and taking the differential of both sides (because the determinate 1 has a null differential) $0 = xdz + zdx$ and $dz = \frac{-zdx}{x} = \frac{-dx}{xx}$. Q. e. d.

Of $\frac{xx}{a}$ the different. $= \frac{2xdx}{a}$. The demonstration of the preceding is similar to the different. of $\frac{x}{y}$ being $\frac{ydx-xdy}{yy}$. Subtr. $\frac{x}{y}$ from $\frac{x+e}{y+f}$. The remainder is $\frac{ey-fx}{yy+fy}$ = by postul. 1 $\frac{ey-fx}{yy} = \frac{ydx-xdy}{yy}$. Q. E. D. Otherwise supp. $\frac{x}{y} = z$ then $x = yz$ and $dx = ydz + zdy = ydz + \frac{x}{y}dy$ and $dx - \frac{x}{y}dy = ydz$ and $\frac{ydx-xdy}{yy} = dz$. Q. e. d.

[4] Again, a Rule is formed from these:

Rule 4. The differential of any fraction is the product of the Denominator by the differential of the Numerator, minus of the product Numerator by the differential of the Denominator, divided by the square of the Denominator.

So that of $\frac{x}{a+x}$ the different. $= \frac{adx}{aa+2ax+xx}$. Thus the differential of the quantity $\frac{xy+yz}{u+t}$ is =[9]

$$\frac{\begin{array}{l} +ux\,dy + uy\,dx - xy\,du \\ +uz\,dy + ty\,dx - yz\,du \\ +tx\,dy + uy\,dz - xy\,dt \\ +tz\,dy + ty\,dz - yz\,dt \end{array}}{uu + 2ut + tt}$$

[7] Compare to §5 on p. 4.

[8] Compare to §6 on p. 4.

[9] In the expression that follows, the numerator consists of all twelve terms in the first four lines.

and the different. of $\frac{x-y}{u-t}$

$$= \frac{+u\,dx - t\,dx - u\,dy + t\,dy - x\,du + y\,du + x\,dt - y\,dt}{uu - 2ut + xx}$$

On the Differentials of Surd Quantities[10]

The differential of certain quantities contained under radical signs are found as follows: e.g. let the given quantity be $\sqrt{ax+xx}$, which I call $= z$ then $ax + xx = zz$ and $a\,dx + 2x\,dx = 2z\,dz = 2\,dz\sqrt{ax+xx}$. Therefore, $\dfrac{a\,dx+2x\,dx}{2\sqrt{ax+xx}} = dz =$ the differential of $\sqrt{ax+xx}$.

In the same way the differential of $\sqrt[3]{ax+xx}$, is found, which is[11] $\dfrac{a\,dx+2x\,dx}{3\Box\sqrt[3]{ax+xx}}$. In a like manner also the differential of[12] $\sqrt[4]{yx+xx}$ is $= \dfrac{y\,dx+x\,dy+2x\,dx}{4\mathrm{C}\sqrt[4]{yx+xx}}$. In the same way, the different. of $\sqrt[5]{ayx+x^3+zyx}$ is[13]

$$\frac{ay\,dx + 3xx\,dx + yz\,dx + ax\,dy + zx\,dy + xy\,dz}{5\mathrm{QQ}\sqrt[5]{ayx+x^3+zyx}}.$$

[5] In fact, these same Differentials are found in another way from the generation of series, where they are quantities in geometrical Proportion and the Powers are in Arithmetic Proportion.[14]

$x^{(4}$	E.g. To find the differential of $\sqrt{ax+xx}$,
$x^{(3}$	consider the quantity $ax + xx$ as x raised to
$x^{(2}$	the power $\frac{1}{2}$
$x^{(1}$	which is the mean proportional between $x^{(1}$ and $x^{(0} = 1$.
$x^{(0} = 1$	And using Rule 2 its differential is
$x^{(-1} = \dfrac{1}{x}$	$\frac{1}{2}, \overline{ax+xx}^{(-\frac{1}{2}}, a\,dx + 2x\,dx = \dfrac{a\,dx+2x\,dx}{2\sqrt{ax+xx}}$.[15] So it is for x

[10]Compare to §7 on p. 5.

[11]The symbol □ is used to denote the square of the term that follows it.

[12]The symbol C is used to denote the cube of the term that follows it.

[13]The symbol QQ is used to denote the fourth power of the term that follows it.

[14]Compare the discussion that follows with L'Hôpital's discussion of geometric and arithmetic progressions on p. 5. In this paragraph, Bernoulli denotes exponents with a left parenthesis in the superscript, perhaps as a form of emphasis. This notation is rarely used elsewhere in the *Lectiones*.

$x^{(-2} = \frac{1}{xx}$ raised to the power $-\frac{1}{2}$, which is the mean proportional

$x^{(-3} = \frac{1}{x^3}$ between $x^{(0} = 1$ and $x^{(-1} = \frac{1}{x}$. For that reason, $x^{(-\frac{1}{2}} = \frac{1}{\sqrt{x}}$.

$x^{(-4} = \frac{1}{x^4}$ Therefore, together, $\overline{ax + xx}^{(-\frac{1}{2}} = \frac{1}{\sqrt{ax + xx}}$. The half of

which multiplied by $a\,dx + 2x\,dx$ makes $\frac{a\,dx + 2x\,dx}{2\sqrt{ax + xx}}$. That is the differential of $\sqrt{ax + xx}$. The differential of $\sqrt[3]{ax + yy}$ is found in the same way, considering it to be $\sqrt[3]{x} = x^{(\frac{1}{3}}$, which is the first of two proportional means between $x^{(0} = 1$ and $x^{(1}$, and I find the differential of it by Rule 2

$$\frac{1}{3}, \overline{ax + yy}^{(-\frac{2}{3}}, a\,dx + 2y\,dy = \frac{a\,dx + 2y\,dy}{3\square\sqrt[3]{ax + yy}}.$$

In addition, we find the differential of surd quantities by Analogy with the others. E.g. different. $\sqrt[3]{x^3}$ is dx, which can be found by dividing the differential $3xx\,dx$ of the Cube x^3, by $3xx$, three times the □ of the root. Also of $\sqrt{xx}$ differ. $= dx$, which can be found by dividing the differential $2x\,dx$ of the Square xx by $2x$, double the root.

[6] Different. $\sqrt[4]{x^4}$ is $= dx$, which may be found by dividing the differential $4x^3\,dx$ by $4x^3$, four times the cube of the root. Thus, of $\sqrt[5]{x^5}$ differ. $= dx$, which has the division of the different. $5x^4\,dx$ by $5x^4$, five times the Square-Square of the Root. And so on for the others.

Accordingly, therefore, the basis of the rules to have the differentials of surds can be found. So, e.g., one can likewise find the differential of $\sqrt[5]{x}$, namely by dividing the differential dx of x by five times the square-square of $\sqrt[5]{x}$, which will be $\frac{dx}{5\sqrt[5]{x^4}}$. In the same way, of $\sqrt[4]{x}$ different. $= \frac{dx}{4\sqrt[4]{x^3}}$ and of $\sqrt[3]{x}$ the differential $= \frac{dx}{3\sqrt[3]{xx}}$ and of $\sqrt{x}$ different. $= \frac{dx}{2\sqrt{x}}$.

The following Rule can be deduced from the foregoing to find the differential of any surd quantity, to wit:

Rule 5. For any surd quantity raised to any number of dimensions, diminish by unity that which is contained within the radical sign and taking so many as the same number of dimensions contained within the radical sign, divide the differential by all of this; the quotient is the differential that we wish to find.

In characters, the rule can be expressed as:

$$\sqrt[p]{x} \text{ differ.} = \frac{dx}{p\sqrt[p]{x^{p-1}}}.$$

[15]Here and in the following the comma represents multiplication. In addition, we note that in Schafheitlin (1922), the term before the equal sign was given as $ax + 2x\,dx$.

E.g. the differential of $\sqrt{ax+xx}$ is $= \dfrac{a\,dx+2x\,dx}{2\sqrt{ax+xx}}$, the different. of[16] $\sqrt[3]{ax+xx}$ is $\dfrac{a\,dx+2x\,dx}{3\Box\sqrt[3]{ax+xx}}$.

To find the differential of the quantity[17] $\dfrac{\sqrt[3]{ax+xx}}{\sqrt{yx+yy}}$. The differential of the Numerator $\dfrac{a\,dx+2x\,dx}{3\Box\sqrt[3]{ax+xx}}$ is found, and that of the Denominator [7] $=$ $\dfrac{y\,dx+x\,dy+2y\,dy}{2\sqrt{yx+yy}}$ by Rule 5. Find by Rule 4:

$$\frac{\overline{\frac{a\,dx+2x\,dx}{3\Box\sqrt[3]{ax+xx}}} \text{ in } \sqrt{xy+yy}+\overline{\frac{-y\,dx-x\,dy-2y\,dy}{2\sqrt{xy+yy}}} \text{ in } \sqrt[3]{ax+xx}}{xy+yy}$$

the differential of[18] $\dfrac{\sqrt[3]{ax+xx}}{\sqrt{yx+yy}}$. The differential of the quantity $\sqrt{ax+xx+\sqrt{aay+y^3}}$ is

$$= \frac{\overline{2a\,dx+4x\,dx} \text{ in } \sqrt{aa\,y+y^3}+aa\,dy+3yy\,dy}{2\sqrt{ax+xx+\sqrt{aa\,y+y^3}} \text{ in } 2\sqrt{aa\,y+y^3}}.$$

[8] This page is blank.

[9] The Use of Differential Calculus in Solving Problems

Problem I. To find the Tangent of the Parabola.[19]

From the nature of the Parabola, $ax = yy$. Therefore, in addition $a\,dx = 2y\,dy$. Consequently[20] $a : 2y :: dy : dx$. Moreover, because each Curved line is understood to be of an infinity of straight lines, by Postul. 2, the tangent AD is in a straight line with the infinitely small portion DF of the parabola BDF. Therefore draw (Fig. 11.1) DG parallel to the Diameter AE, $\triangle DGF$ and $\triangle ACD$ will be similar. Wherefore, $FG : GD :: CD : AC$, which is[21] $dy : dx :: y : s$ (subtangent) $:: a : 2y$ (from prev.) thus $s = \frac{2yy}{a} = \frac{2ax}{a} = 2x$. If therefore AC is taken as twice the abscissa BC itself,

---

[16] L'Hôpital gave this example on p. 8.

[17] L'Hôpital gave this example on p. 8.

[18] L'Hôpital gave a similar example on p. 8.

[19] Compare this to Example I, Part 1, on p. 13.

[20] Bernoulli wrote this proportional relationship as: $a \cdot 2y :: dy \cdot dx$.

[21] The relation $dy : dx :: y : s$, concerning the Differential Triangle, is given in Proposition I on p. 11.

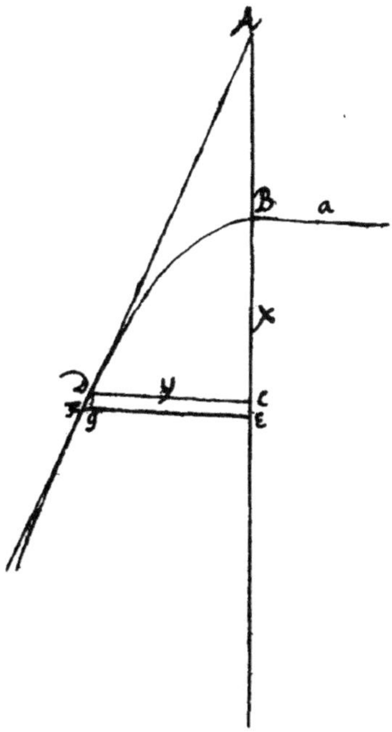

Fig. 11.1 Tangent to the Parabola

and through the point A and the given point D on the Curve the straight line AD is drawn, that will be the Tangent. Q. e. i.[22]

In the same way, to find the tangent in the Cubic Parabola:[23]

1. If its nature is $aa\,x = y^3$. The Subtangent will be $= 3x$. Indeed, $aa\,dx = 3yy\,dy$, and $aa : 3yy :: dy : dx :: y : s$

$$= \frac{3y^3}{aa} = \frac{3aax}{aa} = 3x.$$

2. If $a\,xx = y^3$. Then[24] $s = \frac{3x}{2}$. For[25] $2ax\,dx = 3yy\,dy$. Because of this $2ax : 3yy ::$ $dy : xx :: y : s = \frac{3y^3}{2ax} = \frac{3axx}{2ax} = \frac{3x}{2}$.

[22]This is an abbreviation for *Quod erit inveniendum*; "which was to be found."

[23]Compare what follows to the generalized parabola, where fractional exponents may be used, on p. 13.

[24]In Schafheitlin (1922) the denominator was given as a.

[25]In Schafheitlin (1922), the term $2ax$ was given as $2a$.

In the Biquadratic Parabola

1. If $a^3 x = y^4$. $s = 4x$. For $a^3\,dx = 4y^3\,dy$. Therefore, $a^3 : 4y^3 :: dy : dx :: y : s = \frac{4y^4}{a^3} = \frac{4a^3x}{a^3} = 4x$.
2. [10] If $aa\,xx = y^4$. $s = 2x$. For $2aax\,dx = 4y^3\,dy$ and $2aax : 4y^3 :: dy : dx :: y : s = \frac{4y^4}{2aax} = \frac{4aaxx}{2aax} = 2x$.
3. If $a\,x^3 = y^4$. $s = \frac{4}{3}x$. $3axx\,dx = 4y^3\,dy$ and $3axx : 4y^3 :: dy : dx :: y : s = \frac{4y^4}{3axx} = \frac{4ax^3}{3axx} = \frac{4x}{3}$.

In Other Parabolas

1. If $a^4 x = y^5$. $s = 5x$. $a^4\,dx = 5y^4\,dy$.
2. If $a^3 xx = y^5$. $s = \frac{5}{2}x$.
3. If $aa\,x^3 = y^5$. $s = \frac{5}{3}x$.
4. If $a\,x^4 = y^5$. $s = \frac{5}{4}x$.

And thus for the Others. From which the general Rule can be formed: If the nature of any Parabola is $a^{(c}\,x^{(m} = y^{(n}$.

Its Subtangent is $\dfrac{nx}{m}$.

In fact $a^c\,mx^{m-1}\,dx = ny^{n-1}\,dy$, and

$$a^c\,mx^{m-1} : ny^{n-1} :: dy : dx :: y : s = \frac{ny^n}{a^c\,mx^{m-1}} = \frac{a^c\,x^m\,n}{a^c\,x^{m-1}\,m} = \frac{a^c\,x^{m-1}\,nx}{a^c\,x^{m-1}\,m} = \frac{nx}{m}.$$

Problem II. [11] To Find the Tangent of the Ellipse.[26]

Let the Diameter (Fig. 11.2) be $BJ = b$. The Parameter $= a$, the Abscissa $= x$ the applied ordinate $= y$. The differential of the abscissa $CE = dx$, and the differ. of the applied ordinate $FG = dy$. Here (for the same reason, as noted in the Parabola) $\triangle DFG$ and $\triangle ACD$ will be similar. Therefore, $FG : GD :: DC : AC$ i. e. $dy : dx :: y : s$. Moreover, from the nature of the Ellipse, $b : a :: bx - xx : yy$. Therefore, $abx - axx = byy$, and taking the differentials on both sides $ab\,dx - 2ax\,dx = 2by\,dy$. Thus $ab - 2ax : 2by :: dy : dx :: y : s$. Therefore

$$s = \frac{2byy}{ab - 2ax} = \frac{2abx - 2axx}{ab - 2ax} = \frac{2bx - 2xx}{b - 2x}.$$

[26]Compare this to the more general Example II on p. 13.

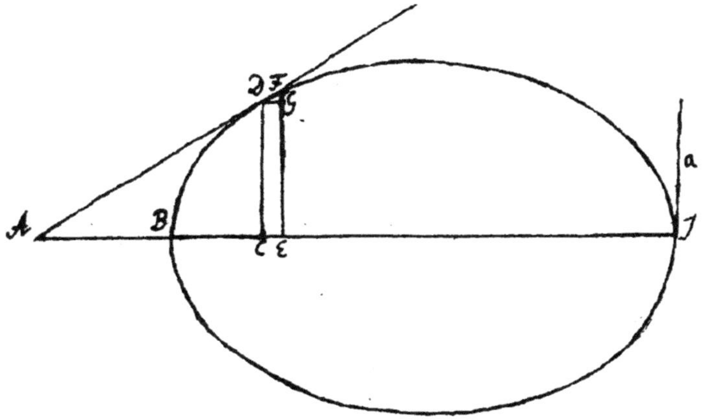

Fig. 11.2 Tangent to the Ellipse

Problem III. To Find the Tangent of the Hyperbola.[27]

With the same set out in accordance with the Ellipse, likewise $dy : dx :: y : s$. And (Fig. 11.3) from the nature of the Hyperbola $b : a :: bx + xx : yy$. For this reason, $abx + axx = byy$ and $ab\,dx + 2ax\,dx = 2by\,dy$. Therefore, $ab + 2ax : 2by :: dy : dx :: y : s =$

$$= \frac{2byy}{ab + 2ax} = \frac{2ab\,x + 2a\,xx}{ab + 2ax} = \frac{2bx + 2xx}{b + 2x}.$$

Even Asymptotes are found from the Tangent, considering them to be Tangents [12] at infinity, and their abscissas x and ordinates y as infinite. And JH and JM will be found, through whose endpoints H and M the Asymptote passes. However, in this case, JH (Fig. 11.4) is the very same as AJ is with regard to any other Tangent, and JM as any JO, evidently $JH = \dfrac{bx}{b + 2x} =$ by postul. 1 (because in this case x is infinitely larger than b) $\dfrac{bx}{2x} = \frac{1}{2}b = \frac{1}{2}$ the transverse Diameter; also $JM = \dfrac{by}{2b + 2x} = \dfrac{\sqrt{abx}}{\sqrt{4b + 4x}} =$ by postul. 1 $\sqrt{\dfrac{abx}{4x}} = \dfrac{\sqrt{ab}}{2} =$ the Conjugate semidiameter.

[27]Compare this to Example II, Part 2, on p. 13 and the more generalized Example III on p. 15.

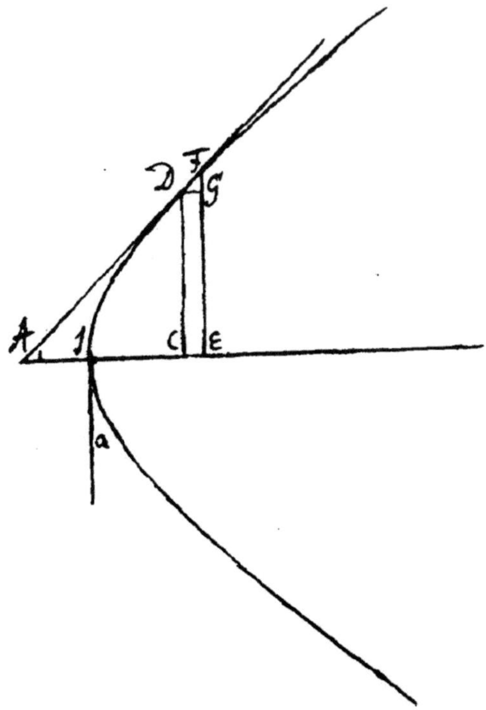

Fig. 11.3 Tangent to the Hyperbola

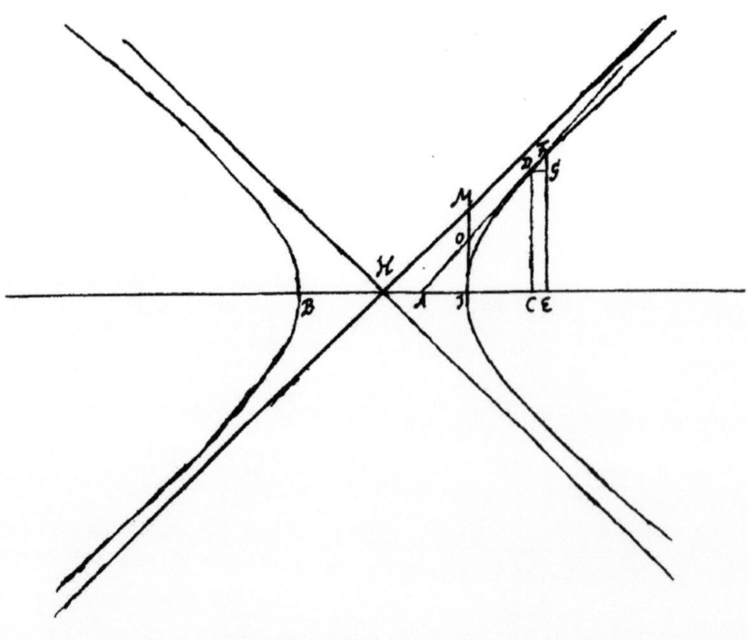

Fig. 11.4 Asymptotes of the Hyperbola

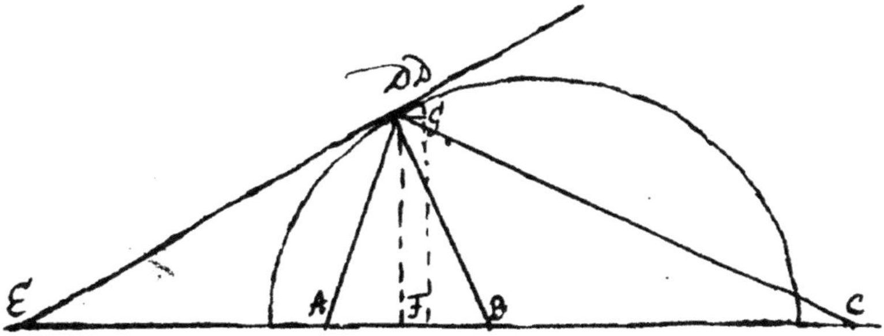

Fig. 11.5 Problem IV

Problem IV. To Find the Tangent of the Curve, which has this property, the sum of the three straight lines drawn from any point on the curve to three given points in any straight line is always equal.

[13] Let (Fig. 11.5) A, B, and C be three given points, whose distances of which are $AB = a, AC = b, AF = x$, and $DF = y$. The Sum of the three lines $AD + BD + CD = c$. Therefore, we will have $AD = \sqrt{xx+yy}$, $BD = \sqrt{aa-2ax+xx+yy}$ and $DC = \sqrt{bb-2bx+xx+yy}$. Therefore

$$\sqrt{xx+yy} + \sqrt{aa-2ax+xx+yy} + \sqrt{bb-2bx+xx+yy} = c$$

and taking differentials of both sides

$$\frac{2x\,dx + 2y\,dy}{2\sqrt{xx+yy}} + \frac{2x\,dx + 2y\,dy - 2a\,dx}{2\sqrt{aa-2ax+xx+yy}} + \frac{2x\,dx + 2y\,dy - 2b\,dx}{2\sqrt{bb-2bx+xx+yy}} = 0.$$

Reducing to quantities in which dx is on one side, and in which dy is on the other, and dividing the entire quantity by $\frac{2}{2}$, we will have

$$\frac{y\,dy}{\sqrt{xx+yy}} + \frac{y\,dy}{\sqrt{aa-2ax+xx+yy}} + \frac{y\,dy}{\sqrt{bb-2bx+xx+yy}}$$
$$= \frac{b\,dx - x\,dx}{\sqrt{bb-2bx+xx+yy}} + \frac{a\,dx - x\,dx}{\sqrt{aa-2ax+xx+yy}} + \frac{-x\,dx}{\sqrt{xx+yy}}.$$

Resolving the equation into a proportion, it is

$$\frac{y}{\sqrt{xx+yy}} + \frac{y}{\sqrt{aa-2ax+xx+yy}} + \frac{y}{\sqrt{bb-2bx+xx+yy}}$$
$$: \frac{b-x}{\sqrt{bb-2bx+xx+yy}} + \frac{a-x}{\sqrt{aa-2ax+xx+yy}} + \frac{-x}{\sqrt{xx+yy}}$$
$$:: dx : dy$$

and $dy : dx = y : s$ (because of similar triangles DDG and EDF), which is therefore

$$\frac{\dfrac{yy}{\sqrt{xx+yy}}+\dfrac{yy}{\sqrt{aa-2ax+xx+yy}}+\dfrac{yy}{\sqrt{bb-2bx+xx+yy}}}{\dfrac{b-x}{\sqrt{bb-2bx+xx+yy}}+\dfrac{a-x}{\sqrt{aa-2ax+xx+yy}}-\dfrac{x}{\sqrt{xx+yy}}}.$$

From this solution it is plainly apparent, that this method is concise, and more succinct than that of Descartes, by which means if this Problem had been solved, it would first be necessary, in order to bring forth that curved line, to find the equation in purely rational terms; a task that is of great labor and tedium.

Problem V. [14] To Find the kind of line whose Subtangents are always equal.[28]

Let (Fig. 11.6) AB be the Tangent, BC the Subtangent $= a$, $AC = y$, $AD = dy$, and cC or $aD = dx$. We will have $dy : dx :: y : a$ and alternately $dy : y :: dx : a$. Because it is true that the ratio $dx : a$ is always constant, likewise $dy : y$ will always be constant, i.e.[29] $\div\!\div\, y : y : y$ makes a geometric progression. And for this reason, this is a Logarithmic line, whose ordinates make a Geometric progression and abscissas an Arithmetic.

Problem VI. To Find The Tangent of the Cycloid.[30]

Let (Fig. 11.7) ABD be a Cycloid, whose Tangent at the point E is to be found, the circle $BHDB$ is drawn in the middle, whose semi-circumference equals half of the base AD or DC, and diameter $= 2a$. EM is drawn further up to be parallel to the base AC, and $BF = x$ is parallel to the same, and to this the ordinate $EF = y = BM$.

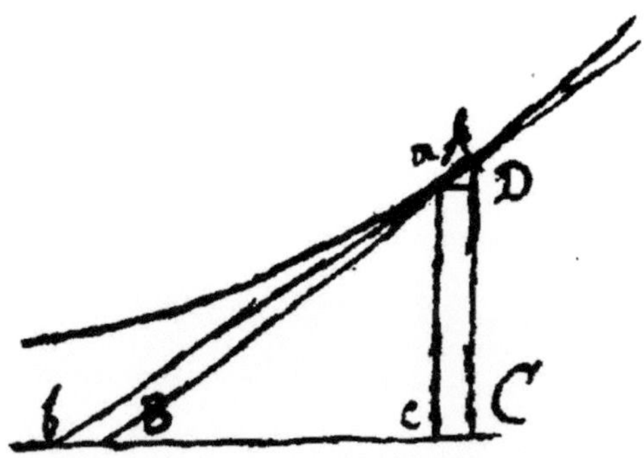

Fig. 11.6 Curve of Constant Subtangent

[28]Compare this to §39 on p. 38.

[29]The notation that follows evidently signifies the progression of y-values.

[30]Compare this §18 on p. 18, which depends on a general proposition in §15.

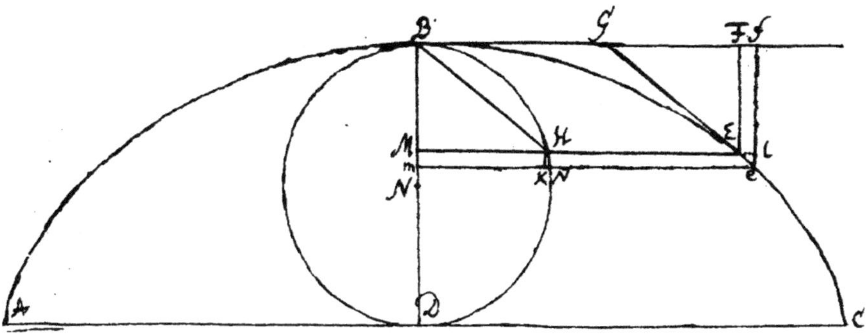

Fig. 11.7 Tangent to the Cycloid

And it is by the nature of this Curve that the straight line EH = the arc $HB = f$. And indeed $x = FB = EH + HM =$

$$f + \sqrt{2ay - yy} \quad \text{and} \quad dx = df + \frac{2a\,dy - 2y\,dy}{2\sqrt{2ay - yy}}.$$

On the other hand [15] $df = HN =$ by postul. 2 the subtense of the Right triangle $HKN = \sqrt{\Box HK + \Box KN} = \dfrac{a\,dy}{\sqrt{2ay - yy}}$. Therefore

$$dx = \frac{2a\,dy - y\,dy}{\sqrt{2ay - yy}} \quad \text{and because} \quad dy : dx :: y : s \quad \text{will likewise have}$$

$$dy : \frac{2a\,dy - y\,dy}{\sqrt{2ay - yy}} :: y : s, \quad \text{that is} \quad 1 : \frac{2a - y}{\sqrt{2ay - yy}} :: y : s$$

which therefore $= \dfrac{2ay - yy}{\sqrt{2ay - yy}} = \sqrt{2ay - yy} = HM$. Because the subtangent $FG = HM$ we will have $FB - FG = EM - HM$ i. e. $GB = EH$ and by consequence the Tangent EG is equal and parallel to the subtense BH.

Problem VII. To Find The Tangent of the Conchoid.[31]

Let (Fig. 11.8) $GL = a$
$CF = AD = b$
$GD = x$
$DE = AB = dx$.

[31]Compare this to §25 on p. 22, which depends on a more general proposition in §24.

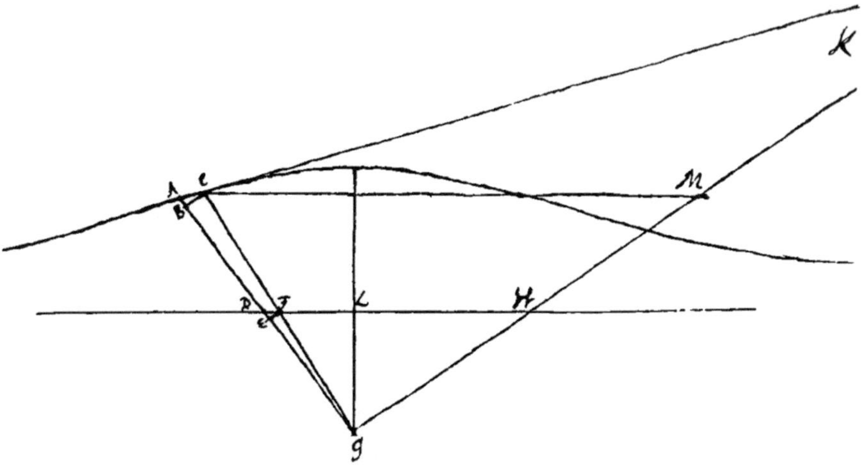

Fig. 11.8 Tangent to the Conchoid of Nichomedes

1. Because of the similar triangles *DEF* and *DLG*, *DL* : *LG* :: *DE* : *EF*

$$\sqrt{xx - aa} : a :: dx : \frac{a\,dx}{\sqrt{xx - aa}}.$$

2. Because of the similar triangles *BGC* and *EGF*, *GF* : *GC* :: *EF* : *BC*

$$x : b + x :: \frac{a\,dx}{\sqrt{xx - aa}} : \frac{ax\,dx + ab\,dx}{x\sqrt{xx - aa}}.$$

3. Because of the similar triangles *ABC* and *AGK*, *AB* : *BC* :: *AG* : *GK* i. e.

$$dx : \frac{ax\,dx + ab\,dx}{x\sqrt{xx - aa}} :: x + b : GK,$$

or if the first and second term are divided by *dx*,

$$1 : \frac{ax + ab}{x\sqrt{xx - aa}} :: x + b : \frac{axx + 2abx + abb}{x\sqrt{xx - aa}} = GK.$$

[16] And by calculation this is solved as follows:[32] Now $GF = GD$, so $DG^2 : (GH \cdot GC) = AG : GK$.

$$AB : BC :: ED : BC \quad \text{(because } ED = AB\text{)} :: ED : EF + EF \cdot BC ::$$
$$DG : GH + GF : GC. :: \square DGF : \square CGH :: AG : GK.$$
$$\text{or } \square DG.$$

Construction. Because of this, the Tangent is easily constructed in this way: *CM* is drawn parallel to *FH* and the points *F* and *M* are connected, let *AK* be made parallel to the line connecting *FM*, this *AK* will be the Tangent that we wish to find.

Proof. Because $CG : GM :: FG : GH$, we will have $\square CGH = \square FGM$. However, $AG : GK :: FG : GM :: \square FG : \square FGM\ (\square CGH)$. Therefore $AG : GK :: \square FG$ or $\square DG : \square CGH$. And therefore by calculation it is discovered that *AK* will be the Tangent of the Conchoid.

Problem VIII. To determine the Tangent of the Cissoid Curve.[33]

Let (Fig. 11.9) *ABC* be a semi-circle, *FB* is erected perpendicular from the Center, *BD* and *BE* are assumed to be any equal arcs, the intersection *H* of the line *AE* that is drawn and the perpendicular *DG* is a point on the Cissoid. Now it is necessary to determine the Tangent at this point. To this end, we wish to find the equation that expresses the nature of this curve, which one does as follows: Let $AF = FC = a$, $AG = x$, and $GH = y$. Therefore *FG* or $KF = a - x$ and *GD* or $KE = \sqrt{2ax - xx}$. Moreover $AK : KE :: AG : GH$ i. e. $2a - x : \sqrt{2ax - xx} :: x : y$ and therefore

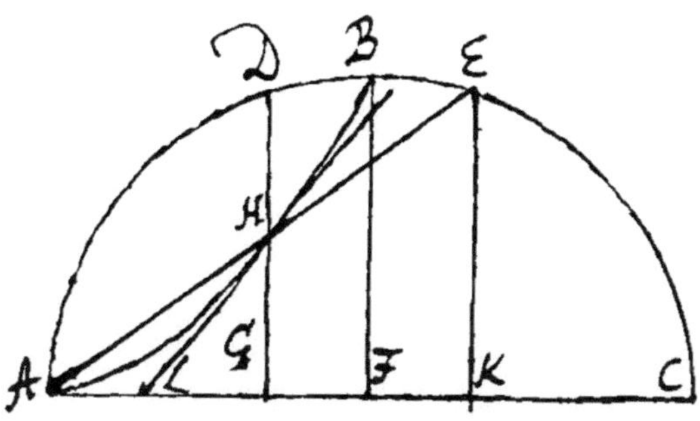

Fig. 11.9 Tangent to the Cissoid

[32]In Schafheitlin (1922), Schafheitlin suggests that we read this argument as: because $ED = AB$, $AG : GK = AB : BC = ED : BC = \frac{ED}{EF} \cdot \frac{EF}{BC} = \frac{DG}{GH} \cdot \frac{GF}{GC}$ Bernoulli uses the rectangle symbol in analogy to the square: $\square$ *DGF* represents the area of the rectangle *DGF*, i.e. the product of *DG* and *GF*.

[33]Compare this to §28 on p. 25, which depends on the more general proposition in §27.

$\sqrt{2a-x} : \sqrt{x} :: x : y$ or $2a - x : x :: xx : yy$. Hence $x^3 = 2ayy - xyy$, the differential [17] of which is $3xx\,dx = 4ay\,dy - 2xy\,dy - yy\,dx$ and $3xx\,dx + yy\,dx = 4ay\,dy - 2xy\,dy$. And for this reason $3xx + yy : 4ay - 2xy :: dy : dx :: y : s$, therefore s or

$$GL = \frac{4ayy - 2xyy}{3xx + yy} = \frac{2x^3}{3xx + yy}$$

or substituting the value of yy, we have $\dfrac{2ax - xx}{3a - x}$.

Problem IX. To Find the Tangent of the Quadratrix.[34]

If (Fig. 11.10) ABC is a Quarter of a Circle, and any arc AD is made, to the portion AE of the radius, as the quarter AB is to the total radius AC, the radius DC and the perpendicular EF are drawn, the point of intersection F will be on the Curve AFG, which is called the Quadratrix. Now it is desired to determine the Tangent at the point F. Let $AC = a$, $AB = b$, $AH = x$, and $AD = f$, then $DH = \sqrt{2ax - xx}$, $AE = \dfrac{af}{b}$, $HC = a - x$, and $EC = a - \dfrac{af}{b}$. Moreover $HC : HD :: EC : EF$, that is $a - x : \sqrt{2ax - xx} :: \dfrac{ab - af}{b} : EF$. Therefore one finds $EF = \dfrac{\overline{ab - af}\sqrt{2ax - xx}}{ab - bx}$. Now, as in the Sixth Problem, we find for the little portion Dd, i.e. for $df = \dfrac{a\,dx}{\sqrt{2a - xx}}$, and for this reason, for the small portion Ee, that is for the differential of the line $AE = \dfrac{aa\,dx}{b\sqrt{2ax - xx}}$, and by Rules 4 and 5 the

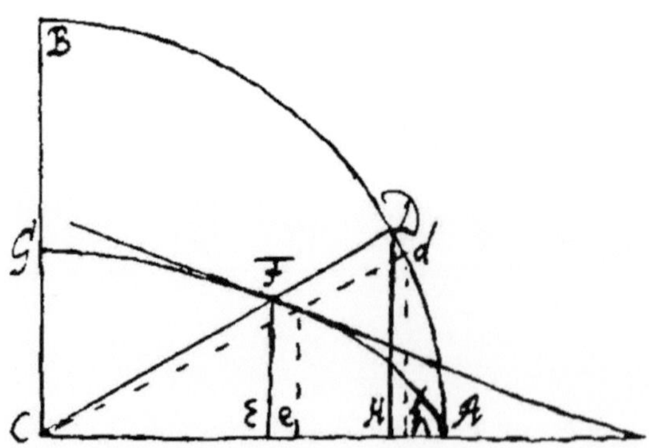

Fig. 11.10 Tangent to the Quadratrix

[34]Compare this to §30 on p. 27, which depends on a more general proposition in §29.

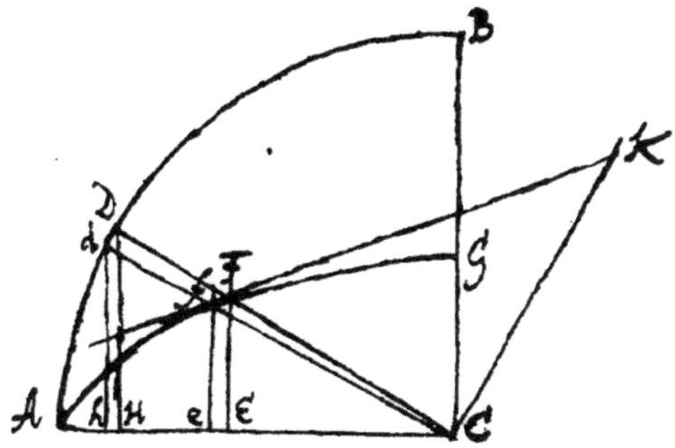

Fig. 11.11 Tangent to the Quadratrix – Alternate

differential of EF (where the value of df is substituted) will be found, because as the differential of EF is to the differential of the line EA, so the line EF is to s.

In Another Way. Let (Fig. 11.11) CK be perpendicular to DC, and retaining the same letters as previously, we wish to find the point K, which if it is joined to the point F, the line FK touches the curve AFG. It is done this way: $HC : DC ::$ $EC : CF$ i. e. $a - x : a :: \dfrac{ab - ef}{b} : CF$.

[18] Therefore $FC = \dfrac{aab - aaf}{ab - bx}$. Likewise $DC : FC :: Dd : Ff$, that is

$$a : \frac{aab - aaf}{ab - bx} :: df = \frac{a\,dx}{\sqrt{2ax - xx}} : Ff.$$

And therefore Ff is equal to $\dfrac{aab\,dx - aaf\,dx}{\overline{ab - bx}\sqrt{2ax - xx}}$. However, the differential of $FC =$

$$\frac{-a^3b\,df + aaxb\,df + aabb\,dx - aafb\,dx}{aabb - 2abbx + bbxx}$$
$$= \frac{-a^4b\,dx + a^3bx\,dx + \overline{aabb\,dx - aafb\,dx}\sqrt{2ax - xx}}{\overline{aabb - 2abbx + bbxx}\sqrt{2ax - xx}}$$

= differ. of FC. However, this differential is to

$$\frac{aab\,dx - aaf\,dx}{\overline{ab - bx}\sqrt{2ax - xx}} \quad \text{as} \quad \frac{-aa + ax + \overline{b - f}\sqrt{2ax - xx}}{a - x} : b - f$$

$$:: \frac{aab - aaf}{ab - bx} = FC : CK = \frac{aabb - 2aabf + aaff}{-aab + abx + \overline{bb - bf}\sqrt{2ax - xx}}.$$

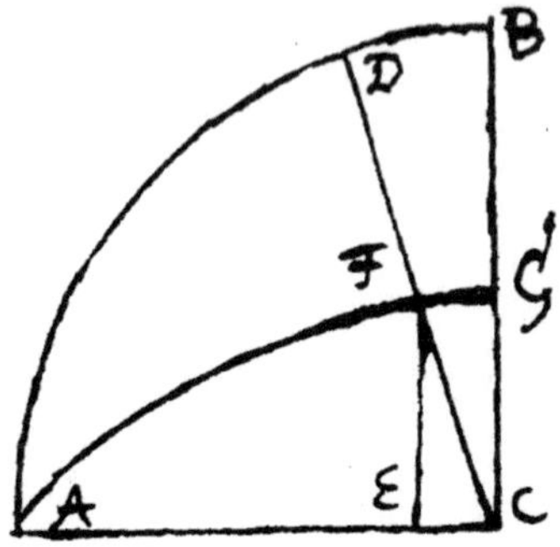

Fig. 11.12 Quadratrix: Point of Intersection

Accordingly, if the tangent is drawn from the point A, $CK = -b$, a quantity which, because it is negative, indicates that CK is to the right, i.e. towards the part AH or x, as one would suppose.

Problem X. To find the point of intersection G (Fig. 11.12) of the Quadratrix curve AG and the perpendicular radius CB.[35]

We imagine a point D, taken near the point B, so that the distance DB is infinitely small, just as [19] the distance CE, and so the radius CD and perpendicular EF are drawn, there will be no doubt the point F is considered to be the same as G, which obviously are not different from one another, owing to an infinitely small interval. This point F is therefore determined: $AB : AC :: AD : AE :: DB : EC :: DB : FG :: CB : CG :: AC : CG$. Therefore CG is the third proportional to the quarter circumference and to the radius. Hence the point G cannot be determined in any other way, without at the same time having the rectification of the circular line.

Problem XI.[36] To Find the Tangent in the Spiral of Archimedes.[37]

That curve is called the Spiral of Archimedes, which is described from a point, which is moved from the center to the circumference of a circle, the radius rotating uniformly in equal durations of time as the point is moved from the center to the circumference.[38] We wish to find the tangent of this curve. Let (Fig. 11.13) the radius $AC = a$, the circumference $DDCD = b$, $AB = x$, and the perpendicular AE to the

[35]Compare this to §31 on p. 28.

[36]In Schafheitlin (1922), it is noted that in the manuscript this is erroneously given as IX.

[37]Compare this to §23 on p. 21, which depends on a more general proposition in §22.

[38]In other words, changes in radial distance are proportional to angular changes of the radius.

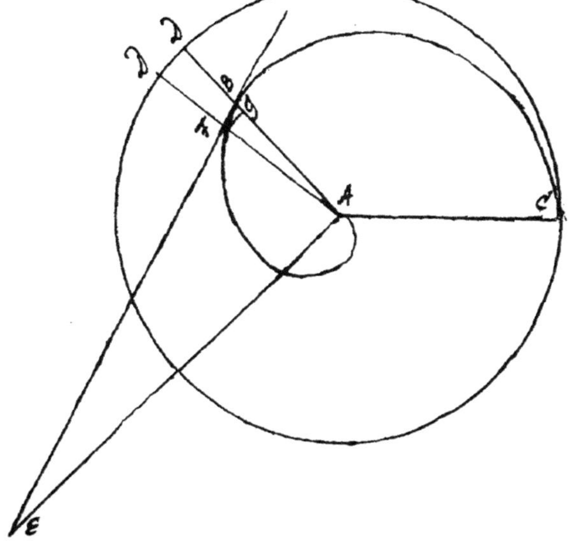

Fig. 11.13 Tangent to the Spiral of Archimedes

line *AB* is drawn. The Radius *AC* will be to the circumference as *AB* is to the arc[39] *CKD*. Also $AD : AF :: DD : FG$, and $a : x :: \frac{b\,dx}{a} : \frac{bx\,dx}{aa}$, again $BG : FG :: AB : AE$, i.e. $dx : \frac{bx\,dx}{aa} :: x : \frac{bxx}{aa} = s$.[40] Thus, if the tangent at *C* is to be drawn, we find that $s = b$, which Archimedes demonstrated by a long discourse.

On Maxima and Minima

Problem XII.[41] [20] To find the largest quantity, quantities are considered as ordinates of any curve concave towards the axis, as in Fig. *ABC* (Fig. 11.14). And *vice versa* to find the least quantity, considered as any ordinates of any curve convex

[39]There is no point *K* in Figure 11.13, although it presumably should be where the line *AE* intersects the circle *DDCD*.

[40]In other words, from the first proportion it follows that $FG = \frac{bx\,dx}{a^2}$ and therefore from the second that $AE = s = \frac{bx^2}{a^2}$.

[41]Compare this to the treatment of maximum and minimum beginning on p. 45.

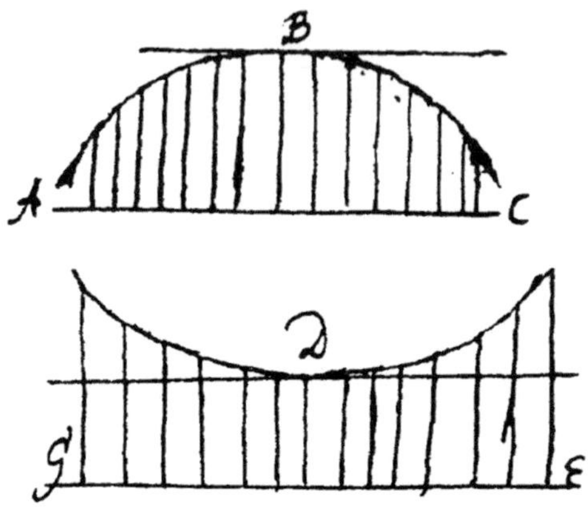

Fig. 11.14 Maximum and Minimum

Fig. 11.15 Product of Three Quantities

towards the axis, as in Fig. *GDE*, with axis *GE*. By this consideration, the tangent at a minimum or maximum point should be drawn parallel to the axis. Because $dy : dx :: y : s$ and y is infinitely smaller than the Subtangent, we will furthermore have that $dy = 0$ with respect to this dx. Therefore, if the maximum is to be found of that rectangle, which is made of two parts x and $a - x$ of a given line a, I consider $ax - xx$ as the ordinate on a certain curve concave towards the axis, and its differential $a\,dx - 2x\,dx = 0$. Therefore, $a\,dx = 2x\,dx$ and $x = \frac{1}{2}a$. Therefore, the rectangle will be maximum if it is assumed that $x = \frac{1}{2}a$.

Problem XIII. To divide a given line into three parts, so that the parts multiplied in turn make the largest solid that it is possible to produce by the three parts of the same line.

Let (Fig. 11.15) the part $AB = x$, the remainder will be cut in two at the point D; however by Probl. XII $\square BD$ or $\square DC$ is the maximum rectangle of the two parts of the line *BC*, therefore the same being multiplied by *AB* gives the maximum [21] solid of the three parts of this line. Thus $aax - 2axx + x^3 =$ Maximum.[42] Its differential $aa\,dx - 4ax\,dx + 3xx\,dx = 0$ and $xx = \frac{4}{3}ax - \frac{1}{3}aa$ and $x = \frac{1}{3}a$.

In the same way, if the line *AC* were to be cut into four parts, so that the parts multiplied in turn make the largest quantity of four dimensions, it is found that

[42]When D bisects BC, the volume of the solid is $x\left(\frac{1}{2}(a - x)\right)^2$, so the quantity being maximized here is in fact a constant multiple of the volume, which achieves its maximum at the same value of x.

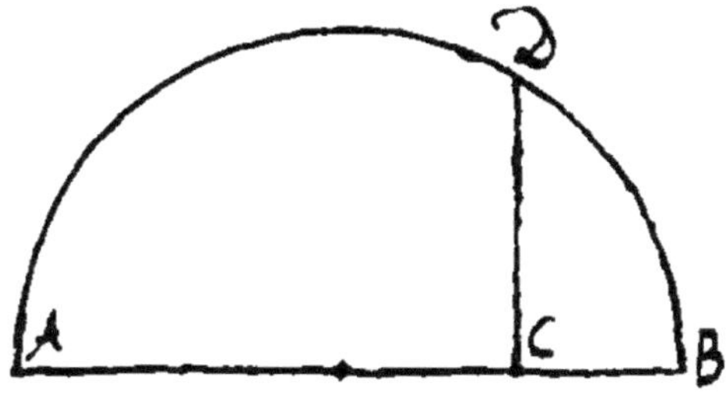

Fig. 11.16 Problem XIV

$x = \frac{1}{4}a$. And if the line AC were to cut in five parts, etc., it is found that $x = \frac{1}{5}a$. And so it is with regard to the others.

Problem XIV. To find the maximum Rectangle, which is described by the abscissa and the ordinate on a Circle.

Let (Fig. 11.16) the Diameter $AB = a$ and $AC = x$, we will have $CB = a - x$, $CD = \sqrt{ax - xx}$, and $\square ACD = \sqrt{ax^3 - x^4}$ = Maximum. Its differential

$$\frac{3axx\,dx - 4x^3\,dx}{2\sqrt{ax^3 - x^4}} = 0.$$

Also, $3axx\,dx = 4x^3\,dx$ and $3a = 4x$, and finally $x = \frac{3}{4}a$. Q. e. i.

Problem XV. To find the maximum Rectangle, which is described from the portions of an ordinate in the quadrant of a Circle that is cut by the subtense of this Quadrant.

Let (Fig. 11.17) $AC = a$ and $DC = x$, we will have $DF = x$ and $DE = \sqrt{2ax - xx}$, so $\square DFE = \sqrt{2ax^3 - x^4} - xx$ = Maximum. Its differential[43]

$$\frac{6axx\,dx - 4x^3\,dx}{2\sqrt{2ax^3 - x^4}} - 2x\,dx = \frac{3ax\,dx - 2xx\,dx - 2x\,dx\sqrt{\cdot\cdot}}{\sqrt{2ax - xx}} = 0.$$

Therefore $3a - 2x = 2\sqrt{2ax - xx}$ and $9aa - 12ax + 4xx = 8ax - 4xx$ and $xx = \dfrac{20ax - 9aa}{8}$ and[44] $x = \frac{5}{4}a \,\text{☋}\, \sqrt{\frac{7}{16}aa}$.

Problem XVI. [22] A traveler[45] A (Fig. 11.18) heading toward E must pass through the flat and well-worn field $AFDB$ and the rough and uneven territory $DBGE$, for any paths that have been walked, in the time a, the distance b in the flat field FDB may be traversed, and in the same time the distance c in the rough place. We wish

[43]In Schafheitlin (1922), the denominator of the second fraction was incorrectly given as $\sqrt{2ax^3 - x^4}$. This was also repeated in the next line.

[44]The symbol ☋ means $+$ or $-$.

[45]Compare to §59 on p. 54.

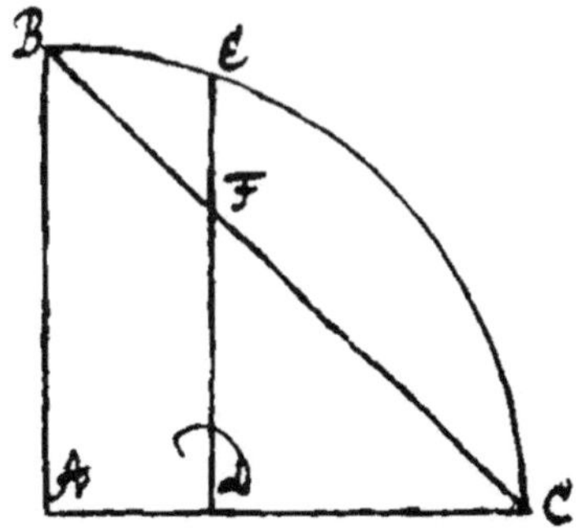

Fig. 11.17 Problem XV

Fig. 11.18 A Traveler in Two Terrains

to find the shortest route from A to E, i. e. how the traveler covers the distance in the smallest time. On the line BD that divides the two different paths, let the perpendiculars $AB = m$ and $ED = n$ be drawn. Let $BC = x$ and $BD = e$, we will have $DC = e - x$, $AC = \sqrt{mm + xx}$ and $CE = \sqrt{ee - 2ex + xx + nn}$. However,

$$b : a :: \sqrt{mm + xx} : \frac{a\sqrt{mm + xx}}{b}$$

= the time taken by the line AC and

$$c : a :: \sqrt{ee - 2ex + xx + nn} : \frac{a\sqrt{ee - 2ex + xx + nn}}{c}$$

= the time taken by the line CE. Therefore, we have

$$\frac{a\sqrt{mm + xx}}{b} + \frac{a\sqrt{ee - 2ex + xx + nn}}{c}$$

= the minimum interval of time. Its differ.

$$\frac{ax\,dx}{b\sqrt{mm+xx}} + \frac{ax\,dx - ae\,dx}{c\sqrt{ee-2ex+xx+nn}} = 0.$$

Therefore

$$\frac{x}{b\sqrt{mm+xx}} = \frac{e-x}{c\sqrt{ee-2ex+xx+nn}}$$

and $cceexx - 2ccex^3 + ccx^4 + ccnnxx = bbeemm + bbeexx - 2bbemmx - 2bbex^3 + bbmmxx + bbx^4$. And

$$\begin{array}{llll} +bb & -2bbe & +bbmm & \\ \quad x^4 & \quad x^3 & & -2bbemm\,x + bbeemm = 0. \\ -cc & +2cce & +bbee & \\ & & & xx \\ & & -ccee & \\ & & -ccnn & \end{array}$$

Problem XVII. [23] On the line *CE* (Fig. 11.19) to find the point D, at which if the lines *DA* and *DB* are drawn from the given points A and B, the sum of them is the minimum of any two lines drawn at the same time from the points A and B to a point of the line *CE*.

The perpendiculars $AC = a$ and $BE = b$ are dropped. Let $CE = c$ and $CD = x$. We will have $DE = c - x$.

$$AD = \sqrt{aa+xx} \quad \text{and} \quad BD = \sqrt{cc-2cx+xx+bb}.$$

And $\sqrt{aa+xx} + \sqrt{cc-2cx+xx+bb}$ = minimum. Its differential

$$\frac{x\,dx}{\sqrt{aa+xx}} + \frac{x\,dx - c\,dx}{\sqrt{cc-2cx+xx+bb}} = 0.$$

Therefore

$$\frac{x}{\sqrt{aa+xx}} = \frac{c-x}{\sqrt{cc-2cx+xx+bb}},$$

$ccxx - 2cx^3 + x^4 + bbxx = aacc - 2aacx + aaxx + ccxx - 2cx^3 + x^4$, $bbxx = aacc - 2aacx + aaxx$, hence $bx = ac - ax$ and $bx + ax = ac$. And finally $x = \dfrac{ac}{a+b}$.

Problem XVIII. On the Radius *AC* (Fig. 11.20) to find the point D, at which if the perpendicular *DE* is drawn to *AC*, the abscissa *FE* contained between the circumference *BEC* and the subtense *BC* is the maximum of all those that can be drawn in the same Quadrant.

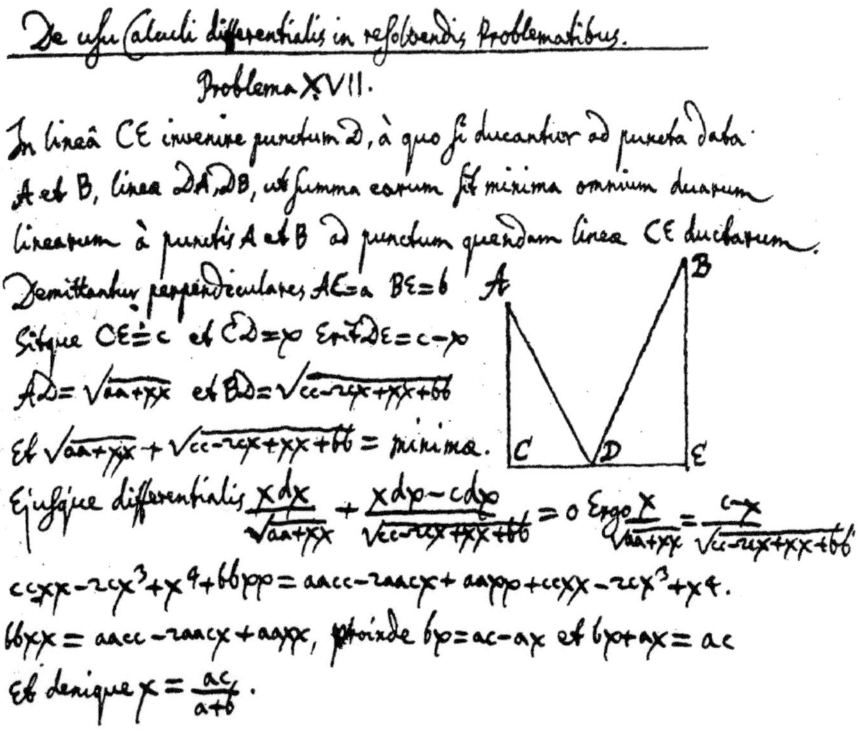

De usu Calculi differentialis in resolvendis Problematibus.

Problema XVII.

In lineâ CE invenire punctum D, à quo si ducantur ad puncta data A et B, lineæ DA, DB, ut summa earum sit minima omnium duarum linearum à punctis A et B ad punctum quoddam lineæ CE ductarum.

Demittantur perpendiculares $AC = a$ $BE = b$

Sitque $CE = c$ et $CD = x$ Erit $DE = c - x$

$AD = \sqrt{aa + xx}$ et $BD = \sqrt{cc - 2cx + xx + bb}$

Et $\sqrt{aa+xx} + \sqrt{cc-2cx+xx+bb} =$ minima.

Ejusque differentialis $\frac{xdx}{\sqrt{aa+xx}} + \frac{xdx - cdx}{\sqrt{cc-2cx+xx+bb}} = 0$ Ergo $\frac{x}{\sqrt{aa+xx}} = \frac{c-x}{\sqrt{cc-2cx+xx+bb}}$

$ccxx - 2cx^3 + x^4 + bbxx = aacc - 2aacx + aaxx + ccxx - 2cx^3 + x^4.$

$bbxx = aacc - 2aacx + aaxx$, proinde $bx = ac - ax$ et $bx + ax = ac$

Et denique $x = \frac{ac}{a+b}$.

Fig. 11.19 Problem XVII

Let $AC = a$ and $DC = x$, we will have $FD = x$ and $DE = \sqrt{2ax - xx}$, $FE = \sqrt{2ax - xx} - x =$ Maximum. Its differential

$$\frac{a\,dx - x\,dx}{\sqrt{2ax - xx}} - dx = 0.$$

Hence we will have $a - x = \sqrt{2ax - xx}$, $aa - 2ax + xx = 2ax - xx$,

$$xx = \frac{4ax - aa}{2} \quad \text{and} \quad x = a \pm \sqrt{\tfrac{1}{2}aa}.$$

Problem XIX. [24] Let the weight *A* (Fig. 11.21) be suspended from a rope *AC* fixed at the point *C* and crossing over a pulley *E*, which hangs freely on a rope affixed at *B*. We wish to find where the pulley *E* and the weight *A* will be at rest.[46]

[46]Compare to §60 on p. 56.

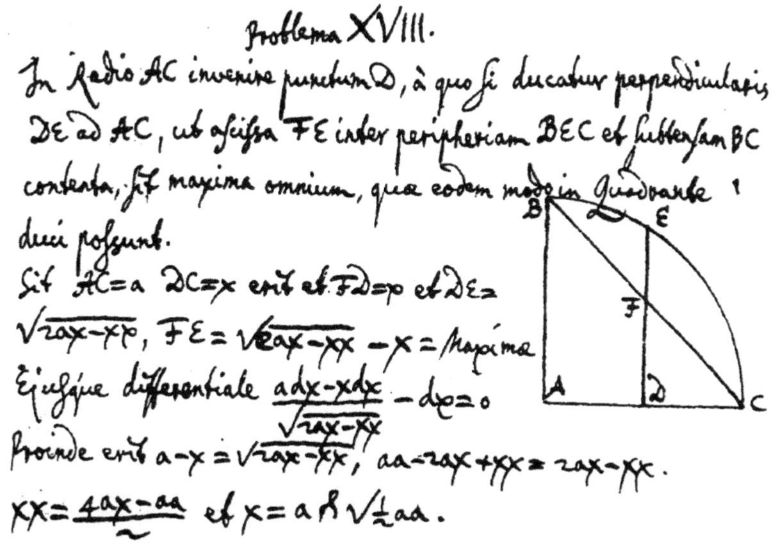

Problema XVIII.

In Radio AC invenire punctum D, à quo si ducatur perpendicularis DE ad AC, ut abscissa FE inter peripheriam BEC et subtensam BC contenta, sit maxima omnium, quae eodem modo in Quadrante duci possunt.

Sit AC = a DC = x erit et FD = x et DE = $\sqrt{2ax - xx}$, FE = $\sqrt{2ax - xx} - x$ = Maxima

Ejusque differentiale $\frac{adx - xdx}{\sqrt{2ax - xx}} - dx = 0$

Proinde erit $a - x = \sqrt{2ax - xx}$, $aa - 2ax + xx = 2ax - xx$.

$xx = \frac{4ax - aa}{2}$ et $x = a \pm \sqrt{\frac{1}{2}aa}$.

Fig. 11.20 Problem XVIII

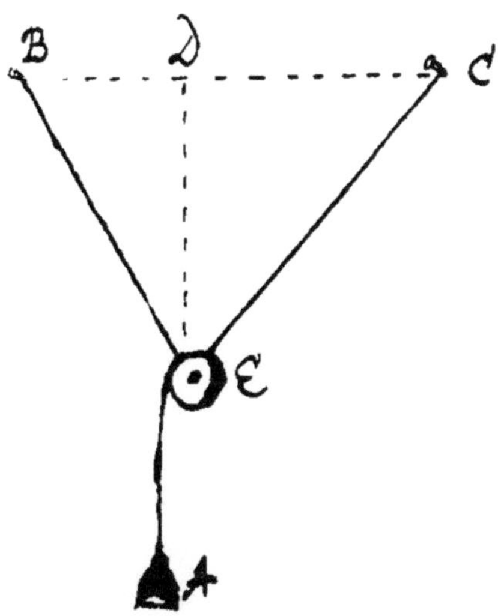

Fig. 11.21 The Pulley Problem

Supposing that the rope and the pulley have no weight, the Pulley and weight rest in that place, where the distance AD of the weight A to the line BC, parallel to the horizon, is a maximum. Therefore, to find the length of the rope, let $AC = a$, $BC = b$, $BE = c$, and $DE = x$. We will have $BD = \sqrt{cc - xx}$, $DC = b - \sqrt{cc - xx}$, $CE = \sqrt{bb + cc - 2b\sqrt{cc - xx}}$, and $AE = a - \sqrt{bb + cc - 2b\sqrt{cc - xx}}$. $AD = x + a - \sqrt{bb + cc - 2b\sqrt{cc - xx}} =$ Maximum. Its different.[47]

$$dx - \frac{bx\,dx}{\sqrt{cc - xx} \text{ times } \sqrt{bb + cc - 2b\sqrt{cc - xx}}} = 0.$$

Hence

$$\sqrt{bb + cc - 2b\sqrt{cc - xx}} = \frac{bx}{\sqrt{cc - xx}}$$

and

$$bb + cc - 2b\sqrt{cc - xx} = \frac{bbxx}{cc - xx},$$

$$bb + cc - \frac{bbxx}{cc - xx} = 2b\sqrt{cc - xx} = \frac{bbcc + c^4 - 2bbxx - ccxx}{cc - xx}.$$

Therefore $b^4c^4 + c^8 + 4b^4x^4 + c^4x^4 + 2bbc^6 - 4b^4ccxx - 6bbc^4xx - 2c^6xx + 4bbccx^4 = 4bbc^6 - 12bbc^4xx + 12bbccx^4 - 4bbx^6$.

In Another Way. Let $BD = x$. We will have $DC = b - x$, $DE = \sqrt{cc - xx}$, $CE = \sqrt{bb - 2bx + cc}$, $AE = a - \sqrt{bb - 2bx + cc}$, $AD = a - \sqrt{bb - 2bx + cc} + \sqrt{cc - xx} =$ Maximum. Its differential $=$

$$\frac{+\,b\,dx}{\sqrt{bb - 2bx + cc}} - \frac{x\,dx}{\sqrt{cc - xx}} = 0.$$

Therefore $x\sqrt{bb - 2bx + cc} = b\sqrt{cc - xx}$. And $bbxx - 2bx^3 + ccxx = bbcc - bbxx$ and

$$x^3 = \frac{2bbxx + ccxx - bbcc}{2b}.$$

Problem XX. [25] To find the shortest Crepuscule.[48]

[47]The Latin preposition *in* was used to mean the multiplication of the two factors. This will be consistently translated as "times."

[48]I.e. the shortest twilight. Compare to §61 on p. 57.

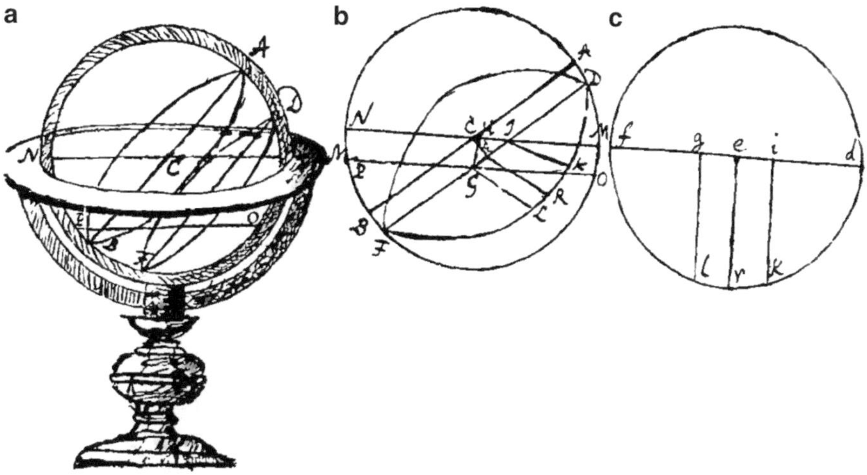

Fig. 11.22 Problem of the Shortest Twilight

Let C (Fig. 11.22a) be the center of the Sphere, *ANB* the meridian, *AB* the Diameter of the Equator, *MN* the diameter of the Horizon, *OP* the diam. of the parallel crepuscular circle, *DF* the diameter of the parallel to the Equator, which we wish to find and is such that, as the Sun rotates along it, the arc *KL*, which is intercepted between the Horizon and the Crepuscular parallel, is to be proportionally the smallest, that is, so that it has the minimum ratio to its circumference, or to its Diameter. Indeed, in this case this is well known to be the Crepuscule that is the smallest of all.

Now let (Fig. 11.22b) the radius be $CA = a$. GH or the sine of the crepuscular arc $PN = b$. GH is to GJ as the sine of the complement of the elevation of the pole is to the total sine. Therefore let $GJ = c$, hence $HJ = \sqrt{cc - bb} = f$. Let CE or the sine of BF, the declination of the sun that we wish to find, be $= x$. Because $GH : HJ :: CE : EJ$ we will have $EJ = \dfrac{fx}{b}$ = the sine of the arc *RK* and $EG = \dfrac{bc - fx}{b}$ = the sine of the arc *RL*. Now separately (Fig. 11.22c), the great circle *drf* is described, with center *e*, diameter *df*, and radius *er* perpendicular to the Diameter. Let it be made that as *DE* is to *EJ*, so *de* is to *ei*, and as *DE* is to *EG*, so *de* is to *eg*, the parallels *ik* and *gl* will be drawn, the arc *kr* similar to[49] *KR* and the arc *rl* [26] similar to the arc *RL* and the whole *kl* similar to the whole *KL*. However, because the arc *KL* must be the minimum proportion, the arc *kl* will be to the constant radius as an absolute minimum and for that reason the subtense *kl* will be the minimum.

[49]In Schafheitlin (1922) the following was written as *kR*.

Consequently by this method, in assuming x, *ie* and *ig* will be found, and from these *ik* and *kl*, from which the subtense *kl* can then be found, granting that this method should give the minimum, as the equation comes forth, which determines x. In fact, because the subtense *kl* can only be found through great skill, this method is exceedingly prolix, and it will produce an equation, for the determination of x, that will have six dimensions, and in the end will contain more than thirty, to such a degree that solving the problem by the Cartesian Method is clearly not possible.

We now consider which of the above things may be obtained by means of the Differential Calculus. In the first place, it is not necessary to find the subtense, or sine, or other straight line, that would determine the arc *kl*, as Cartesian Geometry requires, concerning the measurement, for instance, of those curves that do not turn. It is sufficient to us that the arc *kl* must be the smallest, hence its differential $= 0$, or the differential of the arc kr $=$ different. of the arc *rl*. Consequently, it only remains that the differentials of the arcs *kr* and *rl* be found, which, when equated, will give the equation needed, whose root x will be the sine of the declination of the sun that we wish to find. Moreover, the differentials of the arcs *kr* and *rl* are discovered thusly: because $DE = \sqrt{aa - xx}$ and by Construction $DE : JE :: de : ie$, we will have

$$ie = \frac{afx}{b\sqrt{aa - xx}},$$

in the same way we will have

$$eg = \frac{abc - afx}{b\sqrt{aa - xx}}$$

hence

$$ik = \sqrt{\frac{a^4bb - aabbxx - aaffxx}{aabb - bbxx}} = (\text{because } bb + ff = cc)\ \sqrt{\frac{a^4bb - aaccxx}{aabb - bbxx}}$$

and

$$gl = \sqrt{\frac{a^4bb - aabbxx - aabbcc + 2aabcfx - aaffxx}{aabb - bbxx}} = (\text{because } bb + ff = cc)$$

$$[27] \quad \sqrt{\frac{a^4bb - aaccxx + 2aabcfx - aabbcc}{aabb - bbxx}}.$$

If *ie* is now called m, we will have $ik = \sqrt{aa - mm}$, and the differential of the arc $kr = \dfrac{a\,dm}{\sqrt{aa - mm}}$, therefore because $m = \dfrac{afx}{b\sqrt{aa - xx}}$, we will have

$$dm = \frac{a^3 f\ dx}{\overline{aab - bxx}\sqrt{aa - xx}},$$

and because

$$\sqrt{aa - mm} = \sqrt{\frac{a^4bb - aaccxx}{aabb - bbxx}},$$

we will have

$$\frac{a\,dm}{\sqrt{aa - mm}} = \text{different. of the arc } kr = \frac{a^3 f\ dx}{\overline{aa - xx}\sqrt{aabb - ccxx}}.$$

If eg is now called n, hence $gl = \sqrt{aa - nn}$ we will have diff. of the arc $rl = \frac{a\,dn}{\sqrt{aa - nn}}$. Therefore because $n = \frac{abc - afx}{b\sqrt{aa - xx}}$, we will have

$$dn = \frac{-a^3 f\ dx + abcx\,dx}{\overline{aab - bxx}\sqrt{aa - xx}},$$

and because

$$\sqrt{aa - nn} = \sqrt{\frac{a^4bb - aaccxx + 2aabcfx - aabbcc}{aabb - bbxx}},$$

we will have

$$\frac{a\,dn}{\sqrt{aa - nn}} = \text{different. of the arc } lr = \frac{-a^3 f\ dx + abcx\,dx}{\overline{aa - xx}\sqrt{aabb - ccxx + 2bcfx - bbcc}},$$

and for that reason, because different. of the arc lr = differ. of the arc kr, we have

$$\frac{a^3 f\ dx}{\overline{aa - xx}\sqrt{aabb - ccxx}} = \frac{-a^3 f\ dx + abcx\,dx}{\overline{aa - xx}\sqrt{aabb - ccxx + 2bcfx - bbcc}}$$

dividing each side by $a\,dx$ and multiplic. by $aa - xx$, we will have

$$\frac{aaf}{\sqrt{aabb - ccxx}} = \frac{-aaf + bcx}{\sqrt{aabb - ccxx + 2bcfx - bbcc}},$$

or by cross multipl.

$$aaf\sqrt{aabb - ccxx + 2bcfx - bbcc} = \overline{-aaf + bcx}\sqrt{aabb - ccxx},$$

and taking squares we will have $a^6bbff - a^4ffccxx + 2a^4f^3bcx - a^4ffbbcc = a^6bbff - 2a^4b^3fcx + aab^4ccxx - a^4ffccxx + 2aafbc^3x^3 - bbc^4x^4$. Reducing the equation to zero and dividing by bc, we will have

$$bc^3x^4 - 2aafccx^3 - aab^3cxx + 2a^4bbfx + 2a^4f^3x - a^4ffbc = 0$$

or because $ff + bb = cc$, dividing the equation by c we will have

$$bccx^4 - 2aafcx^3 - aab^3xx + 2a^4fcx - a^4ffb = 0,$$

or substituting the value of $ff = cc - bb$ we will have

$$bccx^4 - 2aafcx^3 - aab^3xx + 2a^4fcx - a^4bcc + a^4b^3 = 0.$$

The equation is divided by $xx - aa$ and we will have

$$bccxx - 2aafcx + aabcc - aab^3 = 0,$$

or (because $cc - bb = ff$) $bccxx - 2aafcx + aabff = 0$, which equation, if [28] solved, will give

$$x = \frac{aaf \pm af\sqrt{aa - bb}}{bc}.$$

I can most easily convert this equation into the simplest proportion in this way: because JG is to JH, or c is to f, as the total sine or a is to the sine of the elevation of the pole, it therefore will be $\frac{af}{c}$. Hence

$$\frac{aaf \pm af\sqrt{aa - bb}}{bc} = \frac{a \pm \sqrt{aa - bb}}{b} \text{ times the sine of the elevation of the pole.}$$

For that reason, $b : a \pm \sqrt{aa - bb}$ :: sine elev. pole : x. Or in terms of Trigonometry we have this proportion: as the right sine of the crepuscular arc is to the versed sine of this arc (for the $-$ sign), or as the versed sine of the complement in two right angles[50] of the same arc (for the $+$ sign), so is the sine of the elevation of the pole to the sine of the declination of the sun towards the South, which we wish to find. If the declination of the sun is known, the location of the same on the Ecliptic can be found by a single operation. But if there should be two minimum crepuscules, and each one is actually frequented, as the sun turns through the Southern Signs, that one which is found with the $-$ sign, generally is at an elevation of the pole that is possible.

[50] I.e., the supplement. In Schafheitlin (1922), the phrase *ad duos rectos* was misprinted as *ad discos rectos*.

The other one with the + sign, however, cannot happen unless it happens very near the Equatorial region. Indeed, at this other one a greater elevation of the pole comes about, as the declination of the sun gets larger, than the maximum declination of the same, which is 23 $\frac{1}{2}$ deg. In fact it sometimes happens, that the sine of the declination that we wish to find gets entirely larger than the total sine, and hence the said minimum Crepuscule is not made only impossible, but clearly imaginary. However, each minimum crepuscule, where it is possible, happens twice in a year, because the sun comes to the same declination twice. The maximum crepuscules happen when the sun appears at the Tropics, which is easy to demonstrate.

[29] NB. I judge that the second of the shortest crepuscules, found by the + sign, by no means satisfies the question, just as happens when only a single root of an equation satisfies a Problem. For if there are two different minimum crepuscules. A maximum crepuscule must necessarily come between them, which is not found by means of our equation, inasmuch as that reached only to two dimensions. Otherwise if there are two minimum crepuscules with a maximum coming between, the equation would rise to at least three dimensions.

On the Discovery of the Inflection Point of Curves

Problem XXI. Let any Curve, which has double curvature, concave toward the axis from the beginning, and afterward convex to the same; or vice versa convex from the beginning, and concave to the end; moreover that point, which separates those two curvatures, or which is the end of the former and the beginning of the latter, is called the point of inflection or of bending back.[51] Therefore as often as the Curve changes its curvature, it has as many inflection points, in whatever way the curve is defined, which we will now find according to one or another method.

First Method

It is clear from the Consideration of the Curve, that, as long as the Curve maintains only one Curvature, the Tangents at successive increasing abscissas recede from the vertex of the curve, but where it first puts on the contrary curvature, the Tangents at increasing abscissas now come near to the vertex.[52] I say this comes from carefully considering the curvature of whatever nature; then from this the infection point is most easily determined. Indeed, because the tangent at the inflection point is farthest from the vertex, [30] the subtangent minus the abscissa, or the abscissa minus the Subtangent, will be the maximum of all possible, that is $t - x = m$ or $x - t = m$,

[51]Compare this to Definition II given on p. 66.

[52]This is clarified in §66, Part 1. on p. 68.

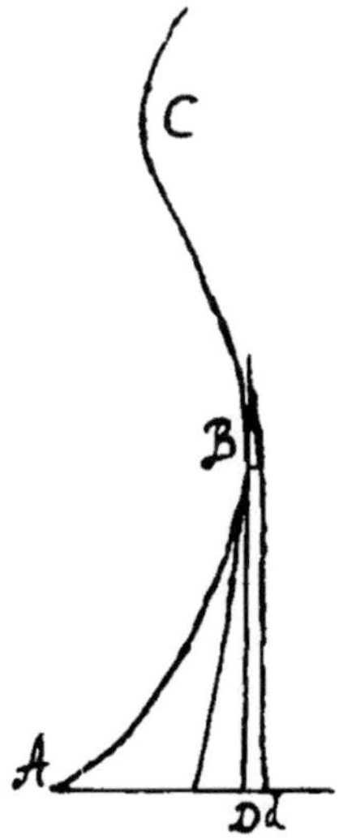

Fig. 11.23 A Curve Related to the "Witch of Agnesi"

hence by the Method of maxima and minima, we will have $dt - dx = 0$ or $dx - dt = 0$. From this equation, the value of the abscissa x emerges, at which the ordinate y determines the inflection point that we wish to find.

Second Method

This same point can be found in a different way as follows: I imagine to be in that place, where the Curve is simultaneously convex and concave, indeed it is equally the one and the other; because it is not possible for both to be true, it is necessary that it should be straight, that is, neither convex nor concave (however, this should not be understood as if a finite portion of the curve were straight, but rather that two infinitely small parts lie in a line). Therefore, whereas in any straight line in which dx is made constant, dy is likewise constant, and hence ddy (the differential of the differential of y) is $= 0$, the inflection point can be found by making $ddy = 0$, from which equation the abscissa x will be determined, and at the same time the inflection point that we wish to find. Examples will to a greater extent illustrate each method of solving the matter.

Therefore let *ABC* (Fig. 11.23) be the given curve,[53] whose nature (setting $AD = x$ and $BD = y$) is expressed by this equation $axx - yxx - aay = 0$, we wish to find the inflection point B. *By the first method.* Differentials of the equation are taken,

[53]Compare this to §68 on p. 71.

and we will have $2ax\,dx - xx\,dy - 2xy\,dx - aa\,dy = 0$ and therefore $2ax\,dx - 2yx\,dx = xx\,dy + aa\,dy$, and hence $2ax - 2yx : xx + aa :: dy : dx :: y : t$. Consequently one finds

$$t = \frac{xxy + aay}{2ax - 2yx} = \frac{axx}{2ax - 2yx} = \frac{ax}{2a - 2y}$$

$$= \left(\text{because } y = \frac{axx}{aa + xx}\right) \frac{aax + x^3}{2aa}.$$

Therefore

$$x - t = \frac{aax - x^3}{2aa} = \text{Maximum.}$$

Hence the differential of the same

$$\frac{aa\,dx - 3xx\,dx}{2aa} = 0.$$

Multipl. by $2aa$ and divis. by dx we will have $aa - 3xx = 0$ or $\frac{1}{3}aa = xx$ and $x = a\sqrt{\frac{1}{3}}$.

[31] *By the second method.* Because

$$y = \frac{axx}{aa + xx}$$

taking differentials we will have[54]

$$dy = \frac{2a^3x\,dx}{\square : \overline{aa + xx}} \quad \text{and} \quad ddy = \frac{2a^7\,dx^2 - 4a^5xx\,dx^2 - 6a^3x^4\,dx^2}{\text{QQ} : \overline{aa + xx}} = 0$$

or multiplic. by QQ : $\overline{aa + xx}$ and divid. by $2a^3\,dx^2$ we will have $a^4 - 2aaxx - 3x^4 = 0$, which equation, if divided by $aa + xx$, will give $aa - 3xx = 0$, as previously.

These two methods succeed no less in Mechanical curves than in Geometric, if the mode is duly applied, and the ratio between dy and dx is obtained.

Let, e.g., *ABC* (Fig. 11.24)[55] be the curve of such a nature that as the semi-circle *AGF* is described on *AF*, and the ordinate *BD* is produced at *G*, *BD* is = the arc *AG*. We wish to find the inflection point *B*.

---

[54]From this point onward, most occurrences $\square, C, QQ$ (for square, cube, and fourth power) are followed by a colon. This is not meant to represent a proportion, but rather the application of the power to the expression that follows it.

[55]The letter C was missing in the figure in the manuscript.

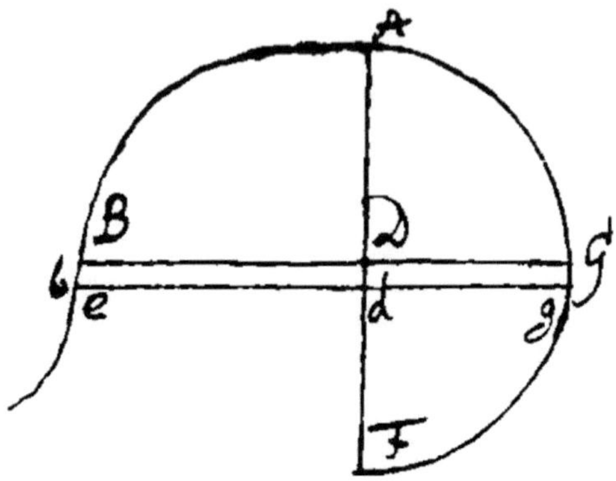

Fig. 11.24 The Inverse Cosine Curve

By method 1. Let $AD = x$, AG or $BD = y$ and $AF = 2a$; we will have

$$dy = (\text{because } be = Gg)\frac{a\,dx}{\sqrt{2ax - xx}}.$$

Also $dy : dx :: y : t$, therefore

$$t = \frac{y\,dx}{dy} = \frac{y\sqrt{2ax - xx}}{a}$$

and $t - x = \dfrac{y\sqrt{2ax - xx}}{a} - x =$ Maximum. Therefore its differential

$$\frac{2ax\,dy - xx\,dy + ay\,dx - yx\,dx}{a\sqrt{2ax - xx}} - dx = 0$$

or $2ax\,dy - xx\,dy = yx\,dx - ay\,dx + a\,dx\sqrt{2ax - xx}$ and

$$dy = \frac{yx\,dx - ay\,dx + a\,dx\sqrt{2ax - xx}}{2ax - xx}.$$

However, it was found above

$$dy = \frac{a\,dx}{\sqrt{2ax - xx}} = \frac{yx\,dx - ay\,dx + a\,dx\sqrt{2ax - xx}}{2ax - xx}.$$

Reducing the equation one has $yx - ay = 0$ and hence $x = a$.

By method 2. Because

$$dy = \frac{a\,dx}{\sqrt{2ax - xx}}, \quad \text{we will have} \quad ddy = \frac{-aa\,dx^2 + ax\,dx^2}{2\,\overline{aa - xx}\sqrt{2ax - xx}} = 0.$$

And for this reason, $-aa + ax = 0$ and $x = a$.

[32] Thus to determine the inflection point of the Conchoid of Nichomedes[56] by the methods we have given, it is necessary to have the nature of the Conchoid by means of an equation, or any relation exhibited between the abscissa and the ordinate.

Therefore let (Fig. 11.25) $AF = a$, $EF = b$, $AD = x$ and $BD = y$; we will have $BG = a$ and $DE = a - x$. Because $DE : EF :: BG : GF$, we will have

$$GF = \frac{ab}{a - x}. \quad \text{Therefore } GE = \frac{\sqrt{2abbx - bbxx}}{a - x}.$$

However $GF : GE :: BF : BD$, that is

$$\frac{ab}{a - x} : \frac{\sqrt{2abbx - bbxx}}{a - x} \text{ or } a : \sqrt{2ax - xx} :: \frac{aa + ab - ax}{a - x} : y$$

and for this reason

$$y = \frac{b}{a - x}\sqrt{2ax - xx} + \sqrt{2ax - xx}$$

and its different.

$$dy = \frac{aab\,dx}{\overline{aa - 2ax + xx}\sqrt{2ax - xx}} + \frac{a\,dx - x\,dx}{\sqrt{2ax - xx}} : dx :: y \left(\overline{\frac{a + b - x}{a - x}}\sqrt{2ax - xx}\right) : t.$$

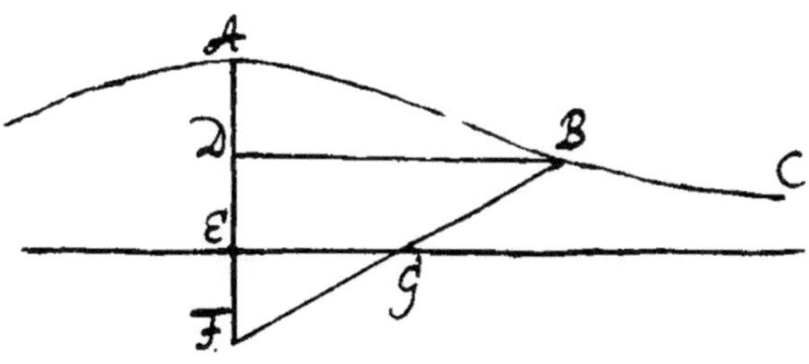

Fig. 11.25 The Conchoid of Nichomedes

[56] Compare this to §71 on p. 72.

Therefore, we will have

$$t = \frac{\overline{ab - bx + aa - 2ax + xx} \quad \text{times} \quad \overline{2ax - xx}}{aab + C : \overline{a - x}}.$$

Now put $a - x = z$ in order to make the Calculation easier, and we will have

$$t = \frac{\overline{bz + zz} \quad \text{times} \quad \overline{aa - zz}}{aab + z^3} = \frac{aabz + aazz - bz^3 - z^4}{aab + z^3}$$

and $t - x = t - a + z = \dfrac{2aabz + aazz - bz^3 - az^3 - a^3b}{aab + z^3} =$ Maximum. And therefore the differential of that will be

$$= \frac{2a^4bb\,dz + 2a^4bz\,dz - 3aabbzz\,dz - 4aabz^3\,dz - aaz^4\,dz}{\Box : aab + z^3} = 0.$$

This is multiplied by $\Box : aab + z^3$ and divided by $aab\,dz + aaz\,dz$, and we will have $2aab - 3bzz - z^3 = 0$ or $z^3 + 3bzz - 2aab = 0$ will be obtained. The Root z of this equation gives the value of $a - x$, or the abscissa ED, the ordinate BD to which goes through the inflection point B that we wished to find. And thus for the first Method.

[33] By the second method, we therefore have:

Because $dy = \dfrac{aab\,dx}{\overline{aa - 2ax + xx}\sqrt{2ax - xx}} + \dfrac{a\,dx - x\,dx}{\sqrt{2ax - xx}}$ substituting z in the place of $a - x$ we will have

$$dy = \frac{aab\,dx}{zz\sqrt{aa - zz}} + \frac{z\,dx}{\sqrt{aa - zz}}$$

therefore its differential[57]

$$\frac{-2a^4bz\,dz\,dx + 3aabz^3\,dz\,dx}{\overline{aaz^4 - z^6}\sqrt{aa - zz}} + \frac{aa\,dz\,dx}{\overline{aa - zz}\sqrt{aa - zz}} = 0.$$

Multiplying the equation by $\overline{aaz^4 - z^6}\sqrt{aa - zz}$ and dividing by $aaz\,dz\,dx$, we have $z^3 + 3bzz - 2aab = 0$, as previously. Noting that if $a = b$ the resulting equation $z^3 + 3azz - 2a^3 = 0$ becomes planar; indeed, it can be divided by $z + a = 0$ and we will have $zz + 2az - 2aa = 0$, hence $z = -a + \sqrt{3aa} = ED$.

[57] In Schafheitlin (1922), the differential dz was missing from the first term of the first numerator.

Now let (Fig. 11.26) *ABC* be another kind of Conchoid,[58] which is such that the Rectangle between *FG* and *GB* is everywhere equal to the Rectangle between *FE* and *EA*. We wish to find the inflection point *B*.

With the same set out as above, the relation between x and y is sought, in this way:

$$DF : EF :: BD : GE \quad \text{therefore} \quad GE = \frac{by}{a+b-x},$$

from which it is found

$$GF = \sqrt{\frac{bbyy}{\square : \overline{a+b-x}} + bb},$$

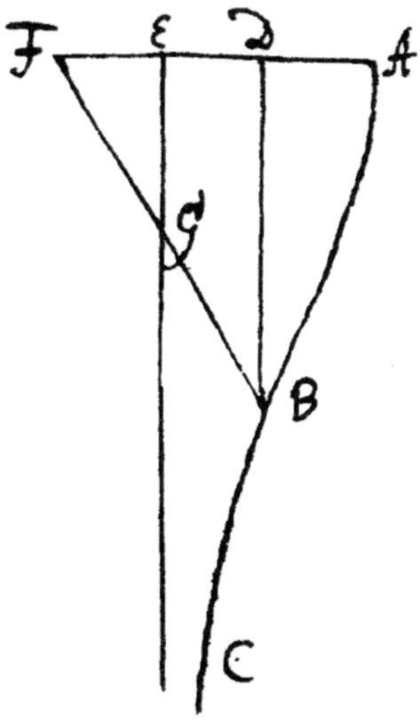

Fig. 11.26 A Different Conchoid

[58]Compare this to §72 on p. 74.

and because

$$EF : ED :: GF : GB \quad \text{we will have} \quad GB = \frac{a-x}{b}\sqrt{\frac{bbyy}{\overline{\Box a - x + b}} + bb}$$

hence

$$\Box FGB = \overline{a-x} \quad \text{times} \quad \overline{\frac{byy}{\overline{\Box a - b + x}} + b} = \Box FEA = ab.$$

To abbreviate the calculation, let $a - x = z$ and the equation that was found will be changed to the following $zyy + z^3 + 2bzz + bbz = azz + 2abz + abb$, and for that reason

$$yy = \frac{azz + 2abz + abb - z^3 - 2bzz - bbz}{z}$$

and

$$y = \overline{z+b}\sqrt{\frac{a-z}{z}} = \sqrt{az - zz} + b\sqrt{\frac{a-z}{z}}$$

therefore

$$dy = \frac{a\,dz - 2z\,dz}{2\sqrt{az - zz}} - \frac{ab\,dz}{2z\sqrt{az - zz}}.$$

Because $dz = -dx$

$$\frac{-a\,dx + 2z\,dx}{2\sqrt{az - zz}} + \frac{ab\,dx}{2z\sqrt{az - zz}} : dx :: y\left(= \overline{z+b}\sqrt{\frac{a-z}{z}}\right) : t$$

therefore

$$t = \frac{2z^3 + 2azz - 2bzz + 2abz}{-az + 2zz + ab}$$

[34] and

$$t - x = t - a + z = \frac{-azz - 2bzz + aaz + 3abz - aab}{-az + 2zz + ab} = \text{Maximum},$$

and for this reason, the differential of this, which is[59]

$$\frac{\overline{-aazz - 4abzz + 2aabz - 4abbz + 3aabb} \quad \text{times} \quad + dz}{\square : \overline{-az + 2zz + ab}} = 0.$$

Multipl. by[60] $\square : \overline{-az + 2zz + ab}$ and divid. by $az + ab$ times dz, we have

$$-az - 4bz + 3ab = 0 \quad \text{hence} \quad z = \frac{3ab}{a + 4b}.$$

By the second method the same is found as follows.

Because

$$dy = \frac{-a\,dx + 2z\,dx}{2\sqrt{az - zz}} + \frac{ab\,dx}{2z\sqrt{az - zz}},$$

therefore its differential

$$\frac{aa\,dz\,dx}{\overline{4az - 4z\sqrt{az - zz}}} + \frac{4abzz\,dz\,dx - 3aabz\,dz\,dx}{\overline{4az^3 - 4z^4\sqrt{az - zz}}} = 0.$$

Multiplying the equation by $\overline{4az^3 - 4z^4}\sqrt{az - zz}$ and dividing by $az\,dz\,dx$ we have $az + 4bz - 3ab = 0$ and hence $z = \dfrac{3ab}{a + 4b} = ED$, as previously.

Third method for finding the inflection point.

The Examples produced will be sufficient to show, that the given methods for finding inflection points can be reduced to the method of Maximum and Minimum. However, it is well known, that it is always necessary, that the relation between x and y is had by means of an equation, if the inflection point is to be found by the proposed method. With the method I now reveal, the inflection point is determined only by means of the generation of the curve, and from this the relation between x and y is obtained. In what follows, however, I go in different directions in understanding the inflection point.

---

[59]In Schafheitlin (1922), the overline extended over the entire numerator.

[60]In Schafheitlin (1922), the minus sign before the az was omitted.

Fig. 11.27 Differentials

[35] I suppose that any curve is composed of an infinity of infinitely small straight lines (Fig. 11.27) *ab*, *bc*, *cd*, etc., and the tangent at any point d to be nothing other than the small line *dc* itself, produced to m. However, it is evident that if the curve is outwardly convex, the tangent of the subsequent little piece *de* falls outside, and makes the infinitely small angle *ldm*. If, however, the curve is outwardly concave, the tangent of the subsequent little piece falls inside. Consequently, the inflection point will be in that place, where the tangent of the subsequent little piece falls neither inside nor outside and hence the preceding little piece coincides with the Tangent, that is, where two adjacent small pieces, such as *de* and *fg*, lie in a straight line.

This is understood of all curves, whose nature does not become known except by generation and by the relation of lines extending from a certain common point to other ones, which, if they have an inflection point, its general equation can be determined.

Indeed, let *ABC* (Fig. 11.28) be any Curve, having an inflection point at B that we wish to find. From the given point F (the line drawn from which to the curve explains the generation or nature of the curve) we understand the lines *FB* and *Fb* to be drawn, making the infinitely small angle *bFB*, and having drawn the perpendiculars *FD* and *Fd* to *FB* and *Fb*, let the tangent *BdD* be drawn from the point B, which (because B is the inflection point) will likewise be the tangent to the point b. Describing the arcs *Be* and *gd* with the center F, let FB or $Fb = z$, FD or $Fd = t$, and $Be = dy$; we will have $be = dz$ and $gD = dt$. Because ang. $BFe = gFd$ we will have $FB : Fd :: Be : gd$, therefore $gd = \dfrac{t\,dy}{z}$ and (because of the similarity of the triangles *beB* and *gdD*) $be : Be :: gd : gD$, which is $dz : dy :: \frac{t\,dy}{z} : dt$, therefore $\dfrac{t\,dy^2}{z} = dz\,dt$, or [36] (because $t : z :: dy : dz$), $\dfrac{dy^3}{dz} = dz\,dt$ and $dy^3 = dz^2\,dt$, from which equation, when dy and dt can be given in terms of dz, z may be extracted, or the line *FB*, hence *FB* is known and also the inflection point B is found.

E.g. let *ABC* (Fig. 11.29) be the first Conchoid, of Nichomedes, of which A is the vertex, F the pole, and *MN* the Asymptote. We wish to find the inflection point

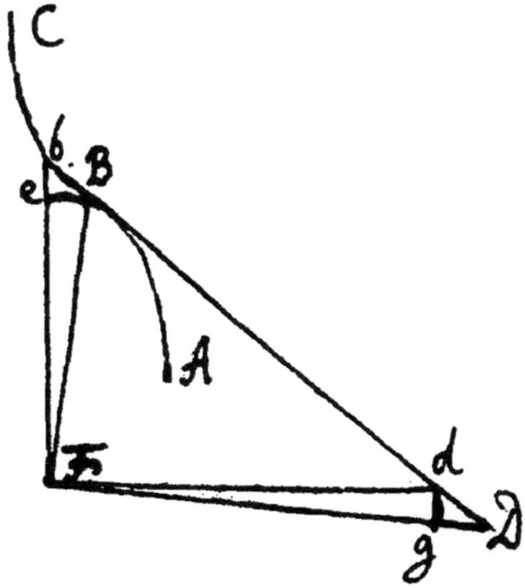

Fig. 11.28 Inflection Point in a Curve Generated from a Pole

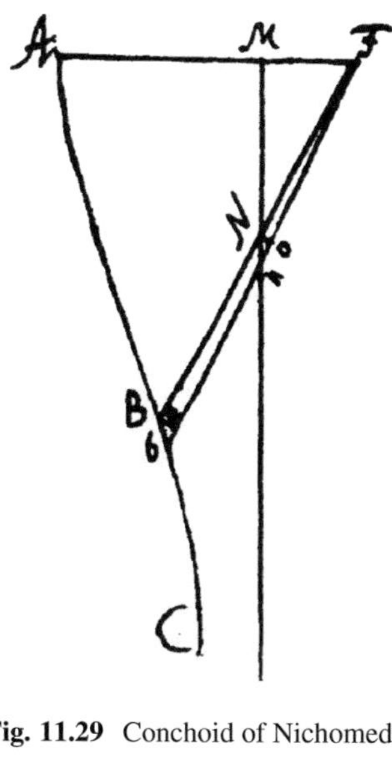

Fig. 11.29 Conchoid of Nichomedes

B, without knowing the relation between abscissas and ordinates, but only from the generation of the Conchoid, for which, of course, however you draw *FB*, the intercepted *NB* is always equal to the constant *AM*.

Given this, make $AM = NB = a$, $FM = b$, *FB* or $Fb = z$, $be = dz$, and $Be = dy$; if *NO* is drawn parallel to *Be*, we will have $FN = z - a$, $no = dz$, and $NM = \sqrt{zz - 2az + aa - bb}$. By simil. of triangles *NMF* and *Non*, $NM : MF :: no :$

oN therefore

$$No = \frac{b\,dz}{\sqrt{zz - 2az + aa - bb}}$$

and because $FN : FB :: No : Be$, we will have[61]

$$Be = \frac{bz\,dz}{\overline{z-a}\sqrt{zz - 2az + aa - bb}} = dy.$$

Likewise

$$be : Be :: bF : t = \frac{bzz}{\overline{z-a}\sqrt{zz - 2az + aa - bb}},$$

and whose differential $-dt$ (NB. $-dt$ is used, because as z increases, t decreases, and hence the different. of t is a negative quantity)[62]

$$= \frac{-2abz^3\,dz + 4aabbzz\,dz - b^3zz\,dz - 2a^3bz\,dz + 2ab^3z\,dz}{\overline{zz - 2az + aa - bb}\sqrt{zz - 2az + aa - bb} \text{ times } \Box : \overline{z-a}}.$$

Consequently the general equation $dy^3 = dz^2\,dt$ is modified into the following[63]

$$\frac{b^3z^3\,dz^3}{\overline{zz - 2az + aa - bb}\sqrt{zz - 2az + aa - bb} \text{ times } C : \overline{z-a}}$$
$$= \frac{2abz^3\,dz^3 - 4aabzz\,dz^3 + b^3zz\,dz^3 + 2a^3bz\,dz^3 - 2ab^3z\,dz^3}{\overline{zz - 2az + aa - bb}\sqrt{zz - 2az + aa - bb} \text{ times } \Box : \overline{z-a}}$$

an equation which multiplied by $\overline{zz - 2az + aa - bb}\sqrt{zz - 2az + aa - bb}$ times C : $\overline{z-a}$ and divided by $bz\,dz^3$, when reduced to zero will give this: $2az^3 - 6aazz + 6a^3z - 3abbz - 2a^4 + 2aabb = 0$. If $a = b$, we have $2zz - 6az + 3aa = 0$ and $z = \frac{3}{2}a + \sqrt{\frac{3}{4}aa} = FB$.

[37] Now ABC (Fig. 11.30) will be the other Conchoid, in which all rectangles FNB are equal to FMA, and setting $FN = x$ as before, we will have $no = dx$, $NM = \sqrt{xx - bb}$,

$$No = \frac{b\,dx}{\sqrt{xx - bb}}, \quad NB = \frac{ab}{x}, \quad FB \text{ or } z = \frac{ab + xx}{x},$$

[61] In Schafheitlin (1922), the overline on the term $z - a$ was omitted.

[62] In Schafheitlin (1922), every occurrence of dz in the numerator was written as dz^3.

[63] In Schafheitlin (1922), every occurrence of dz^3 in the numerator of the second fraction was written as dz.

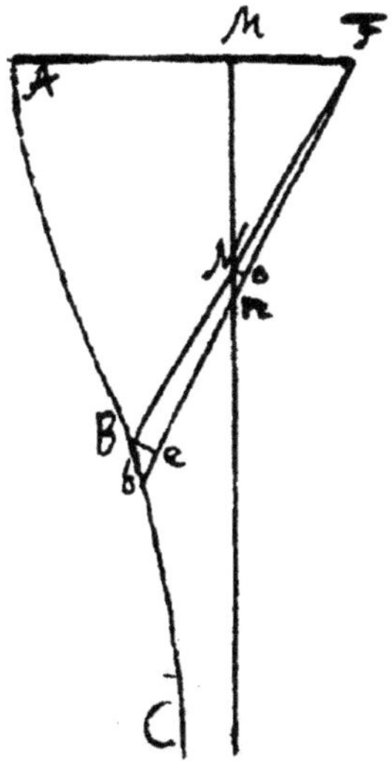

Fig. 11.30 Another Conchoid

and

$$be \;\text{ or }\; dz = \frac{xx\,dx - ab\,dx}{xx},$$

now because $FN : FB :: No : Be$, we will have Be or

$$dy = \frac{abb\,dx + bxx\,dx}{xx\sqrt{xx - bb}}$$

and because $be : eB :: BF : t$, we will have

$$t = \frac{aab^3 + 2abbxx + bx^4}{\overline{x^3 - abx}\sqrt{xx - bb}},$$

taking the differential we will have

$$-dt = -6abbx^6 - b^3x^6 + 5ab^4x^4 - 4aab^3x^4 + 5aab^5xx + 2a^3b^4xx - a^3b^6$$

$$\text{times} \quad \frac{dx}{\overline{xx - bb}\sqrt{xx - bb} \text{ times } \square : \overline{x^3 - abx}}.$$

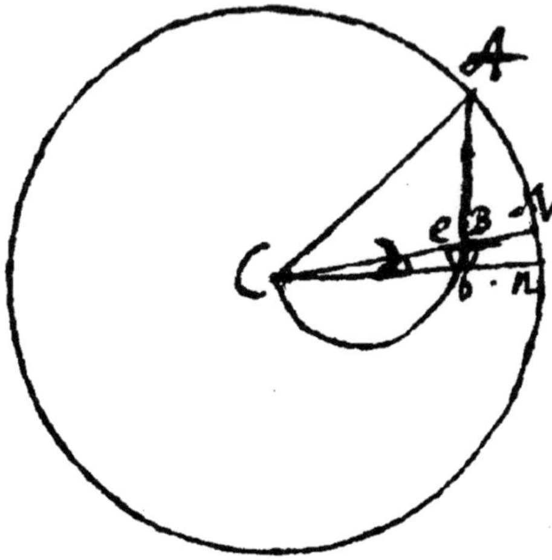

Fig. 11.31 Parabolic Spiral

Therefore if the values we have found are substituted into the general equation $dy^3 = dz^2\, dt$, this equation will come forth

$$\frac{dx^3 \text{ times } C : abb + bxx}{\overline{x^8 - bbx^6}\sqrt{xx - bb}} =$$

$$\frac{6abbx^6 + b^3x^6 - 5ab^4x^4 + 4aab^3x^4 - 5aab^5xx - 2a^3b^4xx + a^3b^6 \text{ times } dx^3}{\overline{x^8 - bbx^6}\sqrt{xx - bb}}.$$

Multiply the equation by $\overline{x^8 - bbx^6}\sqrt{xx - bb}$, and reducing to zero divide by dx^3 and $\overline{abbx^4 + aab^3xx}$, and we will have

$$6xx - 8bb - 2ab = 0 \quad \text{and therefore} \quad x = \sqrt{\frac{4bb + ab}{3}} = FN.$$

Now let *ABbC* (Fig. 11.31) be the Spiral Parabola or Parabolic Spiral,[64] with vertex *A* and Center *C*, whose nature is such that, as from the center *C* through *A* a circle is described, and drawing any radius *CN*, cutting the Curve in *B*, the square of *BN* is equal to the Rectangle between the arc *AN* and a particular constant line, which may be called the Parameter. We wish to find the inflection point *B*.

Let the radius *CA* or $CN = a$, the Parameter $= b$, and $CB = z$, therefore $BN = a - z$, the arc

[64]Compare this to §73 on p. 76.

$$AN = \frac{aa - 2az + zz}{b},$$

hence

$$-Nn = \frac{-2a\,dz + 2z\,dz}{b},$$

$Be = dz$, and because $CN : Ce :: Nn : be$ we will have

$$-be = \frac{+2az\,dz - 2zz\,dz}{ab} = dy.$$

Likewise $Be : be :: BC : t$, we will have

$$t = \frac{-2azz + 2z^3}{ab}$$

[38] therefore

$$-dt = \frac{-4az\,dz + 6zz\,dz}{ab},$$

and for that reason dy^3, that is

$$\frac{+8a^3z^3\,dz^3 - 24aaz^4\,dz^3 + 24az^5\,dz^3 - 8z^6\,dz^3}{a^3b^3} = \frac{4az\,dz^3 - 6zz\,dz^3}{ab}.$$

Multiplying the equation by a^3b^3 and dividing by $2z\,dz^3$, this comes forth on reducing to zero $4z^5 - 12az^4 + 12aaz^3 - 4a^3zz - 3aabbz + 2a^3bb = 0$, the root of which equation gives the quantity CB.

Remark. It is to be remarked in closing, that because in every curve the inflection point maintains the property, that the Tangent at that point simultaneously cuts the Curve such that the angle of cutting is smaller than any given angle, that is, nothing other than a straight line between the tangent (or, if you prefer, the secant) and the curve can be drawn through the inflection point. Indeed, because the inflection point is a jointly concave and convex portion of the curve, and seeing that the tangent is exterior at the convex, but interior at the concave, it is clear the Tangent at the inflection point lies outside this part, however inside the other, that is, cutting the curve itself at this point. However, any angle of cutting is known to be smaller than any you wish to give, since notwithstanding that it cuts the curve, it does not on that account give up the nature of a tangent.

Chapter 12
Selected Letters from the Correspondence Between the Marquis de L'Hôpital and Johann Bernoulli

Letter 5: L'Hôpital to Malebranche
Fougères,[1] October 23, 1690

I cannot express to you, Sir, all the joy that your kind letter has given me, because it assures me of your friendship, which I value more than anything. Since we parted I have received two or three letters from Mr. Huygens concerning the centers of oscillation. In the end, he approved of what I had sent him concerning this and I even believe that he will have it put in the Journals of Holland.[2] I am not happy with the piece that I put in your hands even though you appear to me to approve of it. My plan is to completely change the Arithmetic of the infinities.[3] It seems to me that I have a more general and easier method of demonstrating all of these propositions, which I am sure will not displease you. I have worked on the centers of gravity, which to my mind is the most difficult thing there is in Geometry. I would ask you to be so good as to give me your advice, because this will influence the judgement that I bring to this matter. The wish that I have to be with you will make me hasten my return, which, however, cannot be soon enough, because of domestic affairs that keep me here, where I am very much annoyed not to have the time to apply myself to my studies.

---

[1]The Château of Fougères in Brittany belonged to the estate L'Hôpital's father-in-law.

[2]*Historie des ouvrages des sçavans*, June-August 1690, pp. 440–449.

[3]This manuscript was never published. In Bernoulli (1955, p. 158), the Editor quotes a portion of a letter written by Malebranche to a Fr. Jaquemet at about this time: "Over the holidays, I read the writings of the marquis de l'hopital which occupied two months most agreeably. I believe he will have it printed this winter, if his affairs leave him the time that is necessary for it, and you will see the finest of mathematics, explained with much clarity and few words."

R.E. Bradley et al., *L'Hopital's Analyse des infiniments petits*, Science Networks. Historical Studies 50, DOI 10.1007/978-3-319-17115-9_12

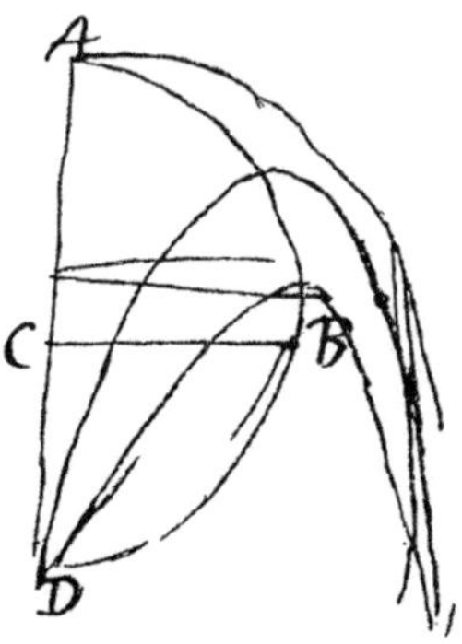

Fig. 12.1 With kind permission of Springer Science+Business Media

I pray you, Sir, to permit me to assure Fr. Bizance[4] of my services. I am also addressing you a letter for Fr. Jaquemet, which you will please take the trouble to read and to give to him. Finally, Sir, I am with all possible respect and esteem your most humble and obedient servant

the Marquis De L'Hôpital

Madame de L'Hôpital thanks you for the honor of your kind regards and sends you her compliments.

Letter 6: L'Hôpital to Bernoulli
Paris, December 8, 1692

I am very much obliged, Sir, with the decency and candor with which your letter[5] is filled. I read the paper that you sent me for which I thank you, although I find the problem of tangents is only half solved, as it seems to me difficult to determine the value of t in terms of x and y, which would nevertheless be necessary to have to find the value of the subtanget, and thus I would wish to have an example. With regard to the curve whose subtangent is $\sqrt{ay+xx}$, it is very easy to remove the incommensurables by supposing that $\sqrt{ay+xx} = m$ (Fig. 12.1). This gives the differential equation

$$2mm\,dm - 2mx\,dx = mm\,dx - xx\,dx,$$

[4]Louis Byzance (ca. 1647–1722) and Claude Jaquement (1651–1729) were fathers in the Oratorian order and members of Malenbranche's circle.

[5]This refers to Johann Bernoulli lost letter of November 1692, which was the first letter in his correspondance with L'Hôpital.

however I have not been able to find the way to take the integrals and I suspect that the curve is not geometric[6] in the sense of Descartes. You would make me very happy by sharing with me what you have discovered concerning this. I would also very much like it if you could send me a general method for solving problems like the following one.[7] Let any semi-Ellipse *ABD* be given, whose semi-axes are *AC* and *CB*. Assuming that there is an infinity of paraboles that all pass through the point *D*, and whose vertices are on the semi-Ellipse; we wish to find the line that touches all of them. Instead of an Ellipse and of parabolas we might assume other lines to infinity. This problem appears to me seems to bear some relationship to what Mr. Leibniz said in the *Acts* of Leipzig in the month of April of this year.[8] Please make a thousand compliments on my part to Mr. your brother,[9] and to thank him for the obliging manner by which he speaks of me in the Journals of Leipzig[10] with regard to what I had previously remarked on the centers of oscillation.[11] I really cannot wait for the continuation of *seriebus infinitis*;[12] because this is a field where it appears to me that there are many things to be clarified, and I believe him to be more capable than anyone. I would be very pleased to know the method by which he reduces the caustics that are formed by refraction to evolutes, and what are his amazing problems of which he spoke to you in his last letter. If he has sent them to Leipzig, as he told you he would, I will wait to see them in the *Acts*.[13] Ask him also what is that proposition that is in his book *de arte conjecturandi* which he would esteem as highly as the quadrature of the circle. You see, Sir, that I continue to ask you to teach me and that I am making use of the freedom that you gave me on this; in return, if I may do anything to render you service in this country, I will set to it with all possible energy, having for you a most particular inclination and esteem,

the Marquis De L'Hôpital

I forgot to tell you that I have not received the *Acts* of Leipzig that Mr. your brother sent me. Tell me by which carriage they will be coming and whom I should address myself here to get them. Please make my compliments to Mr. Stehelin. The secretary sent his to you and to him also.

---

[6]In other words, L'Hôpital believes the curve to be transcendental and not algebraic.

[7]See (L'Hôpital 1696, §146–147).

[8]*Acta Eruditorum*, April 1692, p. 168.

[9]This is how L'Hôpital refers to Jakob Bernoulli in his correspondance with Johann Bernoulli.

[10]*Acta Eruditorum*, July 1691, p. 317.

[11]*Historie des ouvrages des sçavans*, 1690, pp. 440–449; letter from L'Hôpital to Huygens of April 1690.

[12]*Positionum de seriebus infinitis...*, Basel 1692.

[13]Throughout (Bernoulli 1955) this is consistently written as "actes." We have chosen to use the proper formatting for the name of a journal.

Letter 7: L'Hôpital to Bernoulli
Paris, January 2, 1693

I received, Sir, with great pleasure your letter of December 18th,[14] you must never think that I am able to forget a person for whom I have as much esteem and friendship as I have for you and I will demonstrate it to you on all encounters. But let us get down to business.

1. It seems to me that the problem of the curve whose subtangent is $\sqrt{ay + xx}$ is not yet solved; because even though you arrived at the differential equation,

$$2m\,dy = -y\,dm + \frac{a^3\,dy\,dy}{yy\,dm}$$

we find by supposing according to the rule $m = ny^{-2}$, this other equation[15]

$$dn^2 - 2ny^{-2}\,dy\,dn = a^3\,dy^2$$

which still has three terms, and thus we have advanced no further. Therefore, it is necessary, if you please, that you push the problem to the end, that is to say that you give the construction of the curve that satisfies it.

2. The manner by which you find the intersection of two infinitely close parabolas appears very ingenious to me. You are mistaken in calculation because I find[16]

$$z \quad \text{or} \quad EM = \frac{4y^3\,dx^2 - 4yyx\,dy\,dx}{4yy\,dx^2 - 4yx\,dy\,dx + xx\,dy^2}$$

and it becomes easier by assuming that CG is constant, because, it follows that CG, that is to say

$$Ch = \frac{2xx\,dy - 2xy\,dx}{x\,dy - 2y\,dx}.$$

3. The manner by which Mr. your brother determines all quadrable areas of the curve

$$y^4 - 6aayy + 4xxyy + a^4 = 0$$

[14]That letter, in response to l'Hôpital's letter of December 8th, is lost.

[15]As noted in Bernoulli (1955, p. 161) the first term of what follows should be $y^{-1}dn^2$.

[16]See Bernoulli (1955, pp. 161–162) for a detailed explanation of this passage.

is very elegant. Will he not be printing his book *de arte conjecturandi*[17] soon? I have no doubt that it deserves the reputation that it has acquired among scholars. I would simply like to know what it is in this discovery that you have told me he was making such a big deal about.

4. The construction of the curve of Mr. Leibniz, $a\,ddx = dy^2$ is correct; here is the demonstration of it that I have discovered. Let $CG = x$, GH or the arc $AE = y$, $GB = z$, and we will have, because of the logarithm,

$$dx = -\frac{a\,dz}{z}$$

and because of the circle

$$dy = \frac{a\,dz}{\sqrt{aa - zz}}.$$

Now, by the assumption, the differential of

$$dx^2 + dy^2 \quad \text{or} \quad \frac{a^4\,dz^2}{aazz - z^4}$$

must be equal to zero, thus

$$aaz\,d\,dz - z^3\,d\,dz - aa\,dz^2 + 2zz\,dz^2 = 0$$

from which we conclude

$$\frac{aa\,dz^2}{aa - zz} = \frac{aa\,dz^2 - aaz\,d\,dz}{zz},$$

that is to say $dy^2 = a\,d\,dx$. I would very much like to know how you arrived at this construction. Please send me the manner in which Mr. your brother found an infinite series equal to the subtense of an arc of a given circle. You would give me great pleasure by sending me the construction of the curves that you mentioned to me. I do not despair that you will put the method of inverse tangents into perfection and by this method you will solve the curve of descent of Mr. Leibniz.[18] Please allow me to take the liberty that you gave to me by asking you to think about the following questions at your leisure (Fig. 12.2).

1. Let the parabola $AECF$ and the circle BD be given in position in the same plane. From all points B and D of the cirlce the tangents BA, BC, DE, and DF to the

[17] This is a reference to the second part of *Positionum de seriebus infinitis...*, Basel 1692.

[18] That is, the Leibniz Isochrone; see *Acta Eruditorum*, April 1689, p. 195.

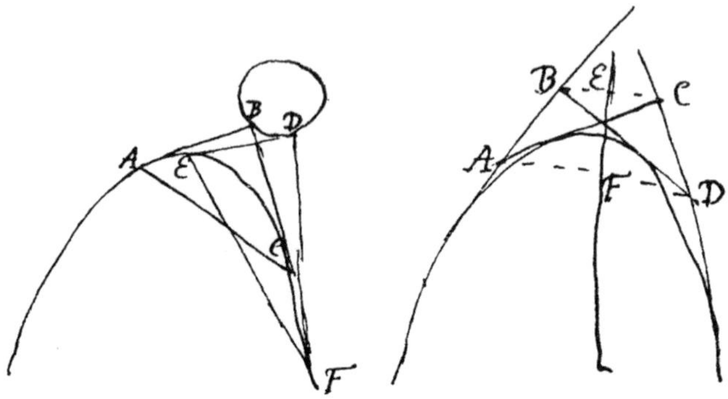

Fig. 12.2 With kind permission of Springer Science+Business Media

parabola are drawn. I wish to find the curve that touches all lines *AC* and *EF* that joins the points of contact.[19]

2. Assuming that the given straight line *AC* slides inside the parabola, we wish to find the curve that it continually touches.[20]
3. The four lines *AB*, *AC*, *DB*, and *DC*, being assumed tangents to a parabola, we wish to prove that the line *EF* that divides the straight lines *BC* and *AD* in the middle is a diameter of the parabola. You know that this theorem was proved by Mr. Newton for the Ellipse and the hyperbola on page 92.[21] However, I do not see that his demonstration extends to the parabola. Please send me the demonstration of this theorem in your reply and if you find something easier than that of Newton for the Ellipse and the hyperbola, by making use of your beloved algebra, you would make me happy by sharing it with me. For the other two questions, that will be at your leisure, because I foresee that the calculations will be long. I finish, Sir, by wishing you a year as fertile in discoveries as the last and by assuring you that I am in the best of my heart entirely yours

the M. De L'Hôpital

Since I wrote my letter I received the Journals of Leipzig by the Strasbourg coach, the last month of which is the month of September. They arrived long ago and it is my fault that I did not look for them earlier, which is why you should send me by the same coach the last three Journals of the year 1692 with the twelfth section of the supplements and the continuation of Mr. your brother's *de seriebus infinitis*, to whom I make a thousand compliments. Tell me, I pray you, to whom I should send money for these journals. Mrs. de L'Hôpital thanks you for remembering her

[19]This problem is the subject of L'Hôpital (1696, §161).

[20]Bernoulli's solution, contained in the lost letter of January, 20, 1693, is the model for L'Hôpital (1696, §158).

[21]This is a reference to Book I, Lemma 25, Corollary 3, of the *Principia*.

and also the little girl who says every day that she wants to go to Basel to eat pies with Mrs. Bernoulli. Before I send her there, it would be good if you let me know if she still likes them so much, as well as all the treats, and whether she has lost the taste for these since your trip to Paris.

Letter 11: L'Hôpital to Bernoulli Paris, June 27, 1693

I am extremely surprised, Sir, that you have made no response to my previous letter and I do not know to what to attribute this silence. I do not wish, however, to imitate this and I would rather make all kinds of overtures to maintain your friendship, to which I attach great importance, rather than to risk seeing it diminished by the discontinuation of the correspondence between us. Mr. the Abbé Bignon came to find me a few days ago and asked me to honor the Academy of sciences (these are his words) by sometimes attending the meetings that they hold, which I accepted, and I went yesterday for the first time. You may be assured that I will not miss any occasion to render service to you and to obtain a solid position for you here. I saw Mr. Varignon who seems to me to be a very good friend of yours, and he gave me a small piece of paper on which had the following question, which he told me you had sent to him. Let the equation

$$\frac{\sqrt{2a^3x - x^4} - a\sqrt[3]{a^2x}}{a - \sqrt[4]{ax^3}} = y$$

express the nature of a curve whose abscissa is x and whose ordinate is y. We wish to know the value of y when x becomes equal to the constant a. Solution. $y = 2a$, because in this case[22]

$$\frac{\sqrt{2a^3x - x^4} - a\sqrt[3]{a^2x}}{a - \sqrt[4]{ax^3}} = \frac{aa - aa}{a - a} = a + a.$$

You have not responded to me about the two curved lines that I proposed to you some time ago, which have the differential equations

$$a\,a\,dy + x\,y\,dy = a\,x\,dx - x\,y\,dx - a\,y\,dx$$

[22]This is the first appearance of the problem of evaluating algebraic expressions that take the form of $\frac{0}{0}$ for certain values of the variable by means of what is now called L'Hôpital 's Rule. This example, which we call Bernoulli's $\frac{0}{0}$ Challenge, appears in L'Hôpital (1696, (see §164)). Bernoulli's letter to Varignon is lost Bernoulli (1955, p. 173, Note 2.). Bernoulli's solution is given in letter 28.

and

$$x^3\,dy = 3\,x^3\,dx + 3\,a\,a\,dx - 2\,x\,y\,y\,dx.$$

These lines are not difficult to describe; however, I will send you the construction if you wish it. Furthermore, Sir, remember that you owe me two responses right away and the only means of repairing your negligence is to send them to me as soon as possible. I await them impatiently because they will tell me your news and I am without reservation entirely yours

The M. De L'Hôpital

The secretary sends his compliments and also complains about your silence.

Since writing my letter I found the solution to your problem, which I send you to be inserted in the *Acts* of Leipzig, if you judge that they are worth the trouble.

PROBLEM.

The curved line *CMN* has the property such that each of its tangents *MT* is always to the part *CT* of the axis, taken between its origin *C* and the intersection *T* of the tangent, in a given ratio of p to q. We wish to know the nature of this line, or the manner of describing it.

[L'Hopital gives his solution to this Reverse Tangent Problem here; see Bernoulli (1955, pp. 174–177). This problem was posed by Bernoulli in *Acta Eruditorum*, May 1693, p. 234. The solution is sometimes called the Bernoulli Curve; this is a frequent topic of discussion in letters 10–23.]

...

If you send this solution to Leipzig please be so kind as to make an exact figure, because you know well enough that I don't have the talent to do so, it will also be necessary to put the discussion in Latin, which you do better than me. Even though my letter is dated the 20th, it is only leaving today, which is the 27th of June.[23]

Letter 12: L'Hôpital to Bernoulli Without Location, July 8, 1693

I have just now received, Sir, your letter from the 2nd of this month, which gave me great great joy, because I feared that some accident had happened to you.

I strongly approve of what you tell me regarding what Mr. Huygens has published in the Journals of Holland and I believe the best thing is not to speak about it. There

[23] *Acta Eruditorum*, September 1693, pp. 398–399.

could be no shorter rule than what you sent[24] me for reducing differential equations of the first degree to quadratures; I strongly doubt that Mr. Leibniz' rule is as short. This makes me want to see the rule in all its extent, I hope that you will share it with me on the 1st occasion. Tell me also if you can use this method to solve the two differential equations that I have proposed to you, and if you are happy with the solution of your problem,[25] apparently because you didn't speak of it in any way in your last letter; because you did not have the time to examine it.

Do not doubt, Sir, that I do everything I can to lure you to this country, I've already spoken about you in a favorable manner, that is to say in rendering you justice in the academy of sciences by sharing with them my solution to the problem you had proposed in the *Acts*,[25] and on the 1st occasion I will speak to Mr. the abbé Bignon about the subject of you and I will give a good account of you, assuring you that I can not have greater pleasure than to be able to entertain you often and to see you face-to-face, that I am entirely yours

the M. De L'Hôpital

Mrs. de L'Hôpital assures you of her services.

Letter 14: L'Hôpital to Bernoulli Paris, August 1, 1693

This letter, Sir, is to let you know that I have received the three months of the *Acts* of Leipzig. In the one for the month of May, I saw your solution to the problem of Mr. de Baune[26] and I found these words:

"Curva autem ipsa *AI* est ex earum numero, quarum rectificationes quidem in abstracto non habentur, longitudines tamen per ipsasmet curvas construi et determinari possunt, quod nob. Dn. Huygens praestitit in nouva sua logarithmica, et ego jam olim in logarithmica vulgari."[27] From these words it is clear that you sent this solution to Leipzig only after having seen the letter from Mr. Huygens,[28] even though you expressly told me the opposite.[29] Because I believe one must speak with

[24]The Editor of Bernoulli (1955) believes this is the rule for integrating homogeneous differential equations.

[25]I.e., the Problem stated in Letter 11.

[26]*Acta Eruditorum*, May 1693, p. 234.

[27]This translates roughly as: "While the curve *AI* is, because of its degree, such that it does not have a rectification in the abstract, nevertheless the lengths of the curves themselves can be constructed and determined, as the noble Mr. Huygens has given in his own new logarithms and in what follows I will proceed in common logarithms."

[28]What is meant here is the open letter of Huygens of February 1693, which prompted Bernoulli to send his second construction of the Curve of de Baune to the *Acta Eruditorum*.

[29]Presumably in the lost Letter (9, 1) in May 1693.

an open heart with one's friends, especially with those that one esteems and with whom one wishes to always maintain friendship, I will tell you frankly that you would have made me happy to have used another method and to have cleared it with me beforehand; because I can truthfully assure you that Mr. Huygens put it in his letter at my insistence and furthermore that my intention had been to send you my construction of the measurement of the logarithm in order to ask you to put in the *Acts* of Leipzig, if you had judged it worthy. It is not that I wish to complain, on the contrary I know that I should thank you because by inserting this solution in the *Acts*, you have done so in the manner that wounds me as little as possible, not having spoken about me. Furthermore, it is a done deal that we must both forget about; although I still can't stop myself from telling you that you could write to Leipzig to have add more or less the following to the end of *my solution to your last problem*: "*ne sinistrae interpretationi locum dent*[30] (I continue in French) what we find in the *Acts* of Leipzig for the month of May and in *l'histore des sçavans* for the month of February, you believe it is appropriate to say that one should not be amazed that two people had found the solution to the same problem and that having intentionally put it without a name in the 34th journal of Paris, it will be attributed to both in what follows." It seems to that in this way you preserve my reputation without hurting your own in anyway, and without going into further detail nor naming anyone, this says all that needs to be said on this occasion. I am sure that if you write to Leipzig as soon as you receive this letter you will still be in time, because things are not printed on the same day that they arrive. No matter what you do in this regard, Sir, you should always count on me as a sincere friend who offers to you with pleasure everything that belongs to him, that is to say his purse and his credit, which is not as great as he might wish. I tell you that if we can attract you to Paris, you will need take no inn other than my house, being with all my heart entirely yours

the M. De L'Hôpital

The secretary sends you many compliments.

When I was about to seal my letter, I received in the post the month of June[31] and one of your letters of July 26th to which I will reply precisely and principally on what concerns you, because I will definitely ask Mr. the abbé Bignon on what we may count on. I forgot to tell you that you would give my pleasure for you to send a request to Leipzig that will include the date of the letter in which I sent you my solution, so that no one might suspect me of having taken it from the letter of Mr. your brother,[32] which as I see is found in the month of June. I ask you to please reply to me about this as soon as possible, because the Leipzig affair causes me some distress.

---

[30]This translates roughly as: "So that the placement should not give over to sinister interpretation."

[31]I.e., the June issue of the *Acta Eruditorum*.

[32]*Acta Eruditorum*, June 1693, p. 255.

Letter 15: L'Hôpital to Bernoulli
Paris, September 2, 1693

[L'hopital discusses a number of issues and gives a construction for the point of intersection of two infinitely close refracted rays.]

...But I ask you how you have found the construction of the tangents of the tractrices that Mr. Varignon communicated to me;[33] and if you are willing I also propose to you at the same time to find the tangents that describe the mean points between the tractrices. I confess that I did not work very hard to solve the equation

$$\frac{\sqrt{2a^3x - x^4} - a\sqrt[3]{a^2x}}{a - \sqrt[4]{ax^3}} = y$$

where $x = a$. Because I see no hope of success, since all the solutions that first present themselves are not correct, I did not want to waste my time unnecessarily, and I'd prefer to learn it from you if you are willing to share it with me. I finish, Sir, by asking you always to love me and to believe me to be entirely yours

The M. De L'Hôpital

Mrs. de L'Hôpital is your servant.

Letter 17: L'Hôpital to Bernoulli
Oucques, October 7, 1693

[L'hopital discusses a number of issues including the Paris Academy prize competition and the caustic by refraction.]

...

Please remember to send me the tangents of the tractrices and those of the curves described by points between tractrices[34] that you have no trouble finding, because things which seem the most difficult to others are easy for you, as I have experienced several times. The method that you use to solve the equation

$$\frac{aa - ax}{a - \sqrt{ax}} = y$$

[33] Bernoulli had already sent this problem of the tangent to the tractrix to Varignon in May, along with his $\frac{0}{0}$ challenge.

[34] L'Hôpital had asked for solutions to both of these problems in letter 15.

where $x = a$, is general.[35] However, it is not practical, as you yourself noted in the proposed equation and so even though I had thought that it was necessary to remove the incommensurables, I did not push this thought any further, seeing that it was impossible to come to the solution in this example.

I have no doubt that your latter method contains something very curious because it extends to all irrationals, which is why I would be most obliged if you would share it with me. I'm afraid of tiring you with all these requests, but chalk it up to your cleverness, and the permission you have given to me. This is a very long letter that might bore you, but I hope that you would be willing to take it as it is and you do not run the risk, because in the end permit me to tell you that if there is not enough space in a letter one may add another page to it. I finish by wishing for perfect health for Mr. your brother and assuring you that I am, as usual, entirely yours

the M. De L'Hôpital

I beg you to send my compliments to Mr. Stehelin and tell him that I thank him for his remembrance, Mrs. L'Hôpital sends you her's as well as Espine and Henriette, who is quite angry that you are not here so that she can make amends for the scratch that she gave you, through a good deed equivalent to the dishonesty with which she received you when you tried to kiss her.

Letter 18: L'Hôpital to Bernoulli Oucques, December 2, 1693

[L'hopital discusses a number of issues including cusps, the radius of the evolute, and the inverse tangent problem.]

...But after that, I beg you to remember your promise on the tangent to the tractrices and the solution to the equation

$$\frac{\sqrt{2a^3x - x^4} - a\sqrt[3]{a^2x}}{a - \sqrt[4]{ax^3}} = y \quad \text{where} \quad x = a$$

where I am sure I will see something quite unique and very beautiful, as in all your inventions.

...

[35] In his response to letter 15, which was lost, Bernoulli presumably suggested this simplified version of his $\frac{0}{0}$ challenge.

Letter 19: L'Hôpital to Bernoulli
Paris, January 16, 1694

I do not know to what to tribute, Sir, the length of time that you are taking to reply to me,[36] however I do not want to imitate you in your silence. You have no doubt heard us speak of the Marquis of Antremonts, my wife's uncle, and whom the Countess of Antremonts had made her sole legity, which obliges us, as you know, to leave Ouques much earlier than we had wished. He died on the 17th of last month, and as my wife is his only heiress and he leaves great wealth with debts and outstanding business, this gives us a lot of trouble for the time being, but it will bring us much later on. We will soon get several properties near Lyon where we will be obliged to spend part of the year, which I think puts us a little closer to you. Please let me know the number of journals that you have sent me and what I owe you for them and where I should send you the money. You would oblige me by sending me the rest of the journals for the year 1693.[37] Please remember to reply to me about everything that I asked you, I would rather wait longer and have it be more extensive. I am sending you my solution to your problem and I am, Sir, essentially yours

the marquis de L'Hôpital

The secretary sends you many compliments, make mine to Mr. the Professor. You showed me while you were in Paris, that the volaria is the same as the (catenary), that is to say that if you assume that the sail is composed of an infinity of small inflexible rectangles equal to one another they will bend themselves in the same manner being pushed by the wind as they would under their own weight. I now ask you if you think the same thing will happen if the number of rectangles is determinate.

Letter 20: L'Hôpital to Bernoulli
Paris, March 17, 1694

I begin, Sir, by making many apologies for having delayed my reply to your letter of January 26 for so long, but the continual problems that the death of Mr. le marquis d'Antremonts give me are the cause of it. There is much property, but we have also found many more debts than we had thought, because there are two hundred thousand pounds, and yet in his will he gives the ladies nearly the same sum of two hundred thousand pounds, and because we are disputing this legacy, this make us enter into a large lawsuit, of which you know that the outcomes are always uncertain. You won't mind that I congratulate you on your fiancé,[38] whom you tell me is so

[36]Bernoulli had not written to l'Hôpital since late September 1693.

[37]The Editor of the Bernoulli (1955) made a correction here; 1694 was written.

[38]In the winter of 1693, Bernoulli became engaged to Dorothea Falkner.

pretty, which I do not doubt because you chose her and you have good taste. For your wedding I would be very happy to find myself there but it does not have the appearance of being possible, but in return I hope that you will come see us on our grounds,[39] because they are but a days journey from Geneva, which is not very far from your house. We will dine well there, because ...[40] it is the best wine in the country and Madame de L'Hôpital fears that if you go there that you will no longer be satisfied with her simple wine that comes from Oucques to Paris.

I will happily to give you a *pension*[41] of three hundred pounds, which will begin the first of January of this present year, and I will send you two hundred pounds for the first half of the year because of the journals you have sent me, and the other half year will be of one hundred fifty pounds, and so on in the future. I promise you to increase this stipend shortly, which I well understand to be very modest, and it will be as soon as my affairs are somewhat straightened out, and I can enjoy the succession, because at present I haven't yet touched anything. I am not so unreasonable as to demand all of your time for this, but I will ask you at intervals to give me a few hours of your time, to work on what I will ask you and also to communicate your discoveries to me, while asking you at the same time not to share any of them with others. I even ask you not to send here to Mr. Varignon, nor to others, any copies of the writings you have left with me; if they should become public I would not be at all pleased. Answer me on all of this and believe me, Sir, to be entirely yours

the M. de L'Hôpital

I have not yet received any of the *Acts* of Leipzig, neither by the Strasbourg coach nor from Mr. Anisson.

Madame de L'Hôpital sends a thousand compliments to you and to your intended spouse, in assuring you that she takes all imaginable joy in everything that is happening to you.

Letter 21: L'Hôpital to Bernoulli
Paris, April 7, 1694

I find you to be very lucky, Sir, to possess a sweetheart so full of charms; the portrait that you have made of her for me makes me curious to judge for myself and I will be very happy should you both be willing to be tempted to come and pass some time in our lands at St. André, my wife takes great pleasure in anticipation and asks you to do so when the time is right (Fig. 12.3).

---

[39]L'Hôpital had just inherited the castle *St. André-de-Briord* in the Department of Ain, near the Swiss border.

[40]This portion is illegible.

[41]In other words, a recurring payment.

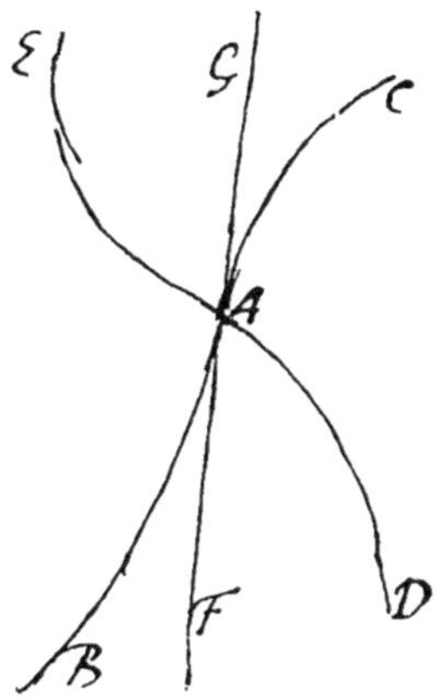

Fig. 12.3 With kind permission of Springer Science+Business Media

With this, I am sending you the letter of exchange for 200 ₶[42] that I promised through Mr. Nicolas Goy, banker, but in the future I will look for somebody else, because he has taken 12 out of 100 for the exchange, which seems too much to me.

Here is a difficulty with regard to evolutes that I ask you to clarify for me. Mr. your brother in the *Acts* of Leipzig[43] for the year 1692, page 116, and Mr. Leibniz,[44] page 443 hold that at an inflection point the radius of the evolute or of the osculating circle always becomes infinitely large. However, I find that it may also be infinitely small or null; for let *BAC* be a curved line, which has an inflection point at *A* and as tangent at this point the straight line *FAG*. It is clear that in beginning to evolve at the point *A*, we describe the curve *AE* by the evolution of the part *CA* and the entire curve *DAE* also has an inflection point at *A*, although at this point the radius of the evolute *BAC* is null. Suppose, for example, that the curve *DAE* is the paraboloid $aax^3 = y^5$; it is easy to show that it has an inflection point *A* and that the radius of its evolute at this point is null or zero. The argument that Mr. Leibniz presents on the same page 443 does nothing against me; it is true that two perpendiculars to a curve that are infinitely close to one another cannot go from convergent to divergent except by first passing through parallelism, supposing that they increase as they approach the inflection point. However, on the contrary suppose that they decrease, they become null or zero, as happens in the examples that I have just given. Now, here is my difficulty. The general expression of the radius of the evolute is

$$\frac{\overline{dx^2 + dy^2}\sqrt{dx^2 + dy^2}}{-dx\,d\,dy}.$$

[42]This currency symbol was used by L'Hôpital for pound.

[43]*Acta Eruditorum*, March 1692.

[44]*Acta Eruditorum*, September 1692.

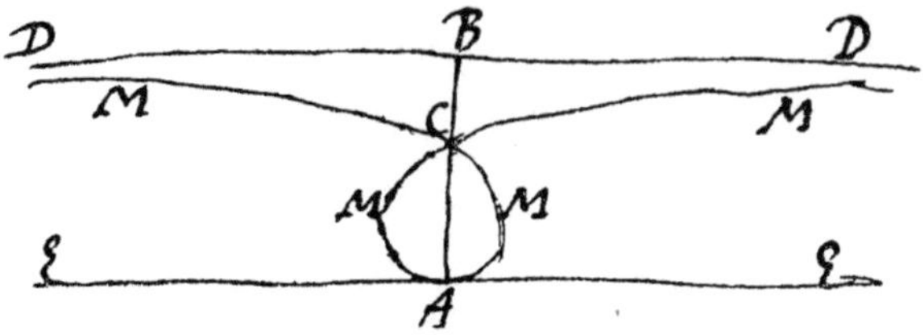

Fig. 12.4 With kind permission of Springer Science+Business Media

Now, at the inflection point we always have $d\,dy = 0$, and consequently this quantity becomes infinite, because the numerator cannot be zero, being the sum of two squares. Furthermore, it is certain that at an inflection point the two infinitely close tangents make no angle between themselves, but that they fall exactly one on the other, from which it follows that their perpendiculars must be parallel and never zero. How then to reconcile this with what we have previously shown.

It seems clear to me that at an inflection point we always have $d\,dy = 0$, but I don't see that this is reciprocal and I believe that their is an infinity of curves that do not change their curvature and which nevertheless have certain points where $d\,dy = 0$. Tell me what you think about this.

I forgot to tell you that Mr. your brother is mistaken in the same place when he holds that in all the paraboloids (except the common parabola) the osculating circle of the vertex is infinitely large, for we may easily show that there are many of these paraboloids where it is infinitely small, as in $ax^3 = y^4$. I have one other thing to ask you (Fig. 12.4):

You told me that your curve for the tangents *AMCM* has as its asymptote a line *BD* parallel to the tangent *AE* when p is greater than q.[45] I would very much like to know the method of determining the points B and C. You would give me pleasure by explaining all this to me and in believing more than ever that I am entirely yours

the M. De L'Hôpital

My sweetheart sends her compliments to yours and she exhorts her to learn mathematics because she lacks only this to be rendered perfect, and for my part I kiss her hand a thousand times.

[45]Bernoulli apparently made this incorrect assertion in letter (20, 1). In Letter 22, he corrects this error and gives a full account of his curve. However, this has been omitted from this volume, because of its length and lack of relevance to L'Hôpital (1696).

Letter 22: Bernoulli to L'Hôpital
Basel, April 22, 1694

Sir[46]

Your generosity extends much further than I deserve, but I see that you take pleasure in making me totally confused, not being in a position to do anything that would match the generous action that you have just extended towards me, because a simple expression of thanks is but a heap of words that gives little satisfaction, and the offer of one's services in return, when they are too small, is more annoying than acceptable. What then can I do on my part not to incur one or the other vice? Being on the one hand entirely ashamed only to give you words when I owe you deeds, and on the other hand to offer you my puny services for which you might care very little. However, I pray you, Sir, to consider well this verse *Si desint vires tamen est laudanda voluntas*,[47] that is to say, although I may not always be able to answer everything you ask of me, or rather to execute your orders, may you nevertheless have the kindness to take my good intentions for the most part (Fig. 12.5). With this conviction, I will satisfy you as much as I can with the difficulties that you had the honor of sharing with me in your last letter.[48]

I have read with great pleasure the great doubts that you have against Mister Leibniz and my brother concerning the inflection point. I have even already sent it to Leipzig to the Editor of the *Acts* to be communicated to Mr. Leibniz, saying simply that you find that the radius of the evolute at an inflection point is not always infinite, despite what he holds on the said page of the *Acts*, and that if he desires your demonstration, I will send it to him. We will now see how he will respond to it. As for me, I find that were right to correct both of them; my brother, having seen your letter, said the same and already retracted what he said in the *Acts* in the cited location, although he adds that he only put it in incidentally, without having

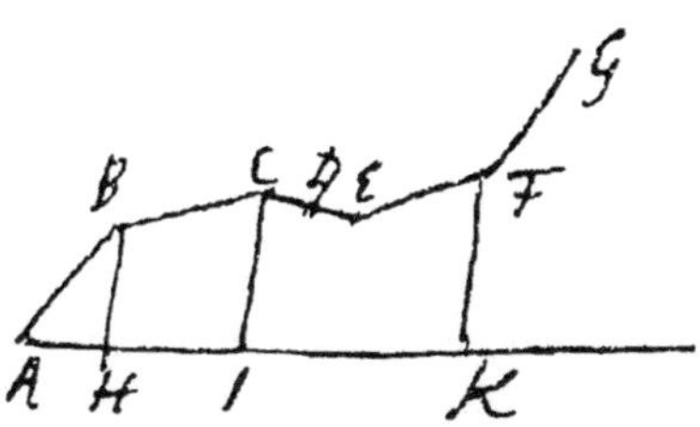

Fig. 12.5 With kind permission of Springer Science+Business Media

[46]This letter, in response to Letter 21, is the first surviving letter from Bernoulli to L'Hôpital. It is a copy of the original, with the note to "Mr. the Marq. De L'Hôpital" in the margin.

[47]This Latin quote translates roughly to: "If strength is lacking, nevertheless the desire is to be praised."

[48]That is, Letter 21.

carefully considered the nature of the point of contour. Indeed, I find that there are as many of even more curves in which the radius of the evolute at the point of contour is infinitely small, as those in which it is infinite, without making any difficulty for the fact that the general expression of this radius is[49]

$$\frac{\overline{dx_q + dy_q}\sqrt{\text{etc.}}}{-dx\,d\,dy},$$

for I do not deny that at an inflection point $d\,dy$ is always $= 0$, far from that it may also be infinite with respect to other $d\,dy$'s; it is this that makes the general expression sometimes also $= 0$. The reason for this is that even though the curve increases, $d\,dy$ increases also, so that when the curvature is infinite, that is to say when the radius of the evolute is infinitely small, $d\,dy$ also becomes infinite with respect to the others. I say with respect, because a differential of whatever kind may be infinitely greater than the others of the same kind and nevertheless infinitely smaller than a differential of the preceding degree, which is easy to demonstrate. This being so, it is a false principle to believe (although I have previously believed it myself) that a curve passing from convexity to concavity must necessarily pass through straightness, which is found between the two different curvatures, and that consequently $d\,dy$ is 0. This does not follow at all, or to say otherwise, this straightness might also sometimes occupy but a differential of the curve, infinitely small with respect to the other differentials, as you see in the present figure, where *ABCFG* denotes a curve, for which *AK* is the axis, and *AB*, *BC*, *CF* and *FG* are the differentials of the curve and *AH*, *HI*, and *IK* are the differentials of the axis. It is clear that the inflection point may be at *D*, in such a way that it is only the two infinitely small parts *CD* and *DE* of the entire differential *CF*, that are set in a straight line and in this way I concede to you that the two perpendiculars at *D* and *E* are parallel, but because of their infinitely small distance they are joined, and it is those at *C* and *F* whose intersection (which without doubt could be infinitely close) makes the radius of the evolute at the inflection point. This is why I say to find the inflection point, we must see if the $d\,dy$ are increasing or decreasing. If they are increasing, $d\,dy$ must be equal to infinity and if decreasing, to zero. It does not appear clear to you that at an inflection point we always have $d\,dy = 0$, but we sometimes have $d\,dy =$ infinity. However, it is true that this is not reciprocal, because there is an infinity of curves that have at a certain point the radius of the evolute either infinite or infinitely small without changing their curvature, which is verified in several paraboloids. In addition, Sir, I must point out to you a considerable thing in such curves, through which we will better see the nature of the point of contour: if we begin to evolve a curve at the inflection point where the radius of the osculating circle is infinite, we will describe another curve which also has an inflection point, but one in which the radius of the osculating circle is infinitely small. If we now evolve this second curve by also beginning at

[49]Here Bernoulli uses the a subscript of q to represent the square of a quantity.

the inflection point, but we describe a third curve which has its inflection point, but for which the radius of the osculating circle is not only infinitely small, but even infinitely smaller than that of the second curve. In evolving the third curve we describe a fourth one, for which the radius of the osculating circle at the inflection point is infinitely smaller than that of the third curve, and so on.[50] This is easily seen when we consider that the first radius of the evolute is equal to the first differential of the same, from which it is clear that there is but one single type of curve for which the radius of the osculating circle at the point of inflection is infinite, but there is an infinity of types where this radius is always infinitely small. On the occasion of all that I have just said, I must tell you of another difficulty that came to mind along time ago on the *maxima et minima*. You know that to find a greatest or least we make dy equal to zero because we have believed until now that the y's for which we wish to find the greatest or the least make a curve whose tangent at the largest or smallest ordinate is parallel to the axis. However, I say that this is not general, because there are curves that I call *bicornes*,[51] in which the greatest or least ordinates pass through a point which I call *de rebroussement*.[52] Now, the tangents at these points are not parallel, but for the most part, perpendicular to the axis as we see in these figures. This is why we make a big mistake when we make $dy = 0$ in these sorts of curves, in order to find the greatest ordinate, it being possible that dy is infinite.[53]

When my Brother holds that in all the paraboloids the osculating circle of the vertex is infinite he means the paraboloids where x is raised to but a single dimension.[54] As for my curve for the tangents that are always in the same ratio to the intercepts, before responding to what you have asked me, it is necessary to put in its entirety the calculation that I made use of to arrive at the general equation that determines the nature of these curves (Fig. 12.6).

...

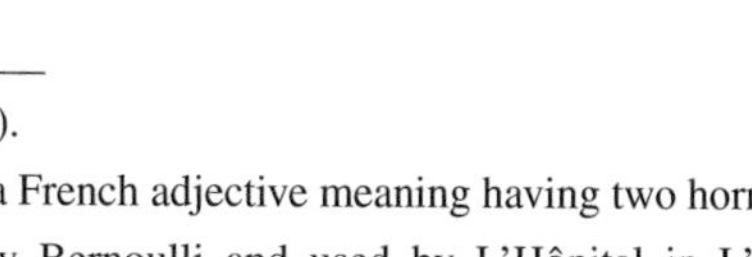

Fig. 12.6 With kind permission of Springer Science+Business Media

[50]See L'Hôpital (1696, §82).

[51]This is Bernoulli's term, a French adjective meaning having two horns.

[52]This term was coined by Bernoulli and used by L'Hôpital in L'Hôpital (1696). In French *rebrousser chemin* means to retrace ones steps. The English word for such a point is cusp, which we will use in what follows. We will also use "bicorn" for a curve with a cusp, even though this is not standard English usage.

[53]The case of an extremum at a cusp was omitted from the *Lectiones*. L'Hôpital included it in §47 (p. 46).

[54]By "paraboloids," Bernoulli means curves whose equation is $x^p = y^q$, where p and q are positive integers. So in this case, Jakob means $p = 1$ and $q > 2$.

[Bernoulli now gives his as yet unpublished solution to the tangent problem stated by l'Hôpital in Letter 11. The details are in Bernoulli (1955, pp. 208–211).]

...

My brother tells me (I do not know if it is true) that he knows a trick for constructing my equation that I found for the curve of descent.

Tell me, if you please, if you have received the *Acts* with the supplements by way of Mr. Anisson. I already have two months of this year with a supplement which I will send to you on the first possible occasion, because I cannot send you anything else by way of Strasbourg coach without passports. However, one of our booksellers sends books to Mr. Anisson from time to time and in the future I will make use of this convenience. I am sending you herewith my inaugural dissertation,[55] I hope that you will also find in it something that is worthy of being read.

My wife and I send our very humble compliment to Madame and thank you for the invitation that you made to us to go and spend some time at your estate. If I can steal away some time for myself, I assure you that we would not miss out on finding ourselves there; if it is impossible for both of us it might be for me alone, for I would be delighted to see you again and to assure you from my own mouth that I am, with a profound respect,

Sir

Your most humble and most obedient and most obliged servant
Bernoulli

Letter 23: L'Hôpital to Bernoulli Without Location, May 1694

I have just received, Sir, your letter of April 22. I take pleasure along with you that your wedding is finished as well as your theses on logic and medicine. I am very pleased that you are happy with my remark on the evolutes, and I was already satisfied in part on the difficulty that I proposed to you, for in considering curves as being composed of an infinity of little arcs which all have their centers in the evolute, it is clear that when the radius is infinitely small, the arc which in this case is but a point is also infinitely small with respect to the other arcs which are the differentials of the curve, and consequently the contiguous perpendiculars fall on one another. Furthermore, in the convex part the value of $d\,dy$ is negative and on the concave part positive, so that at the inflection point we must pass from the one to the other. Now this may happen in two ways, either the $d\,dy$'s increase and therefore it must be that $d\,dy$ becomes infinite, or they decrease and therefore it must be that $d\,dy$ is equal to zero, in the same way that the contiguous perpendiculars may go from convergent to divergent in two ways as I told you my last letter (Fig. 12.7).

[55] *Dissertatio de motu musculorum*, Bernoulli's dissertation for his doctorate in Medicine, 1694.

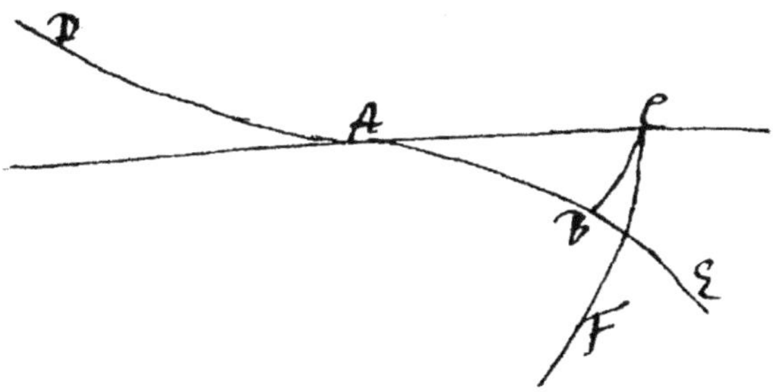

Fig. 12.7 With kind permission of Springer Science+Business Media

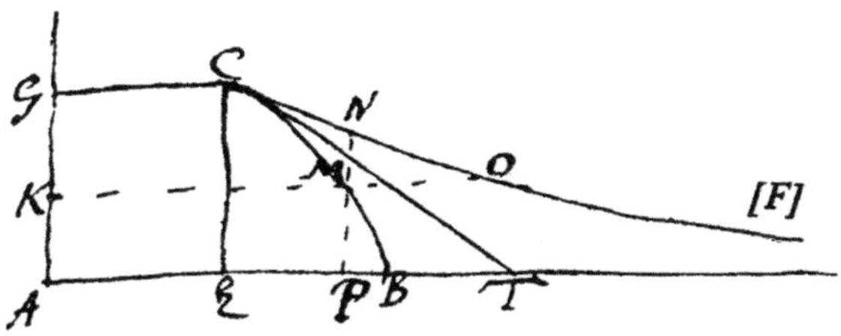

Fig. 12.8 With kind permission of Springer Science+Business Media

With regard to the curves that you call bicorn, and about which I wrote you sometime ago on the occasion of your curve in the *Acts* in the case where q is greater than p, I have several difficulties which I ask you to clarify for me indepth. Secondly, you have told me that at the cusp C the radius of the evolute must necessarily be equal to zero. Now, I find that it may be infinite, which is proven by the example of several paraboloids,[56] and also that it could have a finite or determined magnitude. For let *DAE* be a curve which has the line *AC* as tangent at the inflection point A. If we evolve the part *DAB* beginning at the point D we form the bicorn line *BCF*, and at the cusp C the radius *CA* of the evolute *DAE* has a finite magnitude.[57] You see therefore that the consideration of the evolute does not help us to find this kind of point C. Now, here is how I figure it. Let *BCF* (Fig. 12.8) be a bicorn curve for which the ordinate *PM* or $PN = y$ and the abscissa $AP = x$. It is clear that the dy's increase in the part *CF* as x decreases and on the contrary in the part *CB* it increases as x increases. Therefore, if we imagine a certain curved line that has for its abscissas the dy's, or finite straight lines that are in the same ratio, the question reduces to finding the least of the ordinates in this curve, which gives $d\,dy =$ to infinity or zero. How then to reconcile this with what I have just proven. I have

[56]For example, $y = x^{5/2}$.

[57]The accompanying figure is a picture of a cusp of the second kind appears for the first time, produced through the involution of a curve with an inflection point. See L'Hôpital (1696, §109) and Figure 5.27.

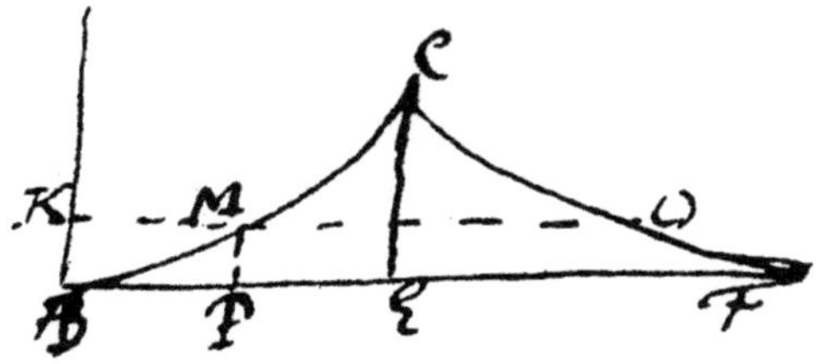

Fig. 12.9 With kind permission of Springer Science+Business Media

pointed out to you at the cusp C (Fig. 12.8) dy must be null with respect to dx as also dx must be null with respect to dy. Because you have not understood my argument, apparently because I explained myself in too obscure a manner, I will make sure I make myself more intelligible. Is it not true that while the abscissa AP (x) remains the same, the ordinate y has two values PM and PN, which become equal to one another when the point P falls on the point E that we wish to find? Therefore if, along with Messrs. Descartes and Hudde,[58] we consider x as known and y as the unknown in the equation that expresses the nature of this curve, and we multiply it as usual by an arithmetic progression, we will form a new equation which we use to determine the cusp C. Now, it is equivalent to taking the differential equation and erasing from it all the terms that are multiplied by dx. From this we see that dx should be null with respect to dy. In the same way we prove that by taking AK as the abscissa and KM or KO as the ordinate, dy must be null with respect to dx. Now, dy is to dx :: $CE : CT$. Therefore, we must have that at the same point C the fraction dy/dx is null, infinite, and determinate. How can this be straightened out? I wait for the clarification of this from you.

Now, let BCF be the bicorn curve (Fig. 12.9) for which the tangent CE at the cusp C is parallel to the ordinates PM. It is clear that in the part BC the dy's are positive and increasing, whereas in the part CF they are negative and decreasing. From this it follows that at the point C, dy must be infinite. I will prove, however, by Messrs. Descartes and Hudde that it must be zero by considering that the abscissas KM and KO (x) become equal. It seems, therefore, that in the case where dy is infinite we also find the same thing as by supposing it equal to zero, so that this consideration would be useless and does nothing but to hinder, and if this were true with regard to dy we would also prove it for ddy. I confess to you that I cannot untangle all of this and you would make me very happy by applying yourself to it with care and to put all the necessary time into it to clarify it entirely.

We know well enough that zero has a constant quantity as its integral. However, I ask what is the integral of infinity, for example, in curves ddy infinite at an inflection point I ask what dy ought to be.

I had forgotten to tell you that it is clear that by evolving a curve at an inflection point for which the radius of the evolute is infinite at that point, we form another

[58]The method of Descartes and Hudde is described in Chapter 10 of L'Hôpital (1696).

curve whose radius is infinitely small at the same point, so that this radius, so to speak, is infinitely small with respect to the first one. However, I do not see that if we form a third curve of the same kind, the radius of its evolute at the inflection point is infinitely smaller still than that of the second one, and so on. This is why I ask you to explain it to me at a little greater length and above all to give as many examples as you can, in order to focus the imagination.

I am writing you this in haste because I am overwhelmed with matters that have left me no spare time up to now. This is why I ask you to pardon me if I made a few mistakes. I send you the best of my heart. My wife has a fever, she sends her compliments as well as mine to your sweetheart.

Finally, Mr. the Abbé Catelan has decided to attack the solution that I put in the memoirs.[59] He holds that the curved line whose nature is expressed by the equation y^4 etc. which I gave does not have the given property; to this end he makes a calculation in which he is mistaken as usual, which I have pointed out in a very short response.[60] I will send you both of them, if you wish, but it is not worth the trouble.

Letter 24: Bernoulli to L'Hôpital Basel, May 21, 1694

Sir, I am very pleased that we are in agreement with regard to the inflection point, but on the matter of the bicorn curves that you have pointed out many difficulties to me, some of which are in truth are rather important. Every time, having put in all the necessary time to clarify them, I apply myself to them with care, so that I believe to be finally brought to the goal, all the ugliness and contradictions that we find on the subject of the cusp are the only things to appear. You will see my reasons for this, however I will reply in order to all the points that you have proposed to me. You were right to say that the radius of the evolute at a cusp may be infinite, but it is similar to looking for the greatest or the least [ordinates], where I have said that the differential of the greatest or the least is not always equal to zero, but also sometimes to infinity. Now you yourself say that in this case this situation does not change, whether we suppose it equal to zero, or to infinity, as I more clearly show below. Nevertheless you propose to me an example where the radius of the evolute at the cusp is neither zero or infinity. You must therefore know, Sir, that you have spoken of curves that turn back on themselves or bicorn curves, I intended only those whose two parts change their convexity, for it is evident that then the radius of the osculating circle must necessarily be either infinite or zero, otherwise the evolute of one part of the curve that turns back on itself cannot pass to that of the other part without making

59 *Memoir Mathematique*, June 1693, pp. 97–101, *Journal des sçavans*, March 1694, p. 146. The reference in Bernoulli (1955, p. 217) was to a different article by l'Hôpital.

60 *Journal des sçavans*, April 1694, pp. 182–183.

an obvious jump, that is to say that this radius is negative and positive at the same time, which is absurd. However, in the case of the other cusps, in which the convex part is opposed to the other concave part, I must confess that I do not know about them. Nevertheless, I say that the consideration of the evolute may also be used to discover the cusp in this second kind of curve.[61] Here is how: let *ABC* be any curve, whose evolute is *GDE*. Also, let two infinitely close radii *CE* and *BD* be drawn and two perpendiculars *EF* and *DF* at the points E and D, meeting at F. As you know, they will form in this way two similar triangles *BDC* and *EFD*. Suppose now that the curve *ABC* has a cusp of the second kind, and that the two radii of the evolute are drawn at the cusp. It is clear that the evolute will always have an inflection point at the point of intersection of these two radii.[62] Now, because, at an inflection point the radius *DF* of the osculating circle is always either infinite or infinitely small, it follows that the ratio of the differential *DE* of the evolute and of the radius *DF* of the osculating circle is either infinitely small or infinitely large with respect to other ratios. Supposing that the differentials of the evolute are equal to one another, or at least that they have a finite ratio, for I can conceive every curve as being divided in whatever manner I wish. This being so, I say that *CB* to *BD* must also have a ratio that is either infinitely small or infinitely large with respect to the others. For this reason, to find the cusp in curves of the second kind, we take the differential of the curve and we divide it by the radius of its osculating circle; what results, must be made equal to zero or to infinity.

...

[What follows is a discussion of other mathematical topics.]

...

There you are, Sir, your difficulties clarified as much as it was possible for me. I am quite surprised that Mr. the Abbé Catelan dares to attack a thing that is quite a bit above his range. Were I to be in your place, I would tell him what Apelles said to a shoemaker: *Ne sutor ultra crepidas*.[63] You would nevertheless give me pleasure by sending me his lovely writing with your reply, if this will not cause too much trouble for the post. Tell me also if you have received the packet of the *Acts* that our bookstore sent to Mr. Anisson. The index that you have asked me for via Mr. Varignon has also arrived; I will send it to you at the first convenience, with the other months that have since arrived (Fig. 12.10).

[61] L'Hôpital discovered this kind of cusp, but Bernoulli recognizes its nature and refers to it as being of the "second kind."

[62] This is incorrect, as l'Hôpital will show in Letter 25.

[63] This translates roughly as "let the shoemaker venture no further." These words were attributed to the painter Apelles (4th century BCE) by Pliny the Elder.

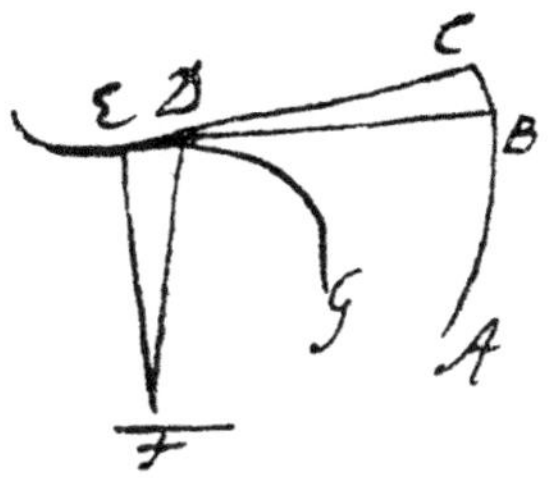

Fig. 12.10 With kind permission of Springer Science+Business Media

I ask you to make my humble compliments to Madame, I am very much affected by her illness. Even so, I would be flattered to hear of her convalescence as soon as you can do me the honor of writing me. My wife also wishes her a happy end to her fever and sends her respects to you.

I am, Sir, forever

Your very humble and very obedient
Servant Bernoulli

Letter 25: L'Hôpital to Bernoulli
Paris, June 7, 1694

The manner in which you conceive the cusp, Sir, appears to me very ingenious. Though, nevertheless, I still find some difficulties there, which I hope that you will be good enough to let me propose to you so that they may be clarified (Fig. 12.11).

The rule that you give for determining the cusp of the second kind appears to me not to be accurate, because it supposes that the evolutes of these sorts of curves

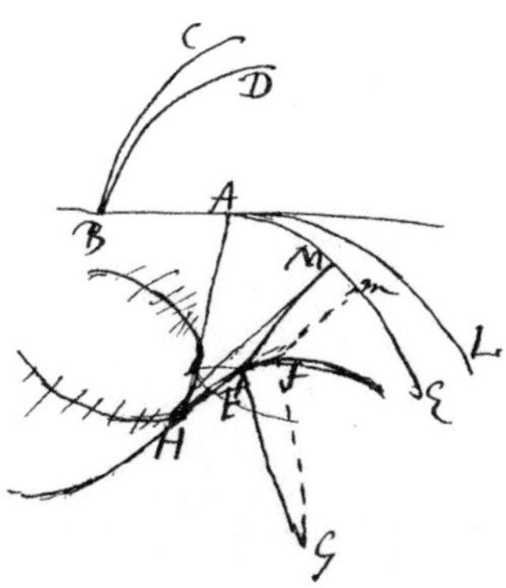

Fig. 12.11 With kind permission of Springer Science+Business Media

always have an inflection point that corresponds to the cusp. Now there is an infinity of cases where this is not so. Let the curve with the cusp of the second kind be *EAL*, which has the straight line *AB* as common tangent at *A*. If we evolve the parts *EA* and *AL* beginning at the point *B*, we form a curve *CBD* with a cusp of the second kind in which the radius of the evolute *BA* at the cusp *B* does not form an inflection point on the curve *BAL* by its extremity *A*, but rather a cusp of the second kind. Furthermore, if we divide the radius of the evolute

$$MF = \frac{\overline{dx^2 + dy^2}\sqrt{dx^2 + dy^2}}{-dx\, d\, dy}$$

by the differential of the curve

$$Mm = \sqrt{dx^2 + dy^2}, \quad \text{we will have} \frac{dx^2 + dy^2}{-dx\, d\, dy}$$

which according to you must be either infinite or zero, from which we conclude that at the cusp *A*, *ddy* must be either infinitely large or null, and consequently that the radius *AH* of the evolute will be either infinite or null, which is contrary to the supposition. In the third case, it seems to me that in the similar triangles *FMm* and *FGf*, we must compose the differentials *Mm* and *Ff* together, as also the radii *MF* and *FG*, because these are magnitudes of the same kind, from which we conclude that because the point *M* falling on the cusp *A*, and the point *F* on the inflection point *H*, the ratio of *Mm*, the differential of the curve *AE*, to *Ff*, the differential of the radius *MF*, must become infinitely large or infinitely small at this point. This gives the general formula

$$dx^2\, d^3y + dy^2\, d^3y - 3\, dy\, ddy^2 = 0 \quad \text{or to infinity}$$

which will serve to determine the cusp *A* in the curves *EAL*. However, I say further that this same formula will also serve to determine the cusp *B* on the curve *CBD*, because it is clear that the radius *BA* being a very small quantity, its differential must be either null or infinite, which gives the same formula.[64] If it is true, as you maintain, that these sorts of cusps are nothing but the touching of an inflection point with the opposing part of the curve, it would follow that the cusp *A* or *B* would still have the nature of an inflection point, and consequently that the radius of its evolute *AH* or *BA* would be infinite or null.

I grant you that at the cusp *dy* is to *dx* in all imaginable ratios but in this, it seems to me, the reason is very different from yours and is founded on a distinction that we have not understood well until now (Fig. 12.12).

I distinguish two kinds of differentials on the ordinates *PM* of curved lines *CBD*, the first sort is the differential *MR* of two ordinates *PM* and *pm* infinitely close

[64] See Bernoulli (1955, p. 224) for a discussion in modern notation.

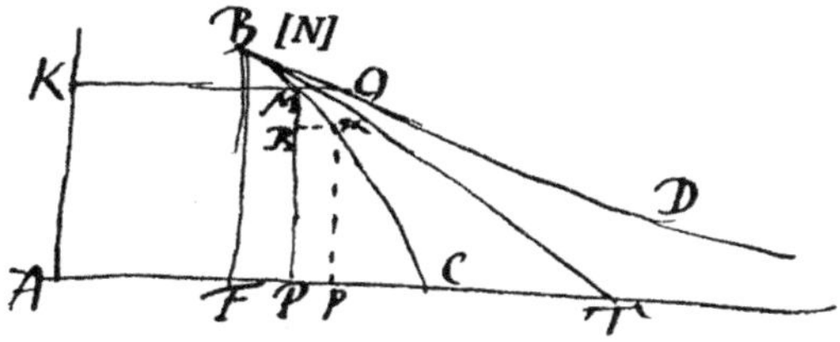

Fig. 12.12 With kind permission of Springer Science+Business Media

to each other, and the other is the differential *MN*, of two ordinates *PM* and *PN* which start at the same point *P*. We must note above all that when we wish to make use of the differentials *MN*, it is necessary that the nature of the curve *CBD* be expressed by an equation in x and y in the manner of Descartes, so that there are no incommensurables, in order that when x remains the same, the unknown y may denote indifferently either of the ordinates *PM* or *PN*, whereas this is not necessary when making use of the differentials *MR*. Given this, it is clear that, if I take the differential of the equation that expresses the nature of the curve *CBD* with a cusp, supposing x to be constant, it is necessary that all the terms that multiply dy, which in this case is *MN*, are together equal to zero, and furthermore that if we suppose *PM* or *AK* (y) to be constant, the terms that multiply dx, which is *MO*, are also equal to zero. From this we see that in taking the differential of the given equation, supposing that x and y are both variable, all the terms that multiply dx on the one hand, and all of those that multiply dy on the other, are equal to zero,[65] which being the ratio of dy to dx may be whatever we might wish, because zero is to zero in whatever ratio we wish. However, when dy denotes *MR*, I say that the ratio of *MR* to *Rm*, which is that of *BF* to *FT* at the cusp *B*, is unique. And this is so true that if we take the differential of the equation $y = \overline{x-b}^{\frac{2}{3}}$, which expresses the nature of the second cubic parabola *EBD*, in which $AB = b$, $AC = x$, and $CD = y$, we have

$$\frac{dy}{dx} = \frac{2}{3\sqrt[3]{x-b}}$$

and making $x = b$, we see that at the cusp *B* the ratio of dy to dx is infinite and not whatever we wish, the reason for which is that dy can only denote the differential of two equally close ordinates because y is only linear in the equation $\overline{x-b}^{\frac{2}{3}}$.

Furthermore, Sir, I submit all of this to your judgement, assuring you that I will ask for nothing more, because I am convinced that having mastered these matters to the depths that you have it can only be very clear. I ask you to send me in your

[65] As the Editor points out in Bernoulli (1955, p. 226) it is incorrect to refer to *MN* as a differential. The distinction is made somewhat clearer in L'Hôpital (1696, §191).

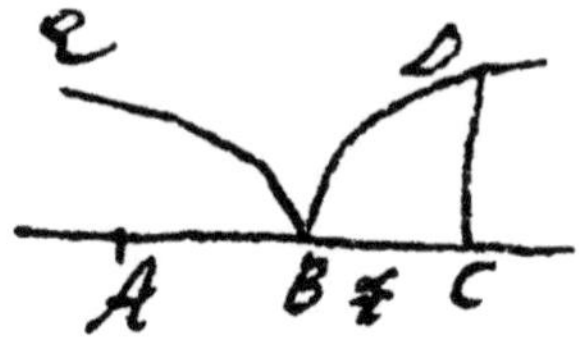

Fig. 12.13 With kind permission of Springer Science+Business Media

response the method of drawing tangents to the tractrix and to the curves described by the mean points between tractrices, as well as the manner of solving the equation

$$y = \frac{\sqrt{2a^3x - x^4} - a\sqrt[3]{aax}}{a - \sqrt[4]{ax^3}}$$

when $x = a$ (Fig. 12.13).

I inquired of Mr. Anisson, who has not received any news from Basel, no parcels nor consequently the *Acts* of Leipzig which you tell me have already been sent; he assures me that if he receives them he will send them to me right away. Mr. Varignon has shown me one of your letters where you say that Mr. Leibniz is writing a book *de scientia infiniti* and that he asks you to share with him your discoveries on these matters. If I dared I would remind you to recall that you have promised me to share them only with me. I will send you the difficulty with the problem of Mr. l'abbé Catelan with my reply, because you have asked for it, however, you will see that it is not worth the trouble. You may count on me, Sir, as on the best of your friends and it will give me real pleasure to let you know this at every encounter.

The M. De L'Hôpital

Mrs. de L'Hôpital is entirely cured from her fever; she sends her compliments without forgetting your dear sweetheart.

Letter 26: Bernoulli to L'Hôpital Basel, June 27, 1694

Sir

The difficulties that you find with the manner in which I conceive of the cusp do not discredit it; all the same, I do not wish to deny that it is subject to several apparent contradictions, which I do not know how to untangle at present. Nevertheless, neither do I wish to doubt that with time I will be able to find the solution, because indeed there are many things that previously seemed inconceivable that we conceive of very clearly today. Who would have led us to believe, for example, this paradox that dy to dx could be at the same time in any imaginable ratio? All the same, this is what causes us the least trouble for the moment. Other than the difficulties that you

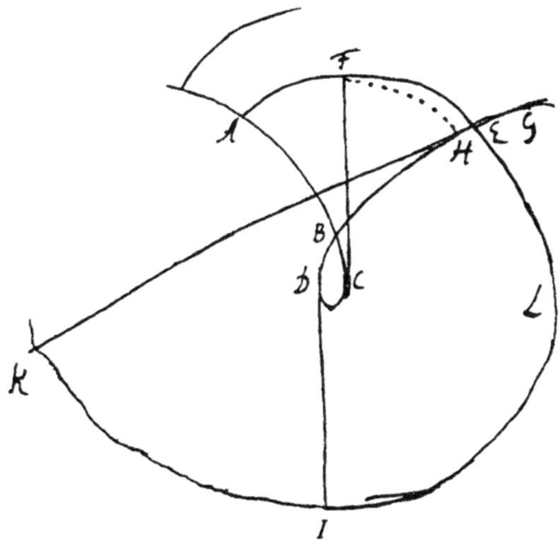

Fig. 12.14 With kind permission of Springer Science+Business Media

have proposed to me, to which I can in some fashion respond, I have others of great consequence, of which here is one. Let *ABCDBHEG* be a curve which intersects itself at *B* and which we evolve beginning at *A*. It is clear that the curve which is generated from the evolution has the form of a spiral *AFELIK*. Now, imagine that the part *BCD* becomes smaller and smaller until it actually vanishes. The curve *ABCDBG* therefore has a cusp, where in this case the parallel radii of the evolute *CF* and *DI* will coincide and be equal, because we suppose *CD* to be of null magnitude, thus the part *FLI* will be a perfect semi-circle. Consequently, the curve that is generated by the evolution of a curve with a cusp is not *AFH*, as we had believed up until now, because *FH* is produced not by the evolution, but by the envelopment of part *BH*. Thus the curve truly generated by the evolution of a curve with a cusp is composed of a curve *AF*, of a semi-circle *FLI* and of a third curve *IK* (Fig. 12.14).

The same thing also happens when we consider the curve with the cusp as an open curve, where the two inflection points *C* and *D* coincide, but the semi-circle *FLI* comes from the other side, as you see represented here with the same letters. We may use the same reasoning on curves with cusps of the second kind. Here then is my difficulty: if every curve describes only a single curve by evolution, how does it happen that the described curve *AFLIK* contains a circular part? I believe to resolve this difficulty we must say that the evolution of cusps may generate various curves; I know several examples well. For example, we know that the evolute of a roulette is another roulette, which has a cusp. If we then evolve this other roulette it reciprocally must generate the first one, but because of what I have just said it will generate three different curves; how then are we to make sense of all of this? The only thing I know to say to this is that I believe that the evolute of a roulette *AFH* is not an entire roulette *ABCDBH*, but rather two parts *AB* and *BH* for which the infinitely small part *BCD*, so to speak, is missing.

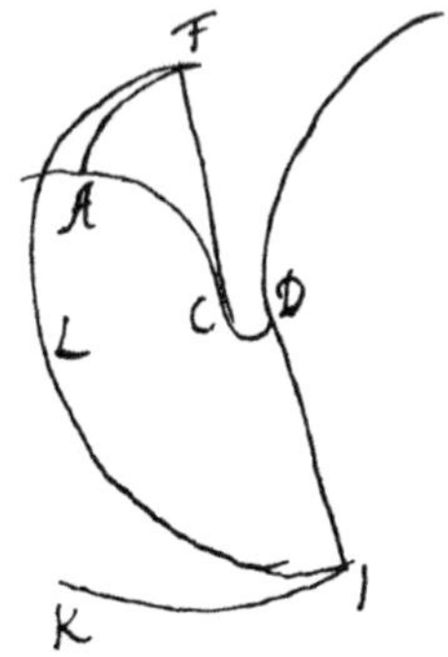

Fig. 12.15 With kind permission of Springer Science+Business Media

This being so it will be easy to respond to your first difficulty, where you believe you have shown me that evolutes of curves with cusps of the second kind do not always have an inflection point, but also some times a cusp, because your demonstration supposes that we evolve a part of this curve and that we envelop the other, which I have said is not a true evolution. For I say generally that *in evolving any curve, the radius always increases towards infinity and never decreases.* In addition, suppose apart from this that a curve with a cusp of the second kind may have a curve with a cusp of the second kind as its evolute; this does not contradict what I have already put forward, because the cusp also includes an inflection point (Fig. 12.15).

As for your second difficulty, I admit, Sir, that $\dfrac{dx^2 + dy^2}{-dx\,ddy}$ must be infinity or zero according to me, but why must one conclude that *ddy* must be either infinitely large or null at a cusp? It suffices that *dx* or *dy* may be either infinity or zero, because these differentials are not free, being determined by those of the evolute. This is what made me say (not having taken magnitudes of the same type) *with respect to the others*, and thus your general formula $dx^2\, d^3y$, etc. follows immediately in the same way. It is nevertheless good that you have noticed this and expressed it in a formula.

The final difficulty that you proposed to me does nothing against what I have said: it is quite true that the radius of the evolute at an inflection point must be either infinity or zero. The reason for this is because the involute of the concave part cannot pass to that of the convex part without the radius being infinite or null, that is to say positive and negative at the same time. Now, cusps of the second kind have only the concave part, because the convex part is infinitely small and consequently it is not necessary that its evolute passes from one side to the other, nor by consequence that its radius be positive and negative at the same time, and thus it can be of a finite magnitude without contradiction.

The reasoning by which you demonstrate[66] that at a cusp, dy is to dx in every imaginable ratio is very ingenious. However, besides the fact that it gives no idea of this point and does not explain its nature, it also gives rise to some doubts:

1. You take *MN* and *MO* for the differentials of *PM* and of *KM*, but it seems to me that *PM* and *PN*, as also *KM* and *KO*, differ by a finite magnitude, and consequently *MN* and *MO* are not differential magnitudes if it is not the case that P falls on F. However, then *MN* and *MO* are nothing but magnitudes that have become infinitely small, for I believe that there is a difference between infinitely small and differential; all differentials are infinitely small, but not reciprocally. Therefore, *MN* and *MO*, although infinitely small, are not included in the general equation of differentials that compose the curve *MBNO* etc.
2. Supposing that *MN* is to *MO* as zero to zero, you then conclude that *MN* is to *MO* in whatever ratio we may wish. However, I deny that between zero and zero there is always all imaginable ratios, because I have shown you that in the example you proposed to me

$$\frac{\sqrt{2a^3x - x^4 - \text{etc.}}}{a - \text{etc}}$$

when $x = a$, it becomes $\frac{0}{0}$ = to a determinate magnitude, and consequently the numerator zero to the denominator zero does not have whatever ratio one might wish.

I come back, Sir, at the first opportunity to the method of solving this equation as well as the method of drawing tangents to the tractrix and to the curves described by the mean points between tractrices. I had wished to send them to you at present, but finding myself unable to do so because of the evolutes that enveloped my entire brain, I was not able to put it in writing on the spot, because it has been quite some time since I have thought about these. Also, I am now quite occupied in moving, having lived up until now with my father-in-law, which gives me little rest to gather my animal spirits. In a word, I am not yet in the routine of being married (Fig. 12.16).

If you have not yet received the *Acts*, you will receive them soon. Mr. Varignon will give you another packet containing the month of December 1693, Jan., Feb., and March 1694, the supplement Volume IV sect. III and the general index that I sent to him with his *Acts*.

Mr. the Abbé Catelan is without doubt extremely unremitting against you, because he attempts to obscure the clearest truths that come from you, but he acquires little glory in the opinion of knowledgeable men. Tell me please if he wishes to give a method of inverse tangents: if he finds the quadratrix from the

[66] See Letter 25, Footnote 65.

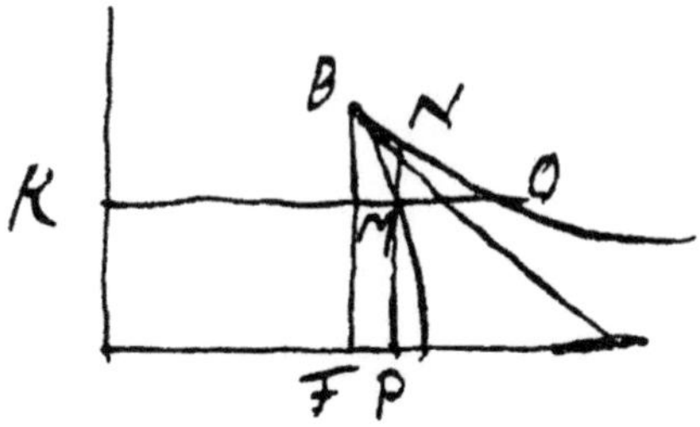

Fig. 12.16 With kind permission of Springer Science+Business Media

quadratrix itself, it seems to me supposing one has what one seeks. Please be so kind as to give the very humble compliments of my wife as well as mine to Madame. We wish her a happy cure from her fever.

I am Sir, etc.

Letter 28: Bernoulli to L'Hôpital Basel, July 22, 1694

Sir, I do not doubt that you have received my last letter, which I gave myself the honor of writing to you almost four weeks ago. This one is to accompany the extract of the month of May of the *Acts* of Leipzig that I sent you by the post. You will see there your solution to my problem with Mr. the Abbé Catelan's difficulty.[67] I am very surprised not to see your clarification, because it is to be feared that this will cause some harm to the differential calculus among people who are not in the know and who might add credence to Mr. Catelan's difficulty without examining the subject. This is what made me believe that it is not one of your friends that communicated to the Editor of the *Acts* your solution accompanied by the difficulty without having attached the response to it. You will also find in that extract the solution to the Florentine Enigma by Mr. Viviani, but it is more or less the same as the one that I gave you in Paris. Here, finally, is the method for drawing tangents to the tractrix and to the curves described by the mean points between tractrices, and also the solution of these sorts of equations

$$y = \frac{\sqrt{2a^3x - x^4} - a\sqrt[3]{aax}}{a - \sqrt[4]{ax^3}}, \quad \text{when} \quad x = a.$$

[67]*Acta Eruditorum*, May 1694, pp. 193–196, 196. These are translations of the articles mentioned by l'Hôpital at the end of Letter 23.

I flatter myself that you will be pleased, having taken all the care that is needed to explain myself clearly.

I received a letter from Mr. Leibniz in the past few days, in which he asks me again to communicate some of my discoveries to be inserted in his treatise *de scientia infiniti*. However, I do not know how I should behave with regard to this, because which ever way I take I find myself quite hindered. It is certainly true that I promised you[68] to share my discoveries with no one but you, but do you counsel me to refuse Mr. Leibniz' request and to entirely cut off correspondance with him? In truth that would be to act with little honesty towards a man to whom I am so indebted. Nevertheless, I believe that in order to keep the promise that I made to you, the best course would be for me would be to communicate to Mr. Leibniz only that which concerns the differential and integral calculus in general and which is already known to my brother and possibly to others as well as to me. However, for that which concerns the things that have gone on between you and me, in particular and the discoveries that I have made on your behalf and that I will make in the future on the opportunities that you give me, I make you a sacred promise, Sir, to always keep them secret and to let nothing at all out. I have already given you proof of this in advance, having always refused under the pretexts fabricated by Mr. Varigion, the principles of the integral calculus that he has solicited so often and so insistently, even though I am much obliged to him for the great services he has rendered me. I do not doubt that he will ask for them again, so please tell me by what means I may withdraw from his requests once and for all. I leave tomorrow for the country 6 leagues from here to drink the waters at Bad Pfäffres, if you wish to take the trouble to write to me you need only use the ordinary address. Having made my very humble respects to Madame and those of my wife, I am, Sir,

Your very humble and
your very obedient servant
Bernoulli

Method for drawing tangents to the tractrix and to the curves described by the mean points between tractrices.[69]

ABCN is a perfectly flexible rope that we drag by one of its extremities *N* along a given curve *NO*, the rope draws with it the attached weights *A*, *B*, *C*, etc., of different magnitudes and of different intervals, as the same number of tractrices, which by the movement of the rope describe the curves *AH*, *BL*, *CM*, etc. We wish to know the method for drawing the tangents.

Sol. It is clear that the curve *AH* has the straight line *AB* itself for a tangent, because *AB* is the direction of the weight *A* that describes the curve *AH*. However, to find the tangent to the curve *BL*, I remark that it cannot be *BC* because it is not

[68]Presumably in the lost reply to Letter 20.

[69]Compare to §45 on p. 41.

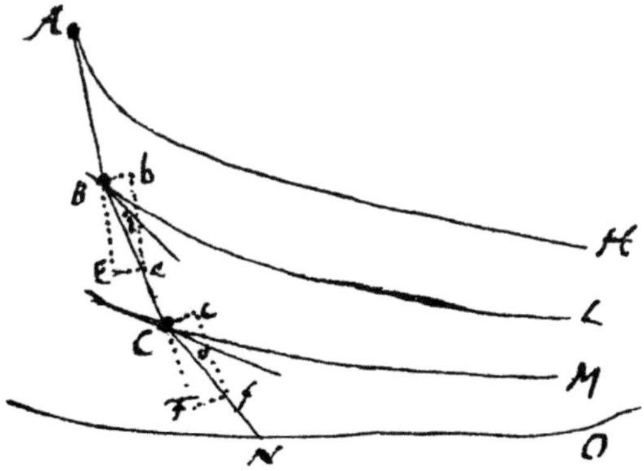

Fig. 12.17 With kind permission of Springer Science+Business Media

the direction of the weight B being impeded and diverted by the weight A. It is therefore a matter of finding the direction of the weight B, which will be the tangent of the curve *BL*, which I do as follows. Having extended the straight line *AB*, I form the rectangle *Eb* on the diagonal *BC* and I consider that if the weight A were not there, the weight B would have the diagonal *Be* as its direction with a quantity of motion denoted by *Be*. Now, this motion is composed of *BE* and of *Bb* and I see that if we now consider the weight A as resisting, it will only be *BE* that will lose its motion, and *Bb* will remain whole, because A does not resist the weight B going along *Bb*, but all the resistance turns towards *BE*. Therefore, if we make the weight $A+$ the weight B to the weight B, as *BE* or *be* is to a fourth quantity *bi*, this *bi* will be the remainder of the motion by the law of statics. This is why having drawn *Bi*, which denotes the motion composed of the whole *Bb* and the remainder *bi*, it will be the direction of the weight B and consequently the tangent of the curve *BL*. In the same way we find the tangent to the curve *CM*: because after having made the rectangle *Fc*, the motion *Cc* remains whole and *CF* is retarded by the weight B and by a part of the weight A, which is to A as *BE* is to *Be*. Therefore, having made

$$\frac{A \times BE}{Be} + B + C : C :: cf : co,$$

the straight line *co* will be the direction of the weight C and consequently the tangent to the curve *CM* (Fig. 12.17). To find a tangent to a fourth curve we must make (using corresponding letters)

$$\frac{A \times BE \times CF}{Be \times Cf} + \frac{B \times CF}{Cf} + C + D : D :: dg : du,$$

the straight line drawn *Du* will be the tangent to the fourth curve; and so on for the others. For the second part of the problem, namely to draw the tangents of the curves

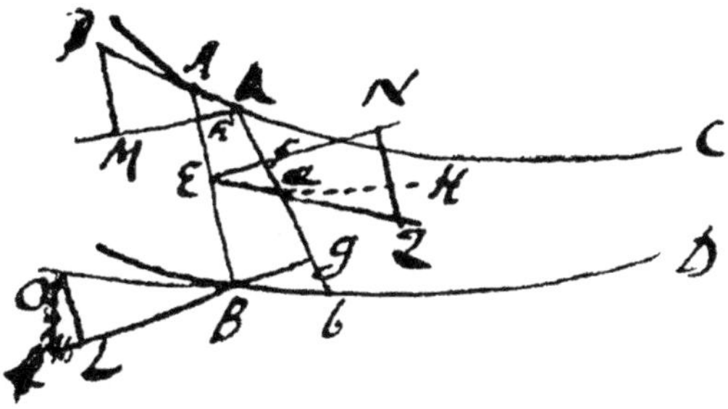

Fig. 12.18 With kind permission of Springer Science+Business Media

described by the mean points between the tractrices, I transform it into another general problem, as follows:[70]

Given two curves *AC* and *BD* to which we apply an infinity of equal lines *AB*, *ab*, etc., all of which we divide in the given ratio of *AE* to *EB*, we wish to know the method for drawing the tangents to the curve *EH* described the divisions E, e, etc.

Sol. Let *AB* and *ab* be two infinitely close equal lines. Having drawn *ah*, *Ef*, and *Bg* perpendicular to *AB* and *ab*, it is clear that *Ah*, *fe*, and *gb* are equal. Let $AE = a$, $EB = b$, *Ah* or *fe* or $gb = dx$, $ah = dy$, and $gB = dz$, then we have

$$fE = \frac{a\,dz + b\,dy}{a+b}, \quad \text{therefore} \quad Ef : fe :: \frac{a\,dz + b\,dy}{a+b} : dx,$$

from which I easily draw the tangent at E, because having drawn three perpendiculars *aM*, *EN* and *BL*, we will draw two tangents *aP* and *BO* to the given curves and we will apply two equal perpendiculars *PM* and *OL* to the angles *OBL* and *PaM*. Thus, by the correspondence of sides of similar triangles, if we take *EN* equal to

$$\frac{AE \times BL + EB \times hM}{AB}$$

and we make *NQ* perpendicular to *EN* and equal to *PM* or *OL*, it is manifest that *EQ* will be the tangent that we wish to find of the curve *EH*. Q.E.I.[71] It is easy to see how this problem includes the second part of the previous one, because the intervals between the weights always remain the same, so the distances of the mean points also do not change (Fig. 12.18).

Probl.[72] Given a curve whose nature is expressed by a fraction equal to y, which in a certain case has the numerator and the denominator equal to zero, we wish to find the value, that is to say the magnitude of the ordinate y.

[70]Compare this to §36 on p. 35.

[71]I.e., That which was to be found.

[72]Compare this to §163 on p. 151.

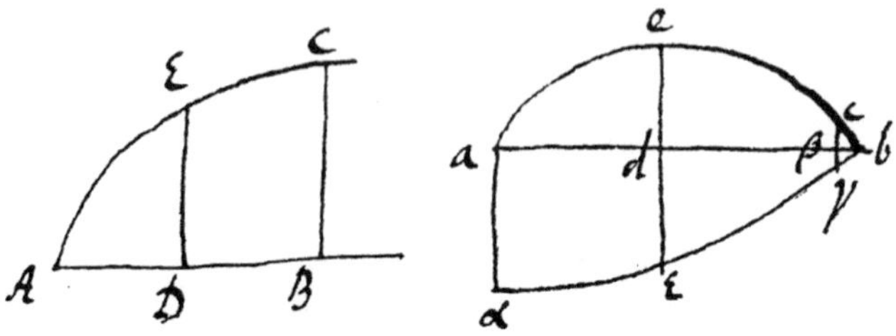

Fig. 12.19 With kind permission of Springer Science+Business Media

Sol. Let *AEC* be the given curve, $AD = x$, $DE = y$, $AB =$ to a constant, such that *BC* becomes equal to a fraction, the denominator and numerator of which are equal to zero. Therefore, to find the magnitude of the ordinate *BC*, I construct on the same axis *adb* two other curves *aeb* and $\alpha\epsilon b$ of such a nature that having taken abscissas equal to *AD* and *ad*, the ordinates *de* are in ratio to the numerator of the general fraction, which expresses the ordinate *DE*, and $d\epsilon$ are in ratio to the denominator of the same fraction. This being done it is clear that *de* divided by $d\epsilon$ may be supposed equal to *DE*. The problem therefore reduces to finding the value of *de* divided by $d\epsilon$ in the case that *ab* is equal to *AB*. Now, I see that in this case, *de* and $d\epsilon$ vanish because the two terms of the fraction vanish, and thus the two curves *aeb* and $\alpha\epsilon b$ intersect at the point b. Therefore, we need only take the differentials[73] βc and $\beta\gamma$, of which the one divided by the other which will tell me the magnitude of *BC* that I seek. This is what gives me the following general rule: *To find the value of the ordinate of the given curve in the given case we must divide the differential of the numerator of the general fraction by the differential of the denominator; the quotient, after having made x equal to the supposed AB, will be the magnitude of BC* (Fig. 12.19).

Example. The curve *ACE* has for its equation

$$\frac{\sqrt{2a^3x - x^4} - a\sqrt[3]{aax}}{a - \sqrt[4]{ax^3}} = y.$$

Thus, if *AB* is $= a$, we have $BC = \frac{0\,a}{0}$, now we wish to know the true value. According to the rule, I take the differential of the numerator $\sqrt{2a^3x - x^4} - a\sqrt[3]{aax}$, which is

$$= \frac{a^3\,dx - 2x^3\,dx}{\sqrt{2a^3x - x^4}} - \frac{a^2\,dx}{3\sqrt[3]{axx}},$$

[73]Bernoullis says *les dernieres differentielles*; literally "the last differentials."

and the differential of the denominator

$$a - \sqrt[4]{ax^3}, \quad \text{which is} \quad \frac{-3a\,dx}{4\sqrt[4]{a^3x}},$$

having now substituted in the place of x the supposed value a, we find $-\frac{4}{3}a\,dx$ for the first differential and $-\frac{3}{4}\,dx$ for the second one. Therefore,

$$\frac{-\frac{4}{3}a\,dx}{-\frac{3}{4}\,dx} \quad \text{or} \quad \frac{16a}{9} = BC.$$

Q.E.I.

To verify this method, we may take a very easy example such as this one

$$\frac{a\sqrt{ax} - xx}{a - \sqrt{ax}} = y,$$

which we may also solve, although with much difficulty, with common geometry by removing the irrationality; for we will find by either method $BC = 3a$.

Letter 29: L'Hôpital to Bernoulli
Paris, August 18, 1694

I received, Sir, the solution of the two problems that I requested of you, for which I am very much obliged to you and with which I am very much pleased. You were very good to send me the *Acts* of Leipzig and I was very much surprised to see the same difficulty with the Abbé C. appear again there, which I thought I had entirely satisfied. However, what you apparently haven't taken note of is that there is a fairly considerable addition that seems to have been taken only from my clarification, as if it had been conceived of in advance. This is what induced me to make the response that I am sending you, and I ask you please to translate it into Latin and to send it as soon as possible to the editor of the *Acts* and to request on your behalf and mine to put it in soon without delay. We're going to St. André at the end of the coming month and therefore, Sir, if you are man of your word prepare yourself to come and see use during this time. We will eat well and drink good wine, for it is thought to be the best in the entire country. I am quite angry that you tell me has happened to your arm,[74] because you tell me that you will take the waters this that makes me

[74]Presumably, Bernoulli mentioned this accident in a postscript to Letter 28 that was not included in the draft. He began Letter 30 on September 9, 1694, with the sentence "It is a few days since I returned from the country; the swelling that I had in my arm from a fall is entirely cured."

apprehensive about your health, and you would do me pleasure by giving me news about it.

I am entirely yours

The M. de L'Hôpital.

My wife sends you her compliments and to Madame your wife, whom she begs to make the journey to St. André.

Letter 31: L'Hôpital to Bernoulli Paris, October 2, 1694

I'm very happy, Sir, that you are cured of the swelling that you had in your arm. Please explain to me at greater length why, in the plenum, the full force, with which the sail *DC* pushes the ship sailing along *BK*, is to that with which it pushes along *BG* as the square *BK* is to the square *BG*, given that you agree in the void it is as *BK* is to *BG*, and it seems that in the first instant the impulse of the wind against the sail should be the same, whether there is nothing to resist it as in the void, or on the contrary, it finds that there is matter which resists the impulse as in plenum. Whether a window, for example, is closed or partially open, it seems that being pushed by the wind, it received the same impulse in the first instant. I am sending you the reply of Mr. Huygens so that you can examine it with care and so that you can tell me your feelings about all of this. I thank you for the trouble you have taken to translate my reply into Latin and to send it to Leipzig. I would be very curious to see what Mr. Leibniz has in the *Acts* for the month of July and, Mr. your brother, in those for the month of June. That is why I ask you to send them to me from Geneva to Lyon, addressed to Mr. Gautier, lumber Merchant at the port of St. Clair, to hold for Mr. the Marquis de L'Hôpital in his château of St. André.

I depart soon to go there, and you will oblige me by sending me as soon possible all the new things that you told me you have found, because this appears very beautiful to me, you should address your letters to Mr. the Count of St. Mesme,[75] the street of the little musk near the arsenal, to be held for Mr. the Marquis de L'Hôpital, if it is necessary to put a Paris address, because if it were not, which you could learn from the post, you need only address your letters to Lyon, where I have told you above.

In any case, we accept no excuses, neither Madame de L'Hôpital or me, for not coming to see us in St. André, and I do not find that there can be any valid reason that keeps you from coming to spend as little time as you wish with us. Madame de L'Hôpital also invites Madame your wife. She hopes soon to receive some of her letters. Be well assured that if you make this voyage, I will take care to pay the costs; I know well that it is not this reason that would keep you from coming, but I

[75]L'Hôpital's father.

also know that this is reasonable and that things must be thus, because you would not take this trip only for the love of me. I have 36 ₶ to give you that Mr. Varignon has asked me to give to you, and in addition the 150 ₶ remaining for the rest of the year. I will give you all this in St. André and when I get there I will tell you how much time it will take you to come find me. I am perfectly yours

The M. de L'Hôpital.

Since writing my letter I have been assured that you need only address your letters to Lyon and they will go without passing through Paris. You should still find out about this at your home to be even more certain.

Letter 39: L'Hôpital to Bernoulli
Lyon, February 3, 1695

You will find, Sir, included here a bill of exchange for two hundred pounds, namely one hundred and fifty pounds for the remainder of last year and thirty six pounds ten *sols*[76] from Mr. Varignon, and the surplus for the *Acts* of Leipzig that you have sent me. You have apparently received my last two letters in which I asked you to make sure that my first draft does not appear in print and to substitute the second one that I have sent in its place.[77] I am sure that you will forget nothing in order to give me the satisfaction that I'm asking of you on this matter. Do not write to me here any further and send your letters to the usual address in Paris, where I hope to be in fifteen or sixteen days starting tomorrow morning.

I am, Sir, entirely yours

The M. De L'Hôpital.

Letter 41: Bernoulli to L'Hôpital
Basel, February 19, 1695

Sir, I am very happy in that I have received two presents all at once. The one comes from you as a result of your generosity, for which I thank you infinitely, promising

[76]The French pound was divided into 20 *sous* or *sols*.

[77]In Letter 33, L'Hôpital had sent Bernoulli his solution to the *Probloma aequilibrii*, a problem posed by Sauveur, to be translated into Latin and sent to the *Acts Eruditorum*. In Letter 36, Bernoulli confirmed that he had done this, but also admitted to adding some remarks of his own, which substantially simplified this solution. In Letter 37, L'Hôpital sent Bernoulli a different solution asking him to sent it to the *Acts Eruditorum* with a request that the first piece be withdrawn. This request came too late and both pieces appeared: in February pp. 56–59 and in Supplement II, 1696 pp. 289–291.

you to put all my efforts into making myself worthy. The other comes from my wife, who has made me the father of a boy, which she delivered eight days ago.[78] I would have been delighted if, in returning to Paris, you had taken the route through here, because I would have taken the liberty of asking you to present him at baptism to be his godfather. I have not failed to constantly give you counsel in such a way that I have strived to give you all possible satisfaction and even more so in what you have asked of me concerning your two solutions. However, I see by your last letter than you had already departed from Lyon when my letter[79] arrived there; for this reason if you would please take the trouble to write to Mr. Gautier, and ask him to send you that letter, for you will see there what I have written to Mr. Menkenius, and also a copy of what I sent him to be inserted in the *Acts*, in case your first article should be printed before the arrival of the countermand. I hope that you will take great pleasure from this precaution that I took, having managed things in such a way that if the counter-order should arrive in Leipzig too late you will still have the biggest part of the new solution, which you will see better in the aforementioned copy. To show you that I hold exactly to my word, see what Mr. Leibniz wrote to me in his last letter. *Quod de infiniti scientiâ cogito opusculum si tuis (ut tuae innuere videntur) auxiliis destitueretur vereor ut mature prodeat in lucem, aut omnino ut prodeat.*[80] He wrote me this way because I had sincerely refused him my writings under the pretext of no longer having them in my hands. However, I am angry that because of this he will entertain thoughts of not making his work current. You must encourage him, for the good of the public, to remain firm in his original design; because I know that he esteems and respects you greatly, calling you *Phoenicem Mathematicorum Galliae*[81] that is to say, the only one in France who understands the inner Geometry. The post that is about to depart is urging me to finish, and begging you to assure Madame of my very humble respects and also those of my wife who is doing well.

I am,

Sir, etc.
Bernoulli.

[78]Nikolaus (II) Bernoulli (1695–1726).

[79]Letter 40.

[80]This may be translated roughly as: "Now concerning *The Science of the Infinite*, I fear that if the work lacks your contributions (as your [letter] seems to hint), then it will not come to light in good time, or else not at all."

[81]The Phoenix of Mathematicians of France.

Letter 44: L'Hôpital to Bernoulli
Paris, February 26, 1695

I have just now received, Sir, your letter of the 19th of this month. I congratulate you on the successful delivery by Mrs. Bernoulli and I wish that the little boy will walk in the footsteps of his father and one day also to be as learned as he.

I am entirely happy with the way that you treat me, and I hope that in the future, when I have my affairs in better order, that you will be even more satisfied with me than in the past. I ask you also to take up some of what I asked of you in my last letter, which would cause you the least trouble, assuring you that I do not wish to impose on you. I ask when you write me that you give me some satisfaction with regards to the questions that I proposed to you in my earlier letters.

My wife sends you and Mrs. Bernoulli a thousand compliments and I am truly, sir, entirely yours

the M. de L'Hôpital.

Letter 47: L'Hôpital to Bernoulli
Paris, March 12, 1695

This morning I received, Sir, your letter of the 5th of this month. I am quite obliged for all the honesty you showed me there, and I assure you that I will not abuse it, and to show you that I consider your career, and not just that which would give me pleasure, I hope to be able to procure you a chair of mathematics in Holland,[82] which is worth twelve hundred pounds of their money, that is to say 1440 ₶ in that of France. If this meets with your wishes, you must make your response as soon as possible, and if this is the case I have two favors to ask of you. One is that you let me know liberally the money that will you will need to make this voyage and I will send it, and the other which is only my concern, would be to go through Paris so that I may provide you with many things. Because if this succeeds, as I have almost no doubt, I fear that I will not see you for a long time, and I confess that if I had only my satisfaction in mind I would have greatly desired to find you a position in this country, but one must forget about oneself when it comes to serving one's friends.

...

[Discussion of roulettes.]

...

I see by what you tell me that some of my letters were lost, and it is precisely that which served as a response to yours, where you sent me your general method for constructing differential equations of the first degree. I have received three letters

[82]In a lost letter, Huygens asked L'Hôpital for a recommendation regarding Bernoulli for a position at the University of Groningen.

from Mr. Leibniz almost at the same time that I have not carefully examined – he shared some of his methods for the inverse of tangents with me. It always appears in my thoughts to have his book on science of the infinite printed. I will send you the location in the memoirs where the theorems of Mr. your brother on evolutes may be found.

Please send me a copy of what you have sent to Leipzig to be added to one of my first solution on the problem of the drawbridge.[83] I looked over the book by Mr. Craig.[84] I find that although he has some intelligence, it is clear that he does not understand these matters as you do, and besides that he does not go straight to the goal.

I ask you to reply to me as soon as possible, and to believe that no one can be more attached to you than I am.

The M. De L'Hôpital.

Madame de L'Hôpital sends you a thousand compliments and to Madame Bernoulli.

Letter 48: Bernoulli to L'Hôpital
Basel, March 26, 1695

Sir, I do not know how to thank you enough for the great care that you have taken in procuring me a chair of mathematics in Holland, I see by this that your affection greatly surpasses the small service that I could do for you. All that I wish for is the continuation of your good graces; if I am not worthy of them, at least I will try to be so. The arrangements that you offer me are beautiful and acceptable and indeed you would have already seen me in Paris to receive this gift myself if I were without a wife and child, because I have the greatest desire in the world to embrace such a favorable occasion. However, because I am married it is necessary that I should bring my wife with me, this is why we must act on this meeting with a bit of forethought, in making use of the council of my friends. Therefore you will be so good as to tell me the circumstances and particulars of this chair, namely in which city in Holland it is, to whom you first spoke about it, the means by which you may procure it for me, and above all, if it is precisely me that they desire; it is on this that I may take the measure of the situation. I particularly hope for your good advice on

[83]I.e., the *Probloma aequilibrii* of Sauveur.

[84]Craig, John. *Methodus figurarum lineis rectis et curvis comprehensarum quadraturas determinandi*, London, 1693. L'Hôpital spelled the name "Graiges."

Fig. 12.20 With kind permission of Springer Science+Business Media

all of this, asking you to tell me as a friend if you think that the annual wage would be sufficient for me to live comfortably, because they tell me it is very expensive to live in Holland particularly in the time of war (Fig. 12.20).

...

[Discussion of roulettes.]

...

Your method for determining the points of caustics by refraction, which you were so good as to send me[85] is the same as what you shared with me a long time ago.[86] Even though it is quite easy, I nevertheless hold that one may go even straighter to the point by seeking only the ratios of the 3 little angles that make up the two incident rays, two refracted rays and two radii of the osculating circle, which can be done without any differential calculus, and by an entirely synthetic means, which consequently is also the most natural. By this method we find that in caustics by reflection the two angles $A + C$ made by the incident rays and by the reflected rays are $=$ to two angles B made by the radii of the osculating circle. From this property we immediately determine the length of the reflected ray DC.[87]

I am sending you a copy of what I sent to Leipzig. By the way, I still owe you a response to your letter of February 28. You tell me that Madame is pregnant, my wife and I wish her a happy delivery; I see you have made her an able arithmetician in teaching her so well the rule of Three.[88] You ask me if my son looks like me, my wife tells me yes, and one must believe her.

I am, Sir,

Bernoulli

[85] *Memoir Mathematique*, August 1693, pp. 129–133.

[86] Letter 15.

[87] For a modern treatment of this argument, see Bernoulli (1955, p. 278).

[88] I.e., cross multiplication. This refers to a joke l'Hôpital made in Letter 32 concerning the birth of his third child.

Letter 49: L'Hôpital to Bernoulli
Paris, April 16, 1695

I did not reply earlier, Sir, to your letter of March 26 because I was awaiting a response from Holland to fully clear up everything that you had asked me on the subject of the chair of mathematics, but as it has not yet arrived and I am afraid that you will accuse me of negligence on that which concerns you, I am writing you this as I wait and as soon as I receive it I will not fail to tell you everything you wish and also to tell you freely my thoughts about it, because you tell me you wish it. I did not reply earlier, Sir, to your letter of March 26 because I was awaiting a response from Holland to fully clear up everything that you had asked me on the subject of the chair of mathematics, but as it has not yet arrived and I am afraid that you will accuse me of negligence on that which concerns you, I am writing you this as I wait and as soon as I receive it I will not fail to tell you everything you wish and also to tell you freely my thoughts about it, because you tell me you wish it.

. . .

[Discussion of roulettes and inflection points.]

. . .

When you reply to me, send me those two pieces by Mr. your brother and by Mr. Huygens and believe me, Sir, to be entirely yours with the best of my heart.

The M. De L'Hôpital.

My wife sends you a thousand compliments and to Mrs. Bernoulli.

Letter 50: Bernoulli to L'Hôpital
Basel, May 3, 1695

Sir

I am eager to learn the news from Holland concerning this chair of mathematics. I would rather you had told me in your last letter, while I was waiting, the city where it is located, and the person that had given you notice. You will greatly increase the large obligation that I have to you, if you tell me everything you know about this.

. . .

[Discussion of roulettes and a fourth order analytic curve.]

. . .

Letter 51: L'Hôpital to Bernoulli
Paris, June 10, 1695

I just received, Sir, your letter of June 5, I have not made a reply to your last letter any sooner because on that same day my father fell into apoplexy from which he has just returned but not without much distress. My wife then delivered a daughter,[89] as you have guessed, and I lost a little time.

Here is the story of the chair of mathematics in Holland. Mr. Huygens, to whom I had already written a long time ago that if there should be a chair of mathematics in Holland it would please me to tell me and to obtain it for you, wrote me a letter, in which he asked me to find out from you if you were still of the same mind, I replied to him that I would have to ask about the salary of this chair and the location of the city so that you would be able to more easily determine your response. He told me that it had a salary of 1,200 ℔ in Dutch money, which is worth 1,440 ℔ in French money, and that the city was Groningen, but he immediately asked me not to tell you the name of the city unless you have positively given your word to accept it. On this matter I wrote you the letter,[90] as you know, and as soon as I had received your response[91] I wrote to Mr. Huygens that I found you to be shaken, but you wished to know if they wanted you in particular, and the city and the other particulars that you mentioned to me in your letter. On this matter, I have had no response from Mr. Huygens even though I have written him two more times since then. However, apparently he made it known to the Curators of the Academy of Groningen, who wrote you the letter of which you told me. Tell me, I pray you, if you will pass through Paris, and with regard to the money that you might need for your voyage, take it, I pray you, from my purse and don't give it another thought, because it always gives me great pleasure to do you a good deed in anything (Fig. 12.21).

...

[Further discussion of the algebraic curve from Letter 50.]

...I finish, my dear Sir, by assuring you of my perfect esteem and of a very sincere friendship on my part, which remoteness can never alter.

The M. De L'Hôpital.

I reopen my letter to you to tell you that I have seen a man this morning who told me that Mr. Huygens was dead.[92] If this is so I am no longer surprised that he has not replied to me. I know well that in his last letter he told me he was very uncomfortable; I would be very angry because he showed me a lot of consideration and I am certain he would have been good friend to you. Here is a little problem which I ask you to think about, it concerns the direct method of tangents.

---

[89] Charolotte Silvia (June 6, 1696-May 5, 1759).

[90] Letter 47.

[91] Letter 48.

[92] This was a mistake. Huygens actually died on July 8, 1695.

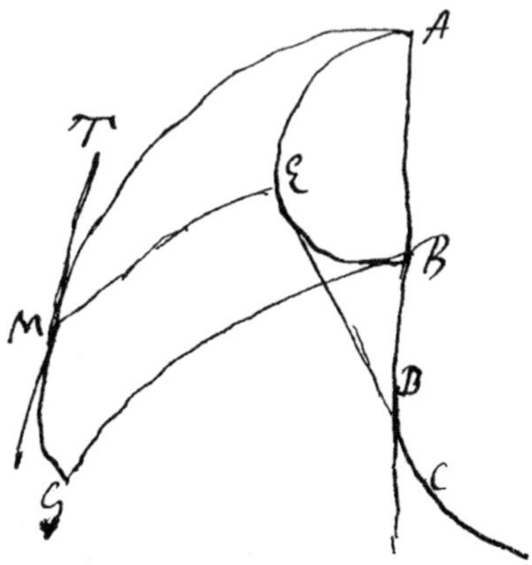

Fig. 12.21 With kind permission of Springer Science+Business Media

The curves *AEB* and *DC* are given and the curve *AMG* has the property that having freely described the arc *EM* by the evolution of the curve *DC*, the relationship of the arcs *AE* and *EM*, and the radius *CE* is expressed by a given equation. We wish to draw the tangent *MT*.

I see by what you tell me about Mr. Leibniz that his construction is not very different from mine. So one might have hoped that it would have appeared sooner. If you write to Mr. Menkenius, he must do it immediately and do not forget to tell him that he should make mention of the date he received my construction so that the public sees than I did not take it from Mr. Leibniz' construction.[93]

Letter 52: Bernoulli to L'Hôpital Basel, June 23, 1695

Sir,

You have told me all kinds of news, both happy and sad. Particularly that of the happy delivery by Madame has given me extraordinary joy because I feared strongly for her, because I had not received any of your letters for a long time. I cannot, therefore help but tell you the joy that this good news caused me. I congratulate you and also Madame with all my heart. You have a special talent for alternating between boys and girls (*amant alterna Camoenae*).[94] The illness of Mr. your Father, and the loss of a lawsuit that you suffered have assuredly distressed me, but the consolation that I find is that Mr. your Father has come back, and that the judgement was small.

[93]Leibniz' solution to the Drawbridge Problem appeared in *Acta Eruditorum* April 1695, page 184. L'Hôpital's second solution appeared two days later in *Acta Eruditorum*, Supp. II, page 372–374.

[94]The Muses Love Alternation.

The most disagreeable news that you tell me is the death of Mr. Huygens; in truth it caused me to be quite dismayed and I have trouble getting over it, because I was already counting very much in advance on his friendship which I might have enjoyed when I was in Holland. Indeed, the desire I had to make the acquaintance of this great man was the first force that draw me to that county. Finally, I have already received three letters from Groningen, in the third of which I am urgently pressed to let my final decision be known so that they know it before the 30th of this month, on which day the Estates of the city gather together to confirm and invite various people that the Curators will have named for the various sciences. Thank God the resolution is taken, because I wrote to Groningen this past Saturday to say that I am ready to obey as soon as they send me a public letter of invitation. However, my wife has not yet decided, it being very difficult for her to leave her country and parents, yet I still hope to persuade her to make this trip with me, or at least to follow me some time after I have taken possession of this chair. You are certainly very generous by wanting to provide me with the fare of this trip; I am in your very humble graces; but I do not know if I can go through Paris, because you may well believe that if I go with my wife, I will also take my child with a governess and a maid; and therefore I must choose the shortest path to avoid spending too much for too many people. But if I travel alone I promise you to take the road to Paris in order to express in person the most humble gratitude that I owe you for all your great kindness.

...

[Further discussion of the fourth degree curve from Letter 50.]

...

Now, to the solution of your problem,[95] I would very much like to know of what use it is, it is pretty, but not very useful (Fig. 12.22).

Lemma. *Let AaaaD be any curve, which we evolve, making two infinitely close arcs Eeeee and Nnnnn. I say that the differential of these two arcs is found by making as EA is to the circular arc ES described by the radius AE and contained between AE and AS parallel to the tangent De, so the distance of these two arcs EN is to the fourth quantity that we wish to find. For imagining that the tangents ae, ae, ae, etc. are drawn, so that they make equal angles EAe, eae, eae, etc. If we now draw nm, nm, nm, etc. parallel to NE, ne, ne, etc., then it is clear that all the triangles mne are similar and equal. Now each part ee exceeds nn by the base me; and consequently the entire curve Eeee exceeds the curve Nnnn by all the me, of which there are as many as there are angles EAe in the angle EAS (because the sum of all the angles eae is = EAS) therefore*

$$EA : Ee + Ee + Ee + \text{ etc. } (ES) :: EN : me + me + me + \text{ etc. } (Eeeee - Nnnnn)$$

Q.E.D.

[95]The following Lemma appears in L'Hôpital (1696, §166). The subsequent proof of the theorem appears in L'Hôpital (1696, §169).

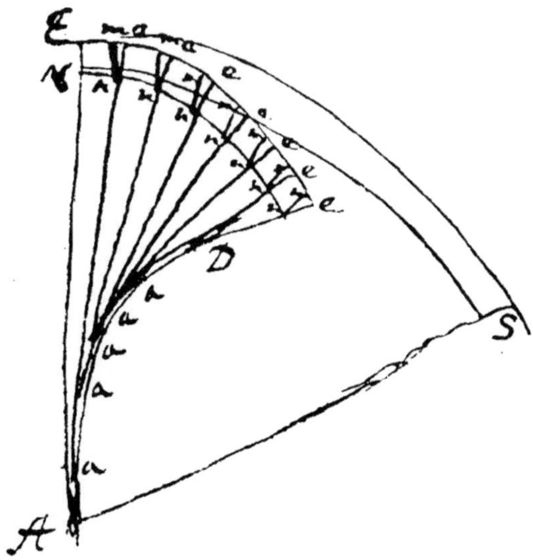

Fig. 12.22 With kind permission of Springer Science+Business Media

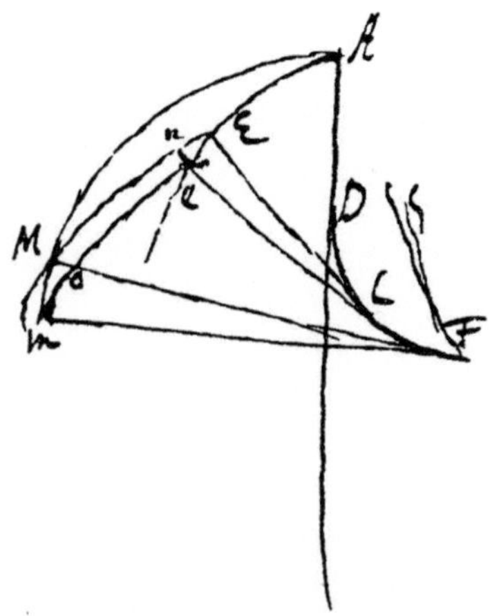

Fig. 12.23 With kind permission of Springer Science+Business Media

We now give the curves *AE* and *DC* [fig. 12.23], and by the evolution of this latter curve, we freely describe *EM* and *em*. The relationship between *AE*, x and *EM*, y is given, and consequently also that between dx and dy. It is required to draw the tangent at M, of the curve that is generated, that is to say to find the ratio between *mo* and *Mo*. I do this as follows. Let $FM = r$, and *Mo* or $ne = dr$, the arc of the circle with radius r contained between *MF* and the parallel *FS* to $CE = s$ $\left(\text{and thus by the lemma } Mn - oe = \frac{s\,dr}{r}\right)$

$$AE = x, \quad EM = y, \quad nE = dz.$$

We therefore have

$$mo = me - ME + ME - Mn + Mn - oe = dy + dz + \frac{s\,dr}{r}.$$

Now the ratio of *En* and *en* to *Ee*, that is to say that of *dz* and *dr* to *dx*, is given, because the curve *AE* is given[96] and we can draw a tangent at E. We have therefore found the ratio of *mo* to *oM*, in *dx* and *dy*, and consequently we have found the way to draw the tangent at M. Q.E.D.

I must tell you on this occasion that having been given any curve, I can construct at the same time another curve, which along with the given curve is equal to an arc of a circle.[97]

Given a curve or any portion of a curve *ABC* [fig. 12.24], whose evolute is *DE*, let *FG* be one of the antivolutes (I call antivolute that which is generated by the evolution of *ED* starting at *E*). Having drawn *CI* parallel to *AG* and equal to *CF* or *AG*, I say that the arc of the circle *FHI* described with radius *CF* will be equal to the given curve *ABC* augmented by its antivolute *FG*, and consequently if the evolute *DE* has its two halves *LD* and *LE* similar, and if we take *EF* = *DA*, it is clear that *FG* will thus be similar and = *ABC*, and so the given curve *ABC* will be = to half of the arc *FHI*. All of this is easily demonstrated through my lemma.

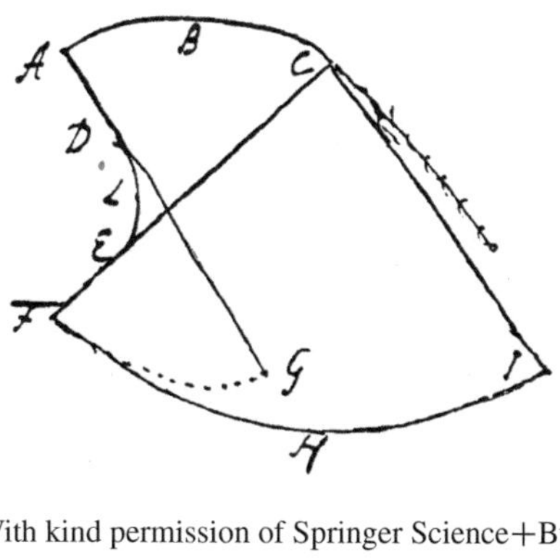

Fig. 12.24 With kind permission of Springer Science+Business Media

[96]As noted in Bernoulli (1955, p. 295), this should say "because the curves *AE* and *DC* are given."

[97]L'Hôpital combined this and the previous lemma in L'Hôpital (1696, §166). See also §168.

I am to you with all respect as I am to Madame, Sir, your very humble and very obedient servant,

Bernoulli

Letter 53: L'Hôpital to Bernoulli
Paris, July 6, 1695

I have received, Sir, your letter of June 23. I am very glad that this chair of mathematics in Holland suits you, and to have contributed in this way to your career. Mr. Huygens is not dead, as I told you,[98] but his condition is not any better because he has a raging headache in such a way that one cannot get any reason out of him; it's a deplorable accident. If you do not bring Mrs. Bernoulli with you and you are willing to travel through Paris, it would give me real pleasure, because otherwise I expect we will not see each other for a long time, however there is no need to disturb you and you should do what suits you best. Please find attached a bill of exchange for 200 ₶.

...

[Discussion of the fourth degree curve from Letter 50 and inflection points.]

...

You have solved my problem on direct tangents very well. I also believe as you do that is not of great utility, but what gave me the opportunity to imagine it was something that Mr. your brother proposed. Regarding that, I sent the solution with the proof of it that I found to Mr. Leibniz with another very simple problem that is the following. You know that you have given the construction of the curve that continually touches the hypotenuse of a right angle by sliding between its sides.[99] I considered the matter more generally in the following way. Given any two lines *AB* and *CD* with a constant straight line *AC*, if one imagines that its extremities *A* and *C* slide between these curves, it is clear that this line *AC* continually touches through this motion a certain curved line *G*, whose points we wish to find.

Solution. I draw the tangents *AE* and *CE*, and dropping the perpendicular *EF* from their point of intersection *E* to *AC*, I take $CG = AF$, and I say that the point *G* will be on the curve we wish to find. I do not send you the proof because you will find it easily.[100]

I should be very glad if you send me all the ways by which you have solved your problem of the centrifugal force under your two different assumptions with the method you have used to reduce them to the rectification of a curved line. You know that I have found two curves that satisfy it and I think are the simplest, but you tell

[98]In Letter 51.

[99]This was something that Bernoulli taught L'Hôpital in 1692 (Bernoulli 1742, pp. 447–448).

[100]This proposition and the full proof are given in L'Hôpital (1696, §158).

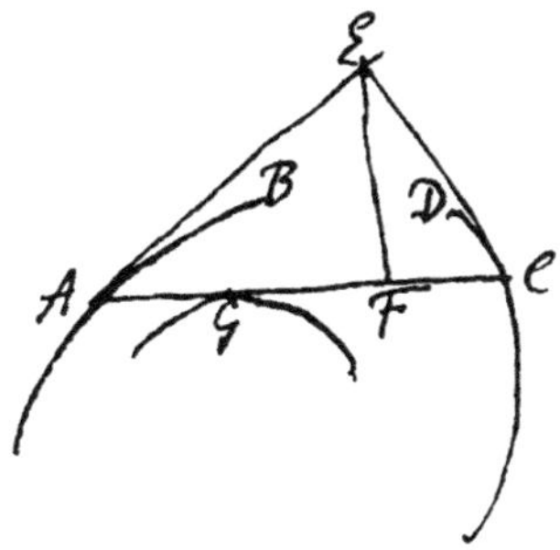

Fig. 12.25 With kind permission of Springer Science+Business Media

me that you can give three under each of these assumptions, which also satisfy it, and this is what I wish to see (Fig. 12.25).

I have a very different method from yours of finding the quadrature of roulettes that have arcs of circles as their bases, whether the describing point falls on, outside, or inside the circumference of the mobile circle. I am sending them to you so that you can to take care of having them put in the *Acts* of Leipzig, if you judge it appropriate.[101]

Please give me notice of the time that you will leave and the party you will take. I hope you will have satisfaction in this new position, and I will always look for opportunities to tell you that I am in my heart, and of my inclination entirely yours

the M. de L'Hôpital.

My wife makes many compliments to you and Mrs. Bernoulli.

As you told me in your penultimate letter that you would be very glad to see what Mr. Leibnitz communicated to me on the inverse of tangents, I am sending them to you and I beg you, when you have read and examined them at your leisure, to return them to me and not to communicate them to anyone, because Mr. Leibniz wrote me in a way to make me believe that he would not be happy if I showed it.

Letter 54: L'Hôpital to Bernoulli Paris, July 20, 1695

Because it has been some time, Sir, since I wrote you and I have received no response, I am at a loss to know if you have received the bill of exchange for the 200 ₶ that I sent you.

I received all the *Acts* of Leipzig you mentioned to me in your last letter, for which I thank you.

[101]This appeared in *Acta Eruditorum*, August 1695, pages 372–374.

A young man from Neuchatel came to see me who says he is a good friend of yours. He told me that Mr. your brother has had the Geometry of Descartes with the commentaries by Schooten printed,[102] and he had added several pages at the end containing remarks of his own. I would be very glad to have these remarks, and so you would make me happy to enclosing them with your letter and sending them to me by post, because from what he said I do not believe that they are very large. I am, Sir, yours with all my heart

the M. de L'Hôpital.

Letter 55: Bernoulli to L'Hôpital Basel, July 26, 1695

Sir

I did indeed receive your letter of the 6th of this month with the bill of exchange, which it pleased your kindness to attach, and for which I thank you infinitely. I am delighted to learn that Mr. Huygens is not dead.[103] I have been told the same thing from Holland, but that he has been extremely ill for four months; I wish him a full recovery of his health, so that I will have the pleasure of enjoying his learned conversations when I will be in Holland – to this end, I wish you would let me know me the city where he lives. Dr. Braun wrote me last week to prepare me for the voyage, because they expect me in Groningen by the month of October the latest, and I will receive the letter of vocation any day now. I am assuredly not a looking forward to such a prompt departure, and what troubles me most is that I do not know yet if my wife will go with me or not, which she has not yet decided, even though I have done my best to persuade her, so I'm entirely at a loss to know how to begin making preparations for the voyage. At very least tell, me for God's sake, if I may go safely and without danger from France to Holland, and if you can procure me a passport or a letter of safe conduct. Messrs. Leibniz, Menkenius, my brother-in-law[104] who is in Amsterdam, and others also wish that I would visit them, but how can I satisfy everyone? My duty compels me, in this case, to leave all by myself and obey you, even if the detour should be three times as larger as it is. You will forgive me, Sir, if I'm short on mathematics at this time, because you can clearly see I am overwhelmed with other thoughts.

...

[Discussion of inflection points, centrifugal force, and other matters.]

...

[102] *Notae et animadversiones tumultuariae in geometriam Cartesii.* Frankfurt 1695.

[103] In letter 51, L'Hôpital told Bernoulli that he had learned that Huygens had died. In letter 53, he reported that this had been in error, but in fact Huygens had died on July 8th.

[104] Emanuel Falkner.

That is everything that I had to tell you or it will be after paying my respects to Madame and also those of my wife.

Sir,

Your very humble and obedient servant
Bernoulli.

Letter 56: L'Hôpital to Bernoulli Paris, August 22, 1695

I have received, Sir, your letter of July 26, by which I see that you will go to Groningen in the month of October at the latest, and that you do not yet know if you will bring Mrs. Bernoulli with you, nor what route you will take. I would be delighted to see you on your way, because I foresee that if I miss this opportunity I will not get another for a long time. Nevertheless, if it is necessary for you to find yourself in Groningen by the month of October and you can only stay here for a few days, I believe it will be better for you to take the shortest route, because the pleasure that I would have in seeing you too soon be followed by the sorrow of losing you, and I had counted on you being able to give me a month of your time that you would spend in Oucques with me, where I am going at the beginning of October. With regard to a passport, I am assured that it will be very easy to get one to bring with you if you bring a certificate with you, because you are from Basel, and the Swiss may pass freely. If you could have given me this time it would have made me very happy, esteeming you as much as I do, but I would be very angry if this should put any obstacles to your career, so take the path that suits you best.

I have no doubt that you know about the death of Mr. Huygens.[105] He was 66 years old and left his inheritance to his nephews to the exclusion of his brother, and named two Dutch mathematicians to edit his writings with care and to have them printed. I am very sorry about this myself, because he was a great friend to me. I am persuaded that had he known you, he would have thought highly of you and rendered you any service that he could have. It was he that named you to the Gentleman of Groningen, who had addressed themselves to him in order to find a mathematician of his stature, having remembered the request that I had made to him on your behalf, as he told me in his letter.

I am about to print my treatise on the conic sections, and being urged to do so by Father Malebranche and several of his friends. I will add to it a small treatise on the differential calculus,[106] in which I will give you all credit that you deserve. I will not speak at all of the integral calculus, leaving this to Mr. Leibniz who intends to write a treatise on it, as you know, with the title *de scientia infiniti*, such that this

[105]Huygens died on the morning of July 8, 1695.

[106]This is the first mention of L'Hôpital (1696) to Bernoulli.

will only be an introduction to his. If you wish to put into order your rules of the logarithmic calculus, that you invented a while ago, I will include them at the end under your name, if you wish, and this will contribute to making this treatise better. Make use of this as you judge appropriate.

...

[A discussion of a question about conic sections and other topics.]

...

With regards to the *Acts* of Leipzig I have asked Mr. the Abbé Bignon if we may send them with his and I will send you the answer. I am, Sir, yours with the best of my heart

The M. de L'Hôpital.

My wife sends you a thousand compliments and to Madame Bernoulli.

Letter 60: The Marquise de L'Hôpital to Bernoulli Oucques, February 1, 1696

I have received, Sir, the letter that you wrote to Mr. de L'Hôpital dated January 4, which I have not given him because he is not in a condition to apply himself to mathematics, having been very sick, although he is beginning to recover a little as soon as he is well again I will give him your letter, which I assure you will give him much pleasure, always having great esteem and friendship for you. I am writing you this letter lest you be worried about not receiving a reply from Mr. de L'Hôpital. His illness delayed our return to Paris, to where I do not think we can return before the end of this month. I am glad that you are adjusting to your new position as well as Mrs. Bernoulli, whose homesickness will pass with time; I send her a thousand compliments. Mr. de L'Hôpital has received from Mr. your brother the Professor the issues of July and August and one of the supplements, but it was at the time that he wrote him and sent him Mr. Menkenius' letter, as he told you, and he even replied not to send him anything until we are in Paris, and we will look for some other way because it costs a lot by this route. Mr. de L'Hôpital will see if he cannot get them by means of the Mr. the abbé Bignon. If he finds that this is not convenient, he will ask you to send them with the help of Mr. your brother the painter;[107] because I know his feelings towards you and you may be assured that they are quite different from those he has for Mr. your brother the professor, whom he respects as one should such an able man, but he has a sincere friendship for you. He[108] has only written to Mr. de L'Hôpital with great decency and offers of service, such as sending him the *Acts* in your place. For myself, I am now sorry not to have cultivated mathematics;

[107] Hieronymus Bernoulli (1669–1760), youngest brother of Jakob and Johann.

[108] That is, Jakob Bernoulli.

perhaps I would now be able enough to propose some questions to you. All that I can do is to assure you, Sir, that you have no truer friend than me.

la marquise de Lhospital

Letter 63: L'Hôpital to Bernoulli
Paris, June 15, 1696

I begin by making many excuses to you, Sir, for not having replied sooner to your letter of April 21,[109] but, I have been impeded by an infinity of matters that have happened to me. At present I am calmer and I will be more regular.

It was very timely that Mr. your brother[110] came to me yesterday, because I would have sent you a bill of exchange for 300 ₶ in today's post through Mr. Compagnie in Amsterdam, because it is easier to find correspondents there than in Basel. As I have seen in your letter of June 5, that he gave me, that you want me to put this money in his hands, I withdrew it from the banker and I will not fail to give it to him prior to his departure and to get back a receipt, as you ask me (Fig. 12.26).

It is true that Mr. Varignon told me a few days ago that you had sent a problem to Leipzig, which consists in determining the nature of the curve in a vertical plane that would be described by a heavy body in traveling from a given point to another given point in the least possible time.[111] This problem seems to me the most curious and the most beautiful that has even been proposed and I would be glad to apply myself to it, but for that it would be necessary for you to send it to me reduced to pure mathematics, because physics puzzles me, and as I have not yet completely

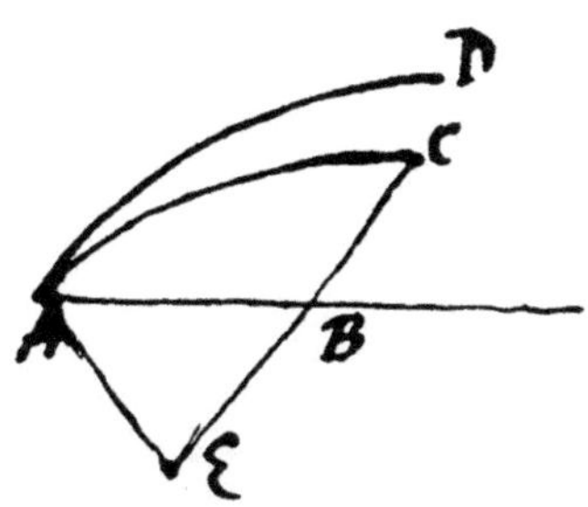

Fig. 12.26 With kind permission of Springer Science+Business Media

[109]The Marquis had recovered from his illness and written to Bernoulli on April 2.

[110]Hieronymus.

[111]This is the beginning of a discussion of the Problem of the Brachistochrone, or line of quickest descent. Bernoulli sent it to Varignon on May 15 and on June 11 it appeared in the *Acta Eruditorum*.

recovered from my illness, I dare not apply myself vigorously to it. Permit me to ask you, however, for the solution of the following question.

Let any two curves *AD* and *AC* be given in a plane, along with a fixed point *E* not on this plane. If we imagine a conic surface *EAC* that has the point *E* as its vertex, and the curve *AC* as its base, and that we make the other curve *AD* turn about its axis *AB*, it is clear that it will describe through this motion a surface that will by its intersection with the conic surface. Now, I wish to find the manner to draw a tangent from a given point on this curved line. My book will appear any day now;[112] I treat only the differential calculus which I attempt to explain in full depth, and my goal has been only to make a proper introduction to what Mr. Leibniz and others may give in what follows. I have no doubt that if you would like to take the trouble, you could give us everything one might want on the integral calculus, and it is this that we ought to get you to work on, because it seems to me that Mr. Leibniz is too busy to be able to explain these things as one would wish. I would have liked to have my conic sections printed at the same time, as I told you, because that has only have barely filled one volume. However, because I did not wish to work on it since my illness it turns out that the first treatise will appear by itself. You will see that I give you the credit that you deserve, and it would be useless to tell me to send you one because you well know that you are the first person to whom I will send it, being with a most sincere friendship entirely yours and to Madame Bernoulli's. My wife wishes both of you a thousand marks of affection.

the M. De L'Hôpital.

I am sending this directly to Groningen, because they assure me at the post that it will go through and it is not necessary to make any detours, so please reply to me as soon as you have received it.

Letter 64: Bernoulli to L'Hôpital
Groningen, June 30, 1696

Sir

I received the honor of you letter from the 15th of this month, in which I saw with pleasure that my brother[113] had come in enough time to get the money that you intended for me; and indeed I learned from a letter I received from him yesterday that you gave him 300 ℔ in return for a note in his hand. Thank you infinitely for your kindness, seeing from this that you still have the same affection for me as you have honored me with in the past; I would consider myself to be the happiest in the

[112]This is l'Hôpital's second mention of his book (L'Hôpital 1696).

[113]Hieronymus.

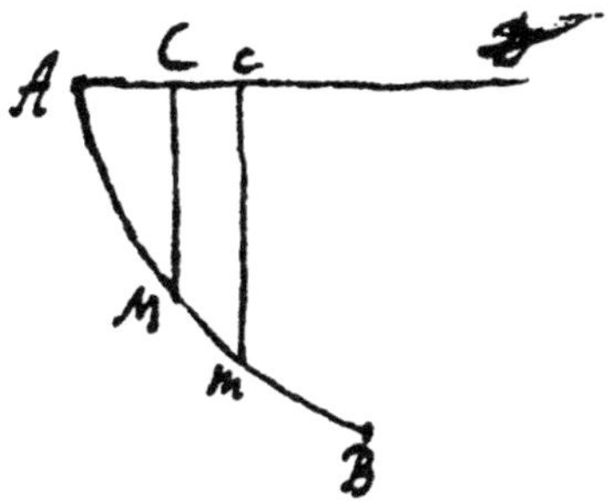

Fig. 12.27 With kind permission of Springer Science+Business Media

world, if you continue to do so. For my part, I will not miss out on any occasion to give you tokens of my obligation and just gratitude that one owes to such a generous benefactor.

It is true, Sir, that I have sent to Leipzig the problem that Mr. Varignon communicated to you on my behalf, but I do not know whether or not if it has been printed. However, having found it very curious as you have also found it, I also sent it to Germany and to England, to be proposed to the Mathematicians of those countries, but I do not know if anybody has yet given the solution. Since I communicated it (which was not long ago) to Mr. Leibniz, I have not yet received a response from him, so I cannot know whether he has solved it – maybe he will solve it, maybe not – because one needs a particular address, which he will not soon know, I am sure (Fig. 12.27).

It is very easy to reduce this problem to pure mathematics, there being nothing else from physics, but the ordinary principle of Galileo that we have also supposed in the isochrone curves, that is knowing that the speeds of falling bodies are in a subdoubled ratio[114] to the perpendicular heights traveled. Given this *let the two points A and B be given, we wish to find the nature of the curve AMB, along which the moving body falling from the point A arrives at the point B in the least time.* Having denoted the horizontal abscissa *AC* by x and the vertical ordinate *CM* by y; the speed of the moving body at *M* will be expressed by $\sqrt{y}$. Now dividing the distance *Mm* traveled by the speed we have that the time it takes to travel along *Mm*, and consequently

$$\frac{\sqrt{dx^2 + dy^2}}{\sqrt{y}}$$

[114]In other words, the speeds of falling bodies are to one another in the same ratios as the square roots of the perpendicular distance traveled.

expresses this time. And thus taking all of the

$$\frac{\sqrt{dx^2 + dy^2}}{\sqrt{y}}$$

from A to B, what results will be the time that the moving object takes to go from A to B.

So now this is how the problem is reduced to pure mathematics. *Among all the lines AMB joining two points A and B we wish to find the nature of the one for which*

$$\int \frac{\sqrt{dx^2 + dy^2}}{\sqrt{y}}$$

is the smallest. I would wish that some of your Geometers who boast of possessing such excellent methods of *maximis* and *minimis*, would set themselves to it, because here is an example that would give them some trouble and maybe more than their method could do.

For heavens sake, Sir, why would you want to break your head when you have not fully recovered from your illness, instead you should break the desire you have to solve this problem, and think about getting some time to rest after having completed your book, which I imagine has contributed more than a little to the bad state of your health.

I solved your question that you proposed to me quite easily, and even more generally than what you had asked for. Because in place of a conoid surface made by the revolution of a given curve around a given axis, and in place of the conic surface, the base of which is a given curve, substitute any two other given surfaces, whether convex or concave, regular or irregular, and finally geometric or mechanical. These two surfaces intersect each other in any manner whatsoever; the section will be another curved line called *solid* as distinct from those that are described in a plane and consequently may be called *plane*. Therefore we wish to find the way of drawing a tangent from a given point on this solid curve. I say to solve this problem that having applied two planes that touch the two given surfaces at the given point, the common section of these two planes, which will be a straight line, will be the tangent to this *solid* curve at the given point; because the demonstration is very clear, I do not put it here.

I must congratulate you, Sir, because your book has begun to appear, for my brother[115] informs me that you gave him several copies for Messrs. Leibniz and Menkenius. Why did you not send them to me? I could have gotten them to these gentlemen much more conveniently than my brother the Professor, because Groningen is more than 3 times closer to Hannover than Basel. I accept with great thanks the offer that you made to me of a copy. It is your usual politeness that you

[115] Hieronymus.

have given a place for my name in your book, for which I am very obliged. I foresee in advance that when it becomes known to the world, this book will provide you with a great reputation worthy of your Illustrious Person. I also believe that Mr. Leibniz will make us wait for a long time for his treatise *de scientia infiniti*, being too busy. You may well imagine the reason why I have not been willing to write anything so far regarding this matter; it is that I have not wanted to do anything without your consent following the promise that I gave you in the past. However, seeing now that you are getting me worked up about it yourself, I might very well get the intent to make a continuation where you have left off, in explaining the integral calculus, provided I have the leisure. My wife and I send our respects to Madame.

I am Sir

Your very humble and very
obedient servant
J. Bernoulli.

Letter 66: L'Hôpital to Bernoulli
Paris, September 10, 1696

I received, Sir, your letter of August 4, which gave me very great pleasure, because it let me know that you have not forgotten me. As for the letter you told me you had written to me beforehand, it must have been lost, because I did not receive it. I am obliged to you for the offer that you made to me to share with me your objections to Mr. Leibniz' system for the estimation of forces and of his responses, but nevertheless I ask you to reserve this goodwill for another time when I will examine this matter in depth, which appears to me of great importance for physics.

I have given the order to my publisher to send two hundred copies of my book to Leers in Rotterdam, and to the aforementioned Leers to send you one as soon as possible. I gave three of them to Mr. your younger brother when he was here, which were the only ones that I had, one of which was destined to Mr. your brother in Basel and the other two for Mssrs. Leibniz and Menkenius. If you have occasion to write to one of them, you will oblige me by asking them if they have received them, because I have received no news. Mr. your brother the professor asked me to have them sent.

Mr. Battier[116] went back to Basel because of a vacant chair he wants to obtain. I am delighted that you are willing to give us the explanation of the integral calculus; it seems that no one can do it better, and that this book will be highly sought after. In it, you will be able to assume what I have explained of the differential calculus in my book, which can be regarded as an introduction to yours. I believe you should waste no time getting to work on this, because curiosity about these matters is very

[116]Samuel Battier (1667–1744) friend of Johann Bernoulli and student of Jakob Bernoulli.

strong. You will find included the receipt for three hundred pounds from Mr. your brother, which I am sending you as you have asked me. I am, Sir, entirely yours and the secretary sends you a thousand compliments as well as to Madame Bernoulli.

The M. De L'Hôpital.

Letter 71: Bernoulli to L'Hôpital Groningen, Mid-February 1697

...

[Discussion of two mathematical problems.]

...

I have finally received a copy of your book and I thank you most humbly for it. You have done me too great an honor in speaking so highly of me in the preface. When I compose something in my turn I will not fail to give you the same in return. You explain things most intelligibly; I also find a beautiful order there and the propositions well organized; everything is admirably well done, and a thousand times better than I could have done. Finally, I desire nothing more than that you had put your name on the cover of the book, which would have given a much greater glory, and lent more Authority to our new method. If not for this, the book would undoubtedly be sought after with greater desire. Now, I must say yet another word with your permission, you seem to me a little too liberal in giving recognition to my brother, as if you had made use of his discoveries, and yet I have noticed nothing so far in your book that might be attributed to my brother with justice. My wife and I pay our compliments to Madame. And I am entirely,

Sir,

Your very humble and very obedient servant
J. Bernoulli.

P.S. For God's sake tell me if Mr. Varignon is dead or alive, and in which corner of the world he might be; he should reply to my very obliging letter that I last wrote to him a long time ago.

Letter 92: The Marquise de L'Hôpital to Bernoulli Paris, December 15, 1707

I am delighted, Sir, that the present that I gave you of the posthumous book of the late Monsieur de L'Hôpital pleases you. Given the esteem I have always had for you, you may well believe that I would not forget you on this occasion; no one can better judge this book than you. I received with pleasure the marks of gratitude that you gave me in your letter, for the memories that I brought back to you by sending you this book. If I could do something more to be at your service from this country, I assure you that I will carry this out with great joy. Mr. Varignon brought Mr. your nephew to me,[117] who asked me on your behalf for my portrait, to put it with the one of the late Mr. de L'Hôpital that you already have. I promised to let him make a copy whenever he wishes. Mr. Varignon tells me that one day he will be a very good painter. I hope he excels in this art as you have done in mathematics, but however skillful he may become, he will be hard pressed to acquire such a great reputation in painting, as you have made in mathematics, and to be as excellent a painter as you are an excellent geometer. I will always be most truly, Sir, your very humble servant.

the Marquise de L'Hôpital.

[117]Niklaus (I) Bernoulli "the Younger" (1687–1769).

Chapter 13
Fontenelle's Eulogy for the Marquis de L'Hôpital

Guillaume François de l'Hôpital,[1] Knight, Marquis of Sainte Mesme, Count of Entremont, Lord of Ouques, la Chaise, le Bréau and other Places, born in 1661 to Anne[2] de l'Hôpital, Lieutenant General of the Armies of the King, first Equerry to His Royal Highness Gaston, Duke of Orleans, and Elisabeth Gobelin, daughter of Claude Gobelin, Quartermaster General of the Armies of the King, and Ordinary Estates Counsel.

The House of l'Hôpital had two branches, to the elder of which Mr. the Marquis de l'Hôpital had joined the name of l'Hôpital to that of Saint Mesme, and the younger, which is presently deceased, produced two Marshals of France, and the Dukes of Vitri. Both of these had as a common stem Adrien de l'Hôpital, Chamberlain of King Charles VIII, Captain of the Hundred men at arms, and Lieutenant General in Bretagne, who commanded the vanguard of the Royal Army in the Battle of St. Aubin in 1488.

Mr. the Marquis de l'Hôpital, whom the Academy of Sciences has lost, being still a child, had a Tutor, who wanted to learn Mathematics in the leisure hours that his job afforded him. The young Pupil, who had little taste, and even, as it would appear, little talent for Latin, and barely an outline of the Elements of Geometry of Circles and Triangles, but when the natural inclination which almost always foreshadows great talents declared itself, he set himself to studying with a passion that which would have frightened anyone other than him at the first sight. He then had another Tutor, who was obliged by his example to apply himself to Geometry, but even though he was a man of intellect and industry, his Student always left him far behind. That which one obtains only through work never measures up to the favors of nature.

One day Mr. the Marquis de l'Hôpital, being only 15 years old, found himself at the house of Mr. the Duke of Roannés, where skilled Geometers, and among

[1]This is a translation of Fontenelle's Eulogy (Fontenelle 1708, pp. 116–145).

[2]Anne-Alexandre.

R.E. Bradley et al., *L'Hopital's Analyse des infiniments petits*, Science Networks. Historical Studies 50, DOI 10.1007/978-3-319-17115-9_13

them Mr. Arnaud, spoke of a Problem from Mr. Pascal on the Roulette, which appeared very difficult. The young Mathematician said that he did not despair of being able to solve it. They found that they could barely forgive this presumption and this temerity, because of his age. Nevertheless, a few days later he sent them the solution to the Problem.

He entered into the services, but without renouncing his more beloved passion. He studied Geometry only in his Tent, it was not only to study that he retired there, it was also to hide his devotion to study. For we must admit that the French Nation, as polite as any Nation, is still in this sort of barbarism, that it doubts if the Sciences, pushed to a certain perfection, do not demean one, and if it is not more noble to know nothing. He was so good at the art of hiding his talents, and being ignorant for the sake of propriety, and he did so well in the craft of war, that those men who were most penetrating on the faults of others never suspected him of being a great Geometer, and I have seen myself some of those who had served at the same time, completely astonished that such a man who had lived like them, and among them, became one of the great Mathematicians of Europe.

He was Cavalry Captain in the Regiment of the Colonel General, but the weakness of his vision, which was so short that he could not see at ten paces, caused him perpetual inconvenience in service, that he had for a long time unsuccessfully tried to overcome. He was finally obliged to give up and leave the career where he might have hoped to match his Ancestors.

As soon as war no longer shared in him, Mathematics benefited. He judged by the Book of *De la Recherche de la Verité* that its Author[3] would be an excellent Guide in the Sciences; he took his counsel, made good use of it, and became friends with him in a friendship which lasted until his death. Soon his knowledge reached the point where it could no longer be hidden; he was only 32 years old, when the Problems derived from the most subline Geometry, selected with great care for their difficulty, and proposed to all Geometers in the *Acts* of Leipzig, tore from him his secret, and forced him to confess to the public that he was able to solve them.

The first was the following, proposed in 1693 by Mr. Bernoulli, Professor of Mathematics at Groningen. *To find a curve such that all its Tangents terminated at the Axis are always in a given ratio with the parts of the axis intercepted between the curve and these Tangents.* It was only solved by Mr. Leibniz in Germany, by Mr. Bernoulli in Switzerland, the brother of the man who had proposed it, by Mr. Huygens in Holland, and by Mr. de l'Hôpital in France.

Mr. Huygens admits in the *Acts* of Leipzig that the difficulty of the Problem had first made him resolve not to think about it, but that such a novel Question had troubled his sleep in spite of himself, had persecuted him relentlessly, and that finally he was no longer able to resist. We readily judge what kind of material in Geometry that this might be, that it appeared so difficult to Mr. Huygens.

All of those who know even a little of the News of the Sciences, have heard tell of the celebrated Problem of the *quickest descent*. Mr. Bernoulli of Groningen

[3]Nicolas Malebranche.

had asked in the *Acts* of Leipzig, *supposing that a heavy body falls obliquely to the Horizon, what would be the Curved line that it should describe to fall as quickly as possible?* For, as he said in the *Historie* of the Academy of Sciences of 1699, p. 67, this most surprising Paradox was demonstrated: that the straight line, although the shortest of all the lines that are drawn between the two given points, was not the path that the Body should take to fall in the least time. He was also certain that Curve in question was not a circle, as Galileo had thought, and the error of so great a man may serve to give a sense of the difficulty of the Problem. Mr. Bernoulli proposed this Enigma in the month of June of 1696, and gave all the Mathematicians of Europe the remainder of the year to think about it. He saw that these six months were not sufficient, he further gave the first four months of 1697, and in those ten months only four Solutions appeared. They were from Mr. Newton, Mr. Leibniz, Mr. Bernoulli of Basel, and Mr. the Marquis de l'Hôpital. England, Germany, Switzerland, and France each gave one Geometer for this Problem.

These same names are found at the top of several similar solutions in the *Acts* of Leipzig, and they seem to be in possession of the rarest and most elevated of knowledge.

An account was even given in the *Hist.* of 1700, p. 78, of a Problem proposed, like almost all the others, by Mr. Bernoulli of Groningen, and which was only solved by Mr. de l'Hôpital. It concerned *Finding in a vertical plane, a curve such that a Body which describes it, descending freely, and by its own weight, always pushes against it at each of its points with a force equal to its absolute weight.* One has aimed to make known the variety of difficulties with this problem, that is to say its beauty. The Geometers of today are not easy to satisfy on these difficulties, and that which made Archimedes leave the Baths to shout through the streets of Syracuse, *I have found it*, would not be a very glorious discovery to them.

The *Hist.* of the Academy of 1699, p. 95, also spoke of a solution of Mr. the Marquis de l'Hôpital, which few others could have attained. Mr. Newton in his excellent Book on the *Mathematical Principles of Natural Philosophy* gave the *figure of the Solid which cleaves water, or other liquid, with as little difficulty as possible.* But he did not let it be seen by what method nor by what route he came to determine this figure. His secret seemed to him worthy of being hidden from the Public. Mr. Fatio, the famous Geometer, fancied that he had discovered it, and sent a printed Analysis of it to Mr. de l'Hôpital. It contained 5 large pages in quarto, almost entirely of calculation. Mr. de l'Hôpital, alarmed by the length and lazy about the new method, believed he would have rather looked for himself for this solution. He had effectively found it within two days, and it was simple and natural. This was one of his great talents. He went not only to Truth, however hidden it was, he went there by the shortest route. A certain kind of fate decrees in all things that the most natural methods or ideas are not those that present themselves most naturally. One almost always puts too great a value on the research that one undertakes, and there are few geniuses, greedy ones fortunately, who expend only what is absolutely necessary. It is not that there is a lack of wealth and abundance to supply to useless expenditures, but there is a greater art in avoiding them, and even a true wealth.

It would take too long to list here all the Masterpieces of Geometry with which Mr. de l'Hôpital, and the small number of his equals, have embellished the Journals of both Germany and of France. One suspects, no doubt, that in order to enter into those Questions which were reserved for them, they must have had, in addition to their natural genius, some particular Key, which was only in their hands. Indeed there was one, and it was the Geometry of the Infinitely small, or the Differential Calculus, invented by Mr. Leibniz, and at the same time also by Mr. Newton, and then continually perfected by them and by Messrs. Bernoulli, and by Mr. de l'Hôpital.

The illustrious Mr. Huygens, who was not the inventor of the differential Calculus, as Mr. Leibniz was, who did not use it in any of his geometrical studies, as Mr. de l'Hôpital and Messrs. Bernoulli, who was led without this help to lofty Theories, and made a most brilliant reputation, who might, in the manner of other men, and perhaps even more legitimately, mistake what he did not know, and considered useless that which had not been necessary to him for these great Works, had nevertheless judged, both by the merit of those who employed this Method, and by the miracles that he saw emerge from it, that it was worthy of his study. He was a great enough man to have admitted that he could still learn something in Geometry, he addressed himself to Mr. de l'Hôpital, who was almost half as old as he, to instruct him in the differential Calculus, and no doubt this feature in the Life of Mr. de l'Hôpital is even more glorious to Mr. Huygens than to him.

It is not that Mr. Huygens did not already know the Land of the Infinite by himself, where one is brought at every moment by the differential Calculus, he had been obliged to go just as far in some of his most subtle researches, especially those he had done for the immortal invention of the Pendulum, because refined Geometry can not go far without penetrating into the infinite. However, there is a great difference between knowing the Map of a County in general, and being familiar, in particular, with all the roads, including the little trails that save so much trouble for Travelers.

Mr. Huygens was then in Holland, where he had retired after leaving Paris and the Academy of Sciences, of which he was one of the principal ornaments. It seems from many of his Letters that were found in the papers of Mr. de l'Hôpital, and above all those of the years of 1692 and 1693, that he consulted Mr. de l'Hôpital concerning his difficulties with the differential Calculus: that when something stopped him, he did not catch on to the Method, but even though he did not possess it well enough, *he saw with surprise and with admiration the extent and the fertility of this Art, that wherever he turned his sight, he discovered new uses, and finally,* in his own words, *he conceived of an infinite progress and contemplation.* He even declared publicly in the publication of the *Acts* of Leipzig, that without a *differential Equation* he would never have come to the goal of finding the Curve whose Tangents, and the portions of the axis always make a given ratio, *and furthermore*, he adds in the same *Acts*, *one must note in this Problem a new and singular Analysis that opens the path to a quantity of things concerning the Theory of Tangents, as the illustrious inventor of the Calculus has well observed, without which we would have great trouble being admitted to such a profound Geometry.* At the same time he wrote to Mr. de l'Hôpital that he owed this differential Equation to his *lessons*, which gave him the solution to the Problem.

Until then the Geometry of the Infinitely small was still nothing but a kind of Mystery, and, so to speak, a Cabalistic Science shared among five or six people. They often gave their Solutions in the Journals without revealing the Method that produced them, and even when one could discover it, it was only a few feeble rays of this Science that had escaped, and the clouds immediately closed again. The Public, or rather the small number of those who aspired to the higher Geometry, were struck with a useless admiration that clarified nothing, and they found a way to attract their applause, while withholding the instruction that they should have given them.

Mr. de l'Hôpital resolved to communicate without restriction the hidden treasures of the new Geometry, and he did it in his famous book on the *Analysis of the Infinitely small*, which he published in 1696. There, all the secrets of the Geometric Infinity were revealed, and the Infinity of the Infinity, and in a word, all the different orders of Infinities, which rise up one above the other, and form the most astonishing and the most daring Edifice that the human Mind has ever dared to imagine.

Just as there are determinate ratios between finite magnitudes, which make up the unique object of Mathematical research, and magnitudes of these different orders of Infinities, one is brought by the path of the infinite to understanding of the finite, which no other Method, that did not have the audacity could ever attain, and at the same time the skill to address the infinite. The Book of the *Infinitely small* was therefore sparkling with truths unknown to the old Geometry, and not only unknown, but often inaccessible to this Geometry. The old truths found themselves lost in a crowd of new truths, and the ease with which we saw them born made one regret the efforts that they had previously cost their inventors. Proofs that by other Methods had demanded an immense detour, where they had even been possible, or which even in the hands of another Geometer instructed in the Method had been long and confused, were of a simplicity and a brevity that almost rendered them suspect.

Such is the effect of the general Methods, once one has sought to discover them. One is at the fountain, and one needs only to follow the peaceful current of the consequences. A single Rule from the Book of Mr. de l'Hôpital gives the Tangents of all imaginable curves; another, all of the greatest and least Ordinates, or all the Inflection points and Cusps, or all of the Evolutes, or all Catoptrics at once, or all Dioptrics. Entire Treatises written by the great Authors sometimes are reduced to a few Corollaries that we meet along the road, and we have trouble distinguishing among the multitude. Everything is related by the kinds of Systems that Mr. de l'Hôpital began to put into Geometry, and which will become widespread in the future.

There are, especially in Mathematics, more good Books than there are ones that are well made, that is to say that we find enough that may teach, but few that teach with a certain method, and, so to speak, with a certain charm. It is easy enough to have good material in one's hands, but to neglect the form, Mr. de l'Hôpital has given us a Book that is as well made as it is good. He had the skill to make a small enough Volume out of an infinity of things, and in it he put this brevity and this clarity, so delicious to the mind, the order and the precision of the ideas so that we might almost dispense with using words. He only wishes to make us think, caring more to bring us to the discoveries of others, than to jealously go about displaying his own.

This work was also received with universal acclaim, because acclaim can be called universal when it's clear how often new and original works fail to be appreciated all over Europe, especially when they require effort to be well understood. Those who pay attention to the events in the History of Science know the eagerness with which the *Analysis of the Infinitely small* was seized by all budding Geometers, who are indifferent to the ancient and the new method, and who have no interest other than to be educated. Because the intention of the Author had been principally to produce Mathematicians, and to sow the seeds of upper Geometry into their minds, he took pleasure to see that they bore fruit from it every day, and that the Problems previously reserved for those who had grown up in the thorns of Mathematics, became first attempts for these young people. Apparently, the revolution became even greater, and with time it found itself with as many disciples as there had Mathematicians.

After seeing how useful his Book on the *Infinitely small* had been, he engaged himself in another work which was also suitable for educating Geometers. In this plan he embraced the Conic Sections, geometric Loci, the Construction of Equations, and a Theory of Mechanical Curves. This was precisely the plan of the *Geometry* of Mr. Descartes, but more extensive and more complete. He made no pretense that this work was as original, nor as sublime, as the former. He might have turned his research to the side of the integral Calculus, which follows and which assumes the Differential Calculus, which has greater, and until now insurmountable difficulties, and which because of this today occupies the greatest Geometers, and which has become the object of their ambition. However, he preferred an enterprise from which the public could draw more general and more necessary instruction, and his Zeal for Geometry had outweighed the interest in his glory. Nevertheless, I am a witness that he could not keep himself from missing the Integral Calculus.

This book was almost finished, when at the beginning of 1704 he was attacked by a Fever, which did not seem particularly dangerous at first, but as one saw that it resisted all the different remedies that were employed, one began to fear it, and the Sick Person faces no greater peril than to contemplate death. He prepared himself in a most edifying manner, and finally fell into an Apoplexy from which he died the next day, on February 2, at 43 years of age.

Some have attributed his death to the excesses that he had done in Mathematics, and what could confirm it, I knew from he himself that often the mornings that he had destined for this study became entire days without him noticing it. He had wanted to renounce this for the sake of his health, but he could never maintain this hardship for more than 4 days. Furthermore, it was natural enough to believe that he should have made great efforts of the mind, when one considers to which point had reached the age of 43 years, and how much time in such a short life had been spent on Mathematics. He had served, he was of a birth that engaged him in a great number of duties, he had a Family, household responsibilities, a very considerable estate to manage, and consequently much business; he was in the commerce of the world, and he lived more or less like those for whom this unproductive occupation was their sole occupation. Still, he was not an enemy of pleasures: there were many distractions, and whatever rare talent we assume that he had for Mathematics, it is impossible that such a prodigious work filled up only a little of his time. Nevertheless, it never seemed that study had altered his health, he had the air of the best and most firm constitution that one might desire. He was not in the least bit somber, nor a day-dreamer; on the contrary, he was happy enough, and he seemed not to have paid any price for his great mathematical genius.

In his most common discourse one senses the justice, the solidity, and, in a word, the Geometry of his soul. He was good company, of perfect integrity, open and sincere, suited to what he was because it was him, and never taking advantage, the true modesty of a great man, quick to declare what he did not know, and to receive instruction, even in matters of Geometry, where it was possible to receive it, not at all jealous, not through knowledge of his superiority, but through his natural fairness, because without this fairness, those who believe themselves and even make themselves superior to others, are nevertheless jealous.

He had married Marie Charlotte de Romilley de la Chesnelaye, Lady of an old noble family of Brittany, and from whom he had received great estates. So close was their union that he shared with her his genius for Mathematics. He left behind one son, and three daughters.

References

Bernoulli, Johann, *Opera Omnia*, vol. 3, Bousquet, Lausanne, 1742.
Bernoulli, Johann, *Der Briefwechsel von Johann I Bernoulli*, vol. 1, ed. O. Spiess, Birkhäuser, Basel, 1955.
Bernoulli, Johann, *Der Briefwechsel von Johann I Bernoulli*, vol. 2, ed. P. Costabel, J. Peiffer, Birkhäuser, Basel, 1988.
Bernoulli, Johann, *Der Briefwechsel von Johann I Bernoulli*, vol. 3, ed. P. Costabel, J. Peiffer, Birkhäuser, Basel, 1992.
Bradley, Robert E., "The Curious Case of the Bird's Beak," *International J. Math. Comp. Sci.*, **1** (2006), pp. 243–268.
Bossut, Charles, *Histoire Générale des Mathématiques*, vol. 2, 2nd ed., F. Louis, Paris, 1810.
Burton, David, *The Hstory of Mathematics: An Introduction*, 6th ed., Mc Graw Hill, Boston, 2007.
Cohen, I. Bernard, *A Guide to Newton's* Principia, in *The Principia*, Newton, Isaac, University of California Press, Berkeley and Los Angeles, 1999.
Descartes, René, trans. , Smith & Latham, *The Geometry of René Descartes*, Dover, New Yrok, 1954.
Eneström, Gustav, "Sur le part de Jean Bernoulli dans la publication de l'*Analyse des infiniment petits*" *Bibliotecha Mathematica*, **8** (1894), pp. 65–72.
Fontenelle, Bernard de, *Histoire du renouvellement de l'Académie royale des sciences*, Boudot, Paris, 1708.
Hahn, Alexander, "Two Historical Applications of Calculus," *The College Mathematics Journal*, **29** (1998), pp. 99–103.
Hall, A. Rupert, *Philosophers at War*, Cambridge U. Press, Cambridge, 1980.
Huygens, Christiaan. *Horologium Oscillatorium sive de motu pendulorum*, F. Muguet, Paris, 1673. English translation by Richard J. Blackwell, *The Pendulum Clock or Geometrical Demonstrations Concerning the Motion of Pendula as Applied to Clocks*, Iowa State University Press, Ames, 1986. Page references are to the 1986 translation of Huygens' *Horologium Oscillatorium.*
Katz, Victor, *History of Mathematics: An Introduction*, 3rd ed., Addison-Wesley, Boston, 2009.
Leibniz, Willhelm G. von, "Nova methodus pro maximis et minimis," *Acta eruditorum*, **3** (1684), p. 467–473.
l'Hôpital, Guillaume F. A. de, "Méthode facile pour déterminer ler points des caustiques ...," *Mémoires de mathématique et de physique, tires des registres de l'Académie Royale des Sciences*, 1693, pp. 129–133.
Anonymous (Guillaume François Antoine, Marquis de l'Hôpital), *Analyse des infiniment petits*, Imprimerie Royale, Paris, 1696.

R.E. Bradley et al., *L'Hopital's Analyse des infiniments petits*, Science Networks. Historical Studies 50, DOI 10.1007/978-3-319-17115-9

l'Hôpital, Guillaume F. A. de, *Traité analytique des sections coniques*, Boudot, Paris, 1707.
l'Hôpital, Guillaume F. A. de, *Analyse des infiniment petits pour l'intellignece des lignes courbes*, 2nd ed., Montalant, Paris, 1715.
l'Hôpital, Guillaume F. A. de, *Analyse des infiniment petits pour l'intellignece des lignes courbes*, 2nd ed. [sic], Montalant, Paris, 1716.
l'Hôpital, Guillaume F. A. de, *Analyse des infiniment petits pour l'intellignece des lignes courbes*, new ed. with a commentary by l'abbé Aimé-Henri Paulian, Didot le jeune, Paris, 1768.
l'Hôpital, Guillaume F. A. de, *Analyse des infiniment petits pour l'intellignece des lignes courbes*, new ed., revised and augmented by Arthur LeFevre, A. Jombert, Paris, 1768.
Lockwood, E. H. (1971). *A Book of Curves*. Cambridge, England: Cambridge University Press.
Montucla, Jean F., *Histoire des Mathématiques*, second ed., vol. 2, Agasse, Paris, 1799.
Schafheitlin, Paul, "Johannis (I) Bernoullii Lectiones de calculo differentialium," *Verhandlungen der Naturforschenden Gesellschaft in Basel*, **34**, pp. 1–32.
Stone, Edmund, *An Analytick Treatise of Conick Sections*, Senex et al, London, 1723.
Stone, Edmund, *The Method of Fluxions, both Direct and Inverse*, Innys, London, 1730.
Suzuki, Jeff "The Lost Calculus (1637–1670): Tangency and Optimization without Limits," *Mathematics Magazine*, **78** (2005), pp. 339–353.
Varignon, Pierre, *Eclaircissemens sur l'analyse des infiniment petits*, Rollin, Paris, 1725.

Index

R.E. Bradley et al., *L'Hopital's Analyse des infiniments petits*, Science Networks. Historical Studies 50, DOI 10.1007/978-3-319-17115-9

D

MIX
Papier aus verantwortungsvollen Quellen
Paper from responsible sources
FSC® C105338

If you have any concerns about our products,
you can contact us on
ProductSafety@springernature.com

In case Publisher is established outside the EU,
the EU authorized representative is:
Springer Nature Customer Service Center GmbH
Europaplatz 3, 69115 Heidelberg, Germany

Printed by Libri Plureos GmbH
in Hamburg, Germany